Proceedings of the 30th International Geological Congress
Volume 1

Origin and History of the Earth

Proceedings of the 30th International Geological Congress

Volume 1 : Origin and History of the Earth
Volume 2 : Geosciences and Human Survival, Environment, and Natural Hazards
Volume 3 : Global Changes and Future Environment
Volume 4 : Structure of the Lithosphere and Deep Processes
Volume 5 : Contemporary Lithospheric Motion / Seismic Geology
Volume 6 : Global Tectonic Zones / Supercontinent Formation and Disposal
Volume 7 : Orogenic Belts / Geological Mapping
Volume 8 : Basin Analysis / Global Sedimentary Geology / Sedimentology
Volume 9 : Energy and Mineral Resources for the 21st Century / Geology of Mineral Deposits / Mineral Economics
Volume 10 : New Technology for Geosciences
Volume 11 : Stratigraphy
Volume 12 : Paleontology and Historical Geology
Volume 13 : Marine Geology and Palaeoceanography
Volume 14 : Structural Geology and Geomechanics
Volume 15 : Igneous Petrology
Volume 16 : Mineralogy
Volume 17 : Precambrian Geology and Metamorphic Petrology
Volume 18.A : Geology of Fossil Fuels - Oil and Gas
Volume 18.B : Geology of Fossil Fuels - Coal
Volume 19 : Geochemistry
Volume 20 : Geophysics
Volume 21 : Quaternary Geology
Volume 22 : Hydrogeology
Volume 23 : Engineering Geology
Volume 24 : Environmental Geology
Volume 25 : Mathematical Geology and Geoinformatics
Volume 26 : Comparative Planetology / Geological Education / History of Geosciences

Proceedings of the

30th International Geological Congress

Beijing, China, 4 - 14 August 1996

VOLUME 1

Origin and History of the Earth

Editors:
Wang Hongzhen
China University of Geosciences, Beijing, China
Borming Jahn
Laboratoire de Géochémie et Géochronologie, Rennes, France
Mei Shilong
China University of Geosciences, Beijing, China

///VSP///

Utrecht, The Netherlands, 1997

VSP BV
P.O. Box 346
3700 AH Zeist
The Netherlands

First published in 1997

ISBN 90-6764-248-7

Printed in The Netherlands by Ridderprint bv, Ridderkerk.

CONTENTS

FOREWORD

During the preparatory period of the Thirtieth International Geological Congress, in the early winter of 1995, it was agreed upon by the Secretariat of the 30th IGC and the VSP International Science Publishers of the Netherlands to publish a series of Proceedings of the 30th IGC. Volume 1 of the series was originally designated to include papers submitted to Special Symposia A—Origin and History of the Earth. The first call for papers for this volume was sent to the participants in the early Spring of 1996, but the number of quality papers received by the deadline was rather limited. It was later decided by the Organizing Committee to include the key-note papers of the Congress in this volume. We deem it an honour to have these important papers included herein, as they encompass some of the most outstanding aspects of geosciences dealt with in this Congress.

Apart from the keynote papers, the present volume contains 15 papers dealing with the different aspects of the symposia, ranging from, in the original order of the sessions, early history of the earth (A-1, jointly organized by IGCP), continental accretion (A-2), core-mantle differentiation (A-3), biological evolution (A-4), palaeoclimate (A-5), to interaction between the lithosphere and the hydro-atmo-biosphere (A-6). Papers within one session are arranged in the alphabetic order of author names.

The title of the Special Symposia, Origin and History of the Earth, is evidently one of the most fundamental issues of earth science research. However, as no vestiges of very ancient rocks (*ca.* 4 Ga) have been preserved in the continental crust, the early history of the earth is also the most controversial and least understood fields of research. For a variety of reasons, we have only assembled a limited number of papers dealing with these subjects. We do not mean to offer an apology, but just wish to note how the edition of this volume has come about. Of course, the authors are solely responsible for the views expressed in their papers. We thank all the contributors for their efforts to make the publication of this volume possible. The cooperation of the VSP Publishers and the assistance of the Secretariat of the 30th IGC are deeply appreciated.

The Editors
March 15, 1997

KEYNOTE PAPERS

Proc. 30th Intern. Geol. Congr. Vol. 1, pp. 1-14
Wang *et al.* (Eds)

The Earth Sciences and Society: The Needs for the 21st Century

W. S. FYFE
President of the International Union of Geological Sciences (1992-1996)
Department of Earth Sciences, University of Western Ontario, London, Ontario, Canada N6A 5B7

Abstract

As world population moves to 10-12 billion next century, there will be, must be, vast development of Earth Resources. If there is to be truly sustainable development, we must use all knowledge about our planet with greater wisdom than in the past. We must integrate all knowledge to solve critical problems like energy supply, food security, waste management and the maintenance of quality air and water for all species who live with us and support us. Knowledge from the Earth Sciences is at the core of intelligent development of our life support systems.

Keywords: resource management, sustainable development, geological education

INTRODUCTION

The 20th century has been remarkable in terms of the development of *Homo sapiens*. Two thousand years ago the human population has been estimated at about 300 million. It then took 1700 years to double the population, social conflict, disease, famine controlled the global population. Then came the birth of modern science and technology and the beginning of new knowledge about our planet. The giants of the period around 1900 (from Lyell to Darwin to Einstein) opened new visions including the understanding of atoms and energy and planetary systems. We now live in the age of observation on all scales. The thoughtless application of the new knowledge led to the present population explosion. For many the quality of life has greatly improved but today we are concerned and following the famous Bruntland report [37]. the new consideration of the concept of "sustainable" development arose.

Today, most agree that the present world human population of about 6 billion will again double in the next 30-40 years. But it will be a world with a new demography. Europe and North America will make a small part of the world population by 2050. The greatest population expansion will occur in the so-called developing (or overdeveloped) nations. Fertility rates tell the story (India, 3.8; Nigeria, 6.5; Egypt, 3.9; Pakistan, 6.2; *cf* Germany, 1.3; Italy, 1.3; Sweden, 2.1).

Recent world data show the rising problems of overpopulation; major areas like Asia and Africa with vast problems of malnutrition [31] and all the problems associated with malnutrition. The expansion of classic diseases like tuberculosis, malaria, *etc.* and with the competition for resources increasing, the often brutal conflicts in some 40 nations.

A new symptom of the present situation is the rise in so-called environmental refugees, "outcasts from Eden". As Fell [7] recently wrote, "The ranks of refugees fleeing floods, drought, desertification and other assaults on the environment could swell to 200 million by 2050. Is there any hope of fending off global catastrophe?"

The surface of our planet will be very different by the year 2100. During the past decades we have seen the growth of environmental science. We have studied in detail changes in environment over the past 100,000 years or so and we have recognized many of the factors from the Sun - atmosphere - hydrosphere - biosphere - solid Earth system that lead to fluctuations on many time scales. Our environment never was, never will be, steady state.

But I think that as Earth Scientists we must be honest. The past will not be the key to understanding the new world. For example, imagine our planet when:

- very few rivers flow freely to the oceans (*e.g.* about 5,300 large dams in 1950, 36,600 in 1985 [1]).

- many nations have no forests

- irrigation on the continents will lead to new patterns of global evaporation - new continental water regimes, salinzation, *etc.*

- global soils become thinner and less bioproductive

- more xenobiotic chemical, pesticides, herbicides, *etc.* cover the planet

- biodiversity is greatly reduced and genetically modified species cover the earth.

In all, our planet will have new outer geosphere chemistry and biology, a new albedo.

Will the average quality of life be greater by 2050? I often think that the lead question which must be addressed by all who are educated must be "can this planet support 10-12 billion humans *well* and leave the planet in good order for all who follow us?" I think the answer to this question is now clear. It may be possible—but not with our present technologies and social morality. I am angry when I read recent reports of the growing income disparities. In 1960, the ratio of the wealth of the top 20% of population to the bottom 20% was a factor of 30, today it has grown to over 60 times. Yes—"globalization" or, as some say, "corporate feudalism" is working.

In what follows I wish to consider some of the new developments which are urgently needed and most of which involve, in part, the Earth Sciences. It is difficult to put such things in any order of priority for all are interrelated.

PRIORITY ONE—UNIVERSAL QUALITY EDUCATION

When we consider the human condition, situations where environments are improving with true change toward sustainable systems, world data are clear (see [38]). In nations with excellent and improving education for all, of all ages, problems are being solved and human society is accepting its place in the biosphere and on the planet. As has been said many times, we require universal literacy, numeracy *and sciency*. All who live on a planet in the solar system, still the only one we have available, must understand how it functions, the factors that control our life support systems. While the need for literacy has been long recognized, it is clear that equal priority must be given to universal science. And it is possible to use the planetary system to introduce the most basic concepts of science (matter-energy dynamics) to people of all ages and stages. We must educate for global responsibility (see [9]). All people must understand that conditions on this planet are never steady state. We must be prepared for surprise and that means we must have surplus.

At advanced levels, universities, the time has come to produce much more integration of the specialized fields. Yes—we must have experts of all kinds. But today we must have experts who can communicate with other experts and not hide behind their jargon. I recently re-read an old report (1963) from UNICEF stating that by 1970, malaria would cease to exist on the planet. Much foreign aid given to developing countries was in the form of DDT. But the experts were wrong, they did not understand ecology or evolution. Under the title of environmental science we need interactions from diverse areas of physics, chemistry, biology, geology, engineering, sociology and economics.

SOME PRIORITIES IN THE EARTH SCIENCES

1. Development of the Geologic Map

As never before, there is need for highly precise geological mapping on scales appropriate to the development problems being considered. Such maps must be precise in describing the timing of events and must be precise in 3 dimensions. For example, if we consider the growth of megacities, the geological knowledge required to prevent-reduce costly engineering mistakes (*e.g.* Kobe, Japan) to provide and protect water, to prevent pollution *etc.*, is of a detail and range far beyond most present mapping systems. Recent studies, as with the German deep drilling experiment (the KTB), clearly show that our techniques for deep remote sensing are far from adequate today. I was recently on a field trip with an excellent group of Portuguese (Lisbon University) structural geologists. They were concerned with mapping a region of some interest in terms of a site for nuclear waste disposal. The region was well known for some major fault structures. But their detailed studies clearly revealed the complexity of the stress patterns and showed that micro-fault systems were present with a frequency distribution of tens of metres. This type of detail is essential to the planning of any major engineering project. At a recent meeting in Norway, Swedish geologists reported on the use of good maps in planning the exact location of new highways resulting in large cost reductions. As urban regions expand, consideration must be given to using the subsurface for storage facilities, factory construction, transport routes, *etc.* The surface of our planet is precious with its role in food and fibre production and bioproductivity in general.

Geologic maps must include the surface chemistry (related to agricultural potential) and the water storage in the top few km. Recent work sponsored by the International Union of Geological Sciences and UNESCO has shown the vast importance of surface geochemical mapping to a host of problems from agriculture, to public health, to waste disposal (see [5]).

2. Quality Control of Materials

Today we have amazing techniques for the description of all materials, inorganic, biological in all phase states, gas, liquid, solid, and for the study of surfaces. Before any material is used and dispersed around our planet, we require exact knowledge of what it is and how it reacts with the outer geospheres. As discussed below, earth materials are used and dispersed on a vast scale and often, their chemistry *etc.* is only studied after problems arise. Some of our classic specializations like mineralogy and petrology, should be considered as the science of natural Earth materials in the broadest sense.

3. Energy

At the present time, the bulk of world energy comes from the combustion of coal, oil and gas; all these resources are based on natural capital and are non-sustainable. The thoughtless waste of such valuable resources is a global disaster. Of the fossil carbon sources, only coal and certain types of carbon rich sediments have reserves of interest for more than a few decades. In a general way, there has been little change in burning technologies—add air—burn—and exhaust to the atmosphere.

There is no need here to discuss the potential future impacts of the climate changes related to the fact that we have rapidly changed the chemistry of the atmosphere. Discussion is normally concerned with CO_2, CH_4, and acid compounds, but, as stressed previously [14], many coals contain significant quantities of all halogens: F, Cl, Br, I, and the steadily increasing ozone depletion catastrophe may be influenced by combustion of these fuels. Also, time after time, the detailed chemistry of coal and coal ash is not well known. Many coals have significant quantities of elements like uranium and arsenic, and an array of heavy metals immobilized in the reducing, sulfur-rich, medium of coal. It is amazing how little is known about the detailed chemistry and phase chemistry of this major world fuel.

Imagine for a moment that we did not use fossil carbon, nuclear, in our energy systems. If we relied only on wind, water, biomass, how many people would live on Earth?

The USA, and nations like China and India will depend on increased use of coal for decades to come. Can the technology be changed at reasonable cost, to reduce the environmental impact of coal combustion? I think the answer is positive. We have been studying the fixation of CO_2 and organics in the cracked, permeable basalts in the caves of Kauai, Hawaii, deep beneath a very heavy forest cover. Every crack is covered with white materials (silica, clays, carbonates), formed by the action of organics with the basalt, a process mediated by ubiquitous bacterial biofilms. Bacteria can live to depths of over 4 km, up to 110°C, in favourable locations [28]. Can such processes be used to fix the exhaust gases of coal combustion? Certainly, some rock types will be better than others, and volcanics with Ca-feldspars and rich in Fe-Mg phases should be ideal, as in Hawaii. Also, it is interesting to note that adding H_2O - CO_2 to appropriate rocks can be a highly exothermic process and the gas disposal could lead to a geothermal energy bonus! For some reactions, the heat produced per carbon atom is about 30% of the original heat of combustion per carbon atom. Recently, on a field trip in China, (East of Beijing), we discussed the possibility of using their rapidly exploited oil-gas fields for disposal of wastes of many types. If a basin can isolate oil-gas for millions of years, it undoubtedly has the capacity to isolate wastes [6]. And, generally, oil field structures and hydrogeologic properties are well known. Interest is growing rapidly in this area (see [15], [18]).

The growing knowledge of the deep biosphere also raises the possibility, with certain types of carbonaceous sediments, of using microorganisms for in-situ methane or hydrogen production. In place of opening deep mines, with all the related water pollution problems, could it be possible to produce bio-gas *in situ*?

There is no shortage of energy sources on this planet. Ultimately, the world must move to solar energy of all types (photovoltaics, wind, tidal) and geothermal energy.. Wind energy use is increasing across the world, and photovoltaic devices are becoming more efficient and cheaper [26].

Geothermal sources are normally associated with regions of high heat flow (volcanic systems) but, for some purposes (city heating, greenhouses and aquaculture systems), the normal geothermal gradient can provide background heating. There are many regions of the ocean floor with impressive potential for geothermal energy. All such potential use requires exact knowledge of deep geologic structures, porosity, permeability and geochemistry. I was interested to read in a recent tissue of the Economist, that there is a world shortage of high purity silicon. Where on the planet is the most pure SiO_2 on the million ton scale. This is the type of problem with which we must be concerned.

4. Water

If the world human population reaches 10 billion, and if these 10 billion are to have adequate nutrition and supplies of clean water, there will be the necessity to manage the global water supply with great care and attention. When we come to consider the water resource question, we see the potential for real limitations and potential environmental disasters [29, 30].

Over the world, precipitation is about 525,100 km^3 (1 $km^3 = 10^{15}g = 10^9$ tonnes). However much of this falls on the oceans (78%). On land, more than half evaporates. The reliable runoff available for *man* is about 14,000 km^3 (much more occurs but only during floods). The reliable usable water is estimated to be about 9000 km^3. At this time, man manipulates about 3500 km^3, almost 40%. And, after man uses water, its chemistry and biology are changed [23].

For any given region, the water supply can be quantified. Perhaps it is best to think of the "island" model (for example, a river may flow from one region to another—but who can guarantee that this inflow—through flow—will be maintained?) In an island model, the water inventory involves: the rainfall, the ground water penetration, the evaporation or evapotranspiration, the runoff to the oceans. A first most important parameter is the reliability , or variations, of the rainfall. Over what time period does one average the availability of the total supply? The numbers to be used will, in part, reflect the technologies that may be available, or possible, to store water over long periods of time.

There are certain realities in preparing a plan of management for the island water supply.
— there must be sufficient runoff to prevent salinization—runoff removes salts and other wastes from the land surface.
— the evaporation processes depend on the nature of the island's surface cover. This will vary, depending on the vegetation, urban developments, transport systems, *etc*.
— the potentially useful ground water resources depend on the geology, the deep porosity, permeability and the chemical nature of the rocks, which will influence the chemistry of the water and its suitability for various purposes, agricultural and industrial.
— surface storage removes land area, and can increase evaporation.
— subsurface storage is only of sustainable use if recharge exceeds withdrawal, and if the cycling does not reduce long term porosity and permeability.

It is remarkable that, in many regions today, such basic data are not considered in the use of water (see [30]). Groundwaters are being mined in many regions.

There is little doubt that, in the near future, there will be development of massive technologies for the massive recycling, re-use of water, particularly in urban and industrial systems. Given an appropriate climate, urban-industrial runoff would be redistilled by abundant solar energy—a sort of Space Ship Earth technology (imagine a colony on Mars!). By use of appropriate rocks, the distilled water could be easily remineralized to required levels of mineral content for human nutrition. For industrial uses, modern filtration systems (inorganic or biological) can allow re-use and reduction of waste discharge.

But for this writer, a great problem with the potential for major problems involves the agricultural-energy uses of major systems of runoff. A major river is dammed for such purposes. The runoff to the oceans now fluctuates over large values, depending on fluctuations in rainfall. Downstream, the aquatic biodiversity and biomass is seriously perturbed. At sites of ocean discharge, the nutritional supply to the ocean biomass is seriously perturbed, and modern satellite pictures show us clearly that the marine biomass is concentrated near continental margins.

But, perhaps most seriously of all, ocean current systems may be perturbed. About 10,000 years ago, the great Gulf Stream current that transports energy to the Atlantic Arctic regions was perturbed. The North went into a little Ice Age, the Younger Dryas event. An explanation by Broecker et al [4] is that, for a period, the Mississippi river system was diverted into the St. Lawrence system. This placed a vast flow of light continental water onto the surface of the N. Atlantic, which perturbed the Gulf Stream and the Northern energy transport. Very rapidly, the North Arctic froze. This model warns us that, if a large outflow is changed, the patterns of ocean currents, ocean mixing, may change effecting local and global climate. But it should be noted that the Broecker et al model has been recently questioned [36]. Such phenomena must be considered when there are plans to modify major components of continental runoff! I am sure that most water engineers have never considered such possibilities.

Finally, we require new technologies for cleaning, remediation, of polluted waters. For inorganic, heavy metal, pollutants, mineral surface reactions can be of great significance. Thus, elements like lead and mercury, can often be sequestered by adsorption on sulphide mineral surfaces. Uranium and many metals can be adsorbed by appropriate microorganisms. Many techniques are being developed for simple applications of solar energy distillation for purification of polluted and saline waters. Where pollution involves toxic organic compounds, again processing via appropriate microorganisms is gaining increasing attention. And the possibility of using subsurface, thermophylic organisms, is of great interest. And in hot regions where evaporation is intense, large area reservoirs should be avoided. We need to explore better strategies for underground storage or even inert cover technology. New observational technologies like Radar satellite technology which can monitor soil moisture, may call on more intelligent planning of the agricultural industries with reduced need for irrigation.

5. Air

Across the world there is growing concern with air quality and public health. For example, it is estimated that in England [17] that "sickness, deaths, lost working days caused by particulate pollution cost the country £17 billion a year." In England the major recognized problems come from the transport technologies. But a recent paper by Nriagu [27] shows the incredible scale of metal emissions to the atmosphere. For example, he shows that four million metric tons of lead are emitted annually to the atmosphere today. As mining of metals has increased, so have the atmospheric emissions.

In many urban areas (see [38]) where particulates are rising, the sources and nature of them is rarely well characterized. There is an urgent need to characterize all particulate and gas emissions to develop technologies for their control The situation in general is similar to that for fossil fuel use. We know that there are possible technologies.

6. Soil

At this time, at least one billion humans do not have an adequate supply of food of well balanced nutritional values [31]. Across the world, wood is becoming an expensive and declining commodity. Despite the electronic revolution, the use of paper products is increasing (per capita, 3x in the last 40 years). In addition, the world's marine resources are declining at an alarming rate. Again, the rich-poor gap is dramatically increasing the nutritional difference in the world's population.

Sustainable food-fibre production depends on climate, climate fluctuations, soil quality and water resources, and knowledge from the geosciences is involved in all these parameters. Given that we are not adequately providing nutrition for the present human population, what are the prospects for the next 5 billion?

All organisms require a large array and balance of the chemical elements (about 50) for efficient production of the organics needed for life [24]. The geochemistry and the mineralogy of soils are critical in estimating the capacity of a soil for sustainable organic productivity. According to the Worldwatch Institute, topsoil loss globally is approaching 1% per year, while natural remediation can take hundreds of years. The technologies exist now for erosion and salinization control, but such technologies are not adequately used, and there is great need for new soil maps which clearly show good soils, soils for forests only, and ones we should leave alone [8]!

Given the chemical and physical properties of a soil, additives may greatly enhance bio-productivity. Often, such additives require the addition of simple mineral materials containing species like K, Mg, Ca, P *etc*., and appropriate trace metals like Co, Mo, *etc*., which may be critical in biofunctions like nitrogen fixation. The types of additives may be closely linked to soil type and climate. For many situations as with the laterite soils of the humid tropics, slow release, mineral fertilizers (K in feldspars, rock phosphates *etc*.) may be more effective and less wasteful than soluble chemical fertilizers (see [20]).

Soil contains a complex array of ultra-fine inorganic and bio-mineral materials, with vast surface areas which control key soil-bio functions. Today, with the modern techniques of surface chemistry (Auger, ESCA...), and the power of modern high resolution transmission electron microscopy [32, 33], we can precisely examine the inorganic-bio-gas-liquid interactions, which was not possible a decade ago. In soils, many of the mineral-forming processes involve reactions with living cells. There is a new world in the science of biomineralization which, as H. Lowenstam showed decades ago, are of vast importance in all aquatic environments to at least 100°C [22].

Today we can greatly reduce soil erosion of all types. Often the massive use of highly soluble chemical fertilizers is not necessary and often our waste products such as sewage, compost, rock dusts, coal ash, *etc*. can be useful soil remediation agents. But, in all such cases of use, there must be strict quality control on the total chemistry and microbiology. And there is increasing evidence that biodiversity does promote bioproductivity (see [35]). Biodiversity can also reduce the impact of climatic fluctuations like drought.

How do we estimate the local and regional health of soil? I would like to suggest that there are simple ways to obtain a useful estimate of the health of soils on a local and regional basis. All organisms, from bacteria to trees, require a wide range of macronutrients and micronutrients. Table 1 shows the level of three key macronutrients in four great rivers of the world. Most of the water in these rivers passes through surface soil and rock. The levels of macronutrients and the total dissolved inorganic material (TDS) speak eloquently about the state of the soils which the rivers drain. The Mississippi and Ganges river systems pass through younger soils that are loaded with rock-forming minerals; the Amazon and Negro systems flow through old laterite terrains. The Amazon water indicates a low capacity to support intense bioproductivity, probably the reason this region has not been heavily populated by the human species, historically.

Table 1. Nutrients in major rivers (from Berner and Berner, [3])

River	Calcium Dissolved	Magnesium Dissolved	Potassium Dissolved	Total Disolved Solids
Mississippi	39	10.7	2.8	265
L. Amazon	5.2	1.0	.8	38
L. Negro	.2	.1	.3	6
Ganges	24.5	5.0	3.1	167

Units are given in parts per million.

7. Materials - Mineral Resources

If we look around us, we notice that we are surrounded by modified materials, mostly derived from the top kilometre or so of the Earth's crust. We live with concrete, glass, bitumen, ceramics, steel, copper, aluminum, stone, zinc *etc*. Our transport machines, our computers, contain a component from half the elements in the periodic table. Advanced societies use about 20 tonnes of rock-derived materials per person, per year. For a population of 10 billion living at an advanced quality of life, this means 2×10^{14} kg of rock per year, or almost 100 km^3 per year. This quantity exceeds the volume of all the volcanism on the planet, on land and submarine, by an order of magnitude. Human actions are now a major component of the processes that modify the planet's surface [11].

Can we supply the necessary materials for all humankind? If we have the energy to drive the machines and transport the materials, the answer is probably yes. But, it is also clear that, for the less common materials, copper, zinc *etc*., with rigorous attention to recycling, the modifications of the environment can be reduced from the scale of today. The new trend to design a product for recycling must become the rule for the future. The savings in total use of materials, use of energy, production of wastes, makes recycling an economic necessity of the future.

A major component of the materials we use today is derived from wood. It is quite clear that the present styles of the use of wood in products for housing, paper, packaging, and the like cannot continue. The environmental impact of removal of forests is too serious to allow present and past careless harvesting to continue. Wood will not be a major material for the 10 billion humans of next century. It is not needed, given the potential of modern systems of building and communication.

Again, I must stress that in the future development of mineral resources quality control must be improved. Before any mine is opened, we must know the total mineralogy of all phases and the total chemistry and chemical siting of *all* elements. And given the new observations on the deep biosphere there are exciting new possibilities for in situ mining using microorganisms (see [12]).

8. Holistic Mining Technologies

As mentioned above, to provide resources advanced societies use about 20 tons of rock per person per year. Vast mine openings operate over a large range of depths, surface to several kilometres. But only recently has any consideration been given to the end use of mining cavities.

First, if the minerals removed in mining are carefully separated and characterized, many may have important use. For example in terrains with laterite soils, common minerals may be useful as soil additives in the local region. And certain types of mines may be useful for waste disposal. Could some very deep mines be useful for the isolation of high and low level nuclear wastes? Could some open cast, near-surface mines be useful for urban waste disposal. And in some cases, given organic wastes, could they be engineered with appropriate clay sealing materials, for useful bio-gas production?

All such potential uses of exhausted mines require holistic planning, beginning to end. And often the key to success is with simple separation of the gangue materials, clays, silicate minerals, carbonates and the like. With care, the consideration of end use, could significantly improve the economics of mining.

9. Wastes

The wastes from the production of energy from fossil fuels are enormous. There is no need here to emphasize carbon dioxide, and the greenhouse, or acid rain from bad coal technology. We use terms like "clean" coal technology. But, unless the carbon dioxide is controlled, a not impossible consideration as many geochemists know, coal combustion is either very dirty, or dirty. The economic and social consequences of even a small climate change or rise in sea level are simply mind boggling. We need holistic economics.

However, while gaseous emissions from fuel burning CO_2, CO, SO_x, No_x *etc.*) are known, another huge problem is ash disposal. We have been working in India, where high ash coal (10-50%) is used, leading to a vast problem with ash disposal. Coal is a material of highly variable chemistry, and can contain significant concentrations of elements like chromium, arsenic, lead, uranium *etc.* But coal can contain useful elements like potassium and phosphorus. The ash being in a reactive, often glassy state, can be environmentally toxic. At present, river dumping is often the disposal method, creating severe problems, such as flooding. Globally, ash disposal is an increasing problem, as at least 1 km^3 is produced annually. Large scale coal mining, often associated with oxidation of sulphides in coal and trace-metal leaching, can perturb surface and groundwater resources.

Decades ago, it appeared possible that nuclear energy could solve the world's energy problems. But today, we are worried. While possibilities exist (see [21]) the nuclear waste disposal problem is still not clearly solved [13]. Hydropower potential still exists on a large scale in some continents, like Africa and South America. We tend to consider hydropower electricity generation as environmentally benign. But recent experience in, for example, the Amazon of Brazil show that this is not always so. River valleys are often great food-forest producers. Unless topography is steep, vast areas may be flooded, forest and farmlands drowned, and a paradise created for tropical water-borne disease. There is increasing interest in bio-fuels (methane, ethanol, methanol), and Brazil leads the world in this potentially renewable resource (sugar $\rightarrow$ ethanol) derived from solar energy. But, again, caution, biofuels use soil. The world has a major problem of inadequate nutrition for all [31]. Unless there is precise control of soil nutrient balance, this technology may not be sustainable. If the most agriculturally productive regions of the world were used for bio-fuels, who would have the food reserve for "the year without summer" [16]?

At a recent Dahlem conference [23], there was discussion on the impact of heavy metals and "xenobiotic" organics that flood modern soil-water-atmosphere systems on ever-increasing scales. At least 50,000 organics, which may have no natural analogues, and of little known environmental behaviour, are used today. Increasingly, there are moves to eliminate dispersion, to containment and, for organics, destruction often by incineration. In the rich world, regulations and controls are much improved (*e.g.* the Rhine river system) but, in much of the developing world, there is still little control. The impacts of strange chemicals on ecology, local and global, are little understood, but recent data from Eastern Europe are alarming (see [38]). The global influence of a group of simple chemicals on ozone destruction nicely illustrates the impact of what were once thought to be harmless chemicals.

There is no need to emphasize the impact on soil, water, air, ecological balance, of overuse of chemical fertilizers, herbicides and pesticides, which is now increasingly recognized. On the other hand, the limits of "organic" farming are also recognized. It has been shown that, if organic farming alone were used, a much greater land area would be needed to feed the present world population. Urbanization has removed any hope for quiet dispersion. Cities of 20 millions will be common next century. Where will the garbage go?

Slowly we are returning to old systems of recycling and reuse. All animal, vegetable waste, with care, makes good fertilizer, or in some systems can be used to generate methane and hydrogen [2]. The key to all the energy saving recycling technologies is front end quality control. If we characterize well all the materials at the input of industry, then it is more economic to effectively recycle the waste products. We probably know more about trace elements on the moon than in city garbage. Landfills should be the last resort—we cannot afford to waste land. And there is growing interest in urban agriculture. We can produce food and reduce wastes in cities (see [25]).

CONCLUSION

World data are hard (see [38]). In nations with a high level of education, and freedom of information, birth rates are falling, people are accepting limits to growth. Some gaseous emissions are also being dramatically reduced (*e.g.* S in Germany, 3743 x 10^3 metric tons of SO_2 equivalent in 1970, reduced to 939 x 10^3 tons, 1990) and, in many countries, it is being shown that recycling is economically positive.

It is also obvious that, for the needed new development, we require new systems to integrate the needed expertise (see [19]. For example, most developing nations need more energy. Who must be involved in planning the new developments? What are the alternatives and the long range economics? For certain, biologists, ecologists, agri-scientists, engineers, hydrogeologists, earth system scientists in general, many from advanced physics, chemistry, materials science, all working with social planners and economists. And such people must be involved in planning from the start! Perhaps Sir Krispin Tickell [34] summarized it all when he said, 'I was recently asked if I was an optimist or a pessimist. The best answer was given by someone else. He said that he had optimism of the intellect but pessimism of the will. In short we have most of the means for cooping with the problems we face, but are distinctly short on our readiness to use them. It is never easy to bring the long term into the short term. Our leaders, whether in politics of business, rarely have a time horizon of more than five years."

There is reason for optimism. In the 40's Aldous Huxley said that we have treated nature with greed, violence and incomprehension. But, today, the incomprehension is no longer excusable. There is need for a totally new approach to the mega-problems, and the key for success is the combination of science with sound economics. The surface of this planet and our hydrosphere, atmosphere, biosphere systems must be managed with great care, if all people are to enjoy their planetary experience [10].

We have many modern studies to show that environmental protection is economically advantageous. ECO-logy and ECO-nomy (Greek, Oikos, my house) are not in conflict when the time scale is decadal. And at our universities, it is time to change many of our programmes and recognize the needs for the world of 2050.

Acknowledgements

Finally I would like to acknowledge the wonderful contribution of the geological community and the government of China in organizing a truly magnificent 30th International Congress in Beijing, 1996. There is no doubt that new ideas and new cooperation, worldwide, will result from this congress, a congress where most of the most urgent problems were discussed.

REFERENCES

[1] J. N. Abramovitz. Imperiled Waters, Impoverished future: The Decline of Freshwater Ecosystems. *Worldwatch Paper* **128**, 80 (1996).
[2] P. Baccini. *The Landfill.* Berlin: Springer-Verlag (1989).
[3] E. K. Berner and R. A. Berner. *The global water cycle.* Prentice Hall Inc., New Jersey (1987).
[4] W. S. Broecker, J. P. Kennett and B. P. Flower. Routing of meltwater from the laurentide ice sheet during the Younger Dryas cold episode. *Nature* **41**, 318-321 (1989).
[5] A. G. Darnley. *A global geochemical database for environmental and resource management.* UNESCO Press, Darnley *et al.*, (Eds.), 122 (1995).
[6] T. M. B. Desseault. Radioactive waste disposal. *Nature* **375**, 625 (1995).
[7] N. Fell. Outcasts from Eden. *New Scientist.* August 31, 24-27 (1996).
[8] W. S. Fyfe. Soil and global change. *Episodes* **12,** 249-254 (1989).
[9] W. S. Fyfe. Education for Global Responsibility. *International Newsletter on Chemical Education* **36**, December 1991(1992).
[10] W. S. Fyfe. The Life Support System in Danger: Challenge for the earth sciences. *Earth Science (Japan)* **47,** 3, 179-201 (1992).
[11] W. S. Fyfe. Global Antrhopogenic Influences. *Encyclopedia of Environmental Biology, Volume 2.* Academic Press Inc., 203-214 (1995).
[12] W. S. Fyfe. The Biosphere is going deep. *Science* **273**, 448 (1996).
[13] W. S. Fyfe. Energy and wastes: from plutonium to carbon dioxide. *Science International Newsletter* **62**, 14-17 (1996).
[14] W. S. Fyfe and M. A. Powell. Halogens in coal and the ozone hole, Letter to the editor. *Chemical Engineering News*, (American Chem. Soc.) April 24, 6 (1995).
[15] W. S. Fyfe, R. Leveille, W. Zang and Y. Chen. Is CO_2 disposal possible? *Am. Chem. Soc. Division of Fuel Chemistry* **41:4**, 1433-35 (1996).
[16] J. M. Grove. *The Little Ice Age.* Methuen, London (1988).
[17] M. Hamer. Clean air strategy fails to tackle traffic. *New Scientist.* August 31, 6 (1996).
[18] B. Hitchon. *Aquifer disposal of carbon dioxide.* Geoscience Publishing Ltd., Alberta, Canada (1996).
[19] A. King and B. Schneider. *The first global revolution.* Simon and Schuster, London (1991).
[20] K. O. Konhauser, W. S. Fyfe, W. Zang, M. I. Bird and B. I. Kronberg. Advances in Amazonian biogeochemistry. In: *Chemistry of the Amazon,* P. R. Seidl, O.R. Gottleib and M.A.C. Kaplan (Eds.) Am. Chem. Soc. Symposium Series, 588 (1995).
[21] K. B. Krauskopf. *Radioactive Waste Disposal and Geology.* London, Chapman and Hall (1988).
[22] H. A. Lowenstam. Minerals formed by organisms. *Science* **211**, 1126-1131 (1981).
[23] D. J. McLaren and B. J. Skinner. *Dahlem conference reports.* John Wiley, New York (1987).
[24] W. Mertz. The essential trace elements. *Science* **213,** 1332-1338 (1981).
[25] T. Nelson. Closing the nutrient loop. *Worldwatch* **9:6**, 10-17 (1996).
[26] *New Scientist.* Cheap solar power, January 14, 11 (1995).
[27] J. O. Nriagu. A history of global metal pollution. *Science* **272**, 223-224 (1996).
[28] K. Pedersen. The deep subterranean biosphere. *Earth Science Reviews* **34**, 243-260 (1994).
[29] S. Postel. *Last oasis-facing water scarcity.* W.W. Norton and Co. New York (1992).
[30] S. Postel. Dividing the Waters: Food Security, Ecosystem Health, and the New Politics of Scarcity. *Worldwatch Paper* **132**, 76 (1996).
[31] N. Sadik. The state of world population, 1989. *United Nations Population Fund, U.N.*. New York. (1989).
[32] K. Tazaki and W. S. Fyfe Observations of primitive clay precursors during microcline weathering. *Contrib. Min. Pet.* **92**, 86-90 (1986).
[33] K. Tazaki, W. S. Fyfe and M. Iwatzaki. Clues to Arctic soil erosion from cryoelectron microscopy of smectite. *Nature* **333**, 245-247 (1987).
[34] S. K. Tickell. The future and its consequences. *The British Association lectures.* The Geological Society, London, 20-24 (1993).

[35] D. Tilman, D. Wedin and J. Knops. Productivity and sustainability influenced by biodiversity in grassland ecosystems. *Nature* **379**, 718-720 (1996).
[36] A. De Vernal, C. Hillaire-Marcel and G. Bilodeau. Reduced meltwater outflow from the Laurentide ice margin during the Younger Dryas. *Nature* **381**, 774-777 (1996).
[37] World Commission on Environment and Development. *Our Common Future*. Oxford University Press, Oxford (1987).
[38] World Resources Institute. *World Resources 1996-97*. Oxford Univ. Press (1996).

Proc. 30th Intern. Geol. Congr., Vol. 1, pp. 15-26
Wang *et al.* (Eds)

Geological Environments in China and Global Change

LIU TUNGSHENG
Institute of Geology, Chinese Academy of Sciences, Beijing100029, P. R. China

Abstract

A large variety of geological evidence are presented and summarized here to address the links between the geological environments in China and global environmental changes in the Cenozoic era. Environmental indicators (coal beds, salt and gypsum deposits, palaeo-vegetation remains *etc.*) from the presently monsoonal regions provide the first terrestrial evidences of the inceptions of the Southeast Asian summer monsoon, the Southwest Asian summer monsoon and the Northwest Asian winter monsoon. The Palaeocene and the Eocene environmental pattern in China was dominated by roughly zonal climates resulted from the planetary wind system. The decrease in the aridity for the southeastern country in the Oligocene indicates the inception of the southeast summer monsoon. Conspicuous changes have occurred for the Miocene when the originally arid south-western part of the country became much more humid, indicating the initiation of the south-west summer monsoon. The south-east summer monsoon was also significantly strengthened at the same time. The geographic location of the arid region in northern China was further close to the present-day one, suggesting that the Himalayans and the Tibetan Plateau may have began to uplift. These were also more or less synchronous with the rapid growth of the Antarctic ice-sheet. Another drastic change occurred around the latest Miocene when extensive aeolian dust began to deposit in the northern China, indicating the emergence of the north-east winter monsoon and an increase in the continental aridity in central Asia. Since then, a stepwise increase of continental aridity in northern China was observed for the last 2.5 Ma as shown by the loess-palaeosol sequences.

Keywords: inception of monsoon, Tibetan Plateau, global change, palaeoenvironment

INTRODUCTION

During the last two decades, a large variety of scientific results have demonstrated that the evolution of the major geologic environments of China, during the Cenozoic, have played an important role in global climate change. The development of the world's highest mountains, Himalayas, and largest plateau, the Qinghai-Xizang (Tibetan) Plateau is thought to have aided late Cenozoic climate deterioration [27, 37], in part by deeply affecting atmospheric circulation patterns. Several key atmospheric components prevailing in the Asian continent include the sub-arctic cold (north-west) winter monsoon, the tropical and subtropical Pacific (south-east) summer monsoon, the cross-equatorial Indian Ocean (south-west) summer monsoon; and the northern hemisphere Westerlies [49, 5]. Furthermore, changes in the size of the South China Sea, located near the modern west Pacific warm pool, in response to glacial-interglacial cycles may have profoundly influenced the thermodynamic role played by the global warm pool [42]. The most extensive mid-latitude deserts in the world are located in north China and central Asia. Recent results (An, Yao, Mayewski and Thompson's verbal comminications) have shown that mineral aerosols in the Greenland ice cores were transported primarily from the arid regions of Asia during the last glacial maximum. The quantity of iron in aeolian dust may be crucial to ocean productivity, and could have significantly affected the concentration of atmospheric CO_2. The modern topography of China consists of a three-step framework: the high western region of the Tibetan Plateau; the central plateau (including the Loess Plateau), and the eastern fluvial plains. This modern topographic pattern developed during the early Cenozoic from an older complex style, and lead to the formation of the modern Yangtze and Yellow Rivers. The formation of both of these rivers has substantially affected the amount of fresh water, and sediment, discharged from the Eurasian continent.

In this presentation, a large variety of prior geologic records of climate change are summarized, along with some new data. Particular emphasis is placed on the aeolian record of aridification which is critical for understanding the link between the oceans, continents, ice-sheets and the atmosphere. This discussion is focused on the following three questions:

(1) What is the Cenozoic history of aridification in China, especially as it pertains to the development of mid-latitude arid and semi-arid zones?

(2) What are the major factors that controlled Cenozoic aridification?

(3) How are these factors linked with global climate change?

ARIDIFICATION IN NORTHERN CHINA IN THE PRE-QUATERNARY PERIOD

The Palaeocene environment of China was inherited from that of the Late Cretaceous (Fig.1a). It was dominated by an approximately west to east zonal climate attributable to the planetary wind system, as reflected in the spatial distribution of environmental proxies such as coal beds, halite and gypsum deposits, animal fossils and palaeo-vegetation indicators. A broad arid zone is evident between about 18-35°N latitude. The present arid region in northern China, located between about 35-40°N latitude, was characterized by a semi-arid, warm temperate to subtropical climate during the Palaeocene. The current monsoon-dominated areas of south China were more arid, leading to widely distributed halite and gypsum deposits. The Tibetan Plateau was not yet developed.

The Eocene pattern is similar to the Palaeocene, however the presence of some coal basins in the southern-most areas, suggest that the climate zones may have moved northwards, perhaps in relation with a warmer global climate background (Fig.1b).

The Oligocene pattern remained zonal, but was marked by a significant northward movement of the southern boundary of the arid zone in the eastern provinces (Fig.1c).

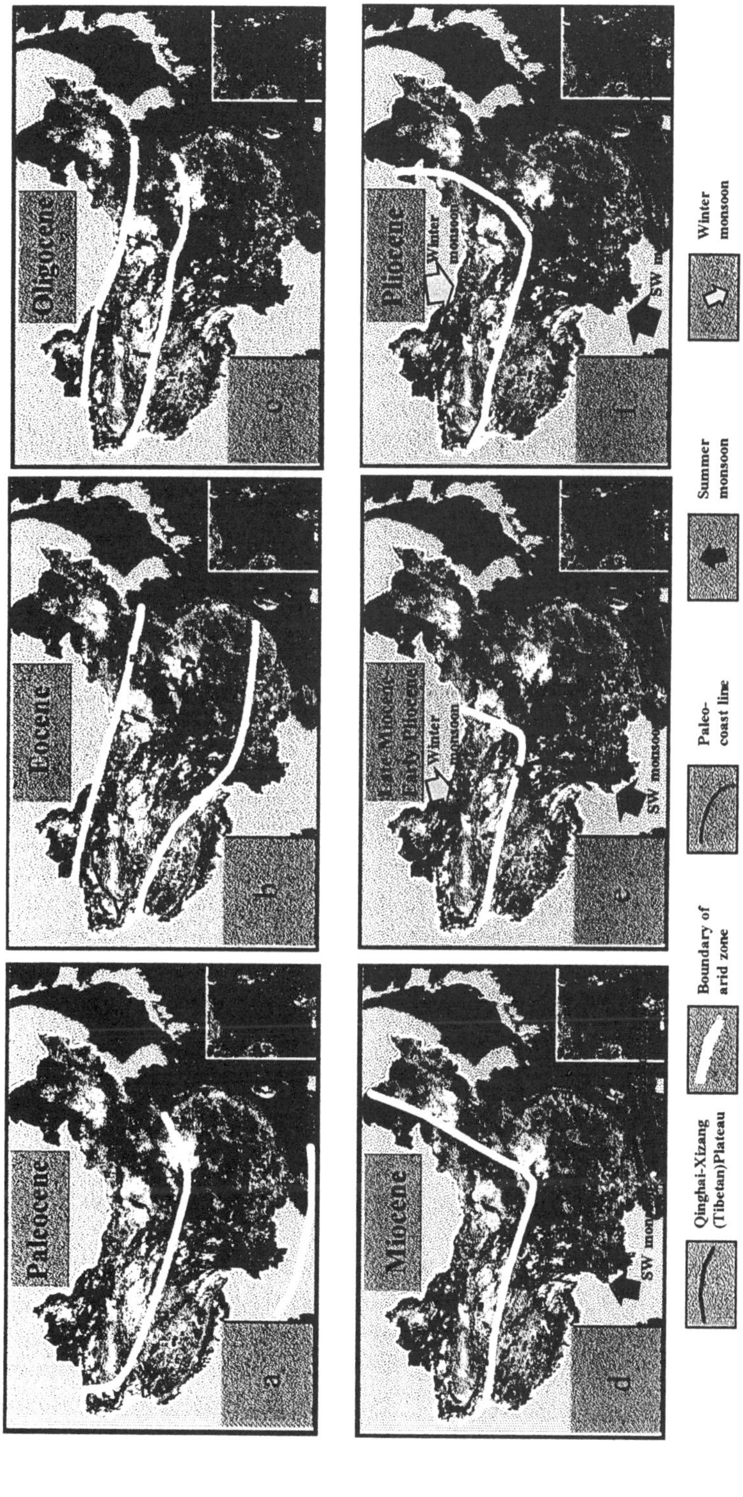

Figure 1. Aridification in northern China from Paleocene to Pliocene (modified after Zheng Mianping).

Several lines of evidence indicate that a forest existed to the south of the present Yellow River. Only limited gypsum deposits were present at this time, primarily near Wuhan. The decrease in aridity of the southeastern areas obviously indicates a moderate increase in precipitation and moisture that may be attributed to the initial stage of the southeast summer monsoon. Thess data suggest an earlier inception for the southeast glaciation occurred in the Ross Sea during the late Oligocene [26]. Since the Tibetan summer monsoon than the Indian monsoon, since the latter appears to have been was initiated during the middle Miocene [35]. There is evidence that an early stage of Plateau had not been uplifted to any considerable height at this time, we speculate that the commencement of the southeast summer monsoon may be linked to the formation of the Antarctic ice cap [6] and/or an alteration of the heating difference between the Eurasian continent and Pacific Ocean.

Conspicuous changes occurred during the Miocene when the south-western part of the country became more humid, as is evidenced by a great number of coal-bearing basins (Fig.1d). The increase in humidity may be attributed to the initiation of, or a further strengthening of the south-west summer monsoon bringing moisture from the Indian Ocean. This is consistent with modeling results [35]. The south-east summer monsoon also strengthened significantly as implied by the absence of arid climate proxies, the presence of abundant coal beds, and forest evidence from southeastern China. The geographic location of the arid region was closer to the present-day position. The Himalayas and Tibetan Plateau may have begun to uplift and were able, to a certain extent, obstruct moisture brought by the south-west summer monsoon. This record of uplift is apparently confirmed by the initiation of the Indus and Bengal fans (*ca.* 20-17 Ma) in the deep Indian Ocean [35, 24], and also by a marked global increase in the ocean $^{87}Sr/^{86}Sr$ ratio [21]. However, coal and forest evidence in southern Tibet suggest a limited height for the southern part of the Plateau. It is likely that the strengthening of the Asian monsoons was synchronous with tectonic changes in the region, and possibly with rapid growth of the Antarctic ice-sheet, resulting in a new state for the ocean-atmosphere system [12]. At the same time, significant aridification may have occurred on the African continent, as indicated by the aeolian dust record of the Indian Ocean [22].

The end of the Miocene marks a period of notable climate change (Fig.1e). The Indian Ocean summer monsoon experienced a significant increase in intensity around 8 Ma [34] consistent with a marked change in the isotope composition of the Pacific Ocean [39]. The initiation of extensive aeolian deposition in northern China near the end of the Miocene resulted in the "Red Clay" formation. This deposit was weathered, but obviously derived from aeolian dust deposition as it shows no significant differences in magnetic anisotropy from Quaternary loess [31] and sedimentologic characteristics (morphological features of coarse grains and clay mineralogy *etc.*)[15]. We attribute the initiation of dust deflation and transportation to the emergence of the north-east winter monsoon and an increase in continental aridity in central Asia. During early to mid-Pliocene time a further northwest retreat of the arid zone is evident. One of the most remarkable changes during the late Pliocene is the appearance of alpine vegetation in the Tibetan Plateau region, suggesting a significant altitude for the plateau (Fig.1f).

The “Red Clay” formation documents the environmental history of northern China at least from the Latest Miocene or early Pliocene to about 2.5 Ma. Magnetostratigraphic results from several sites date the lower boundary at about 5-6 Ma [51, 47], although an older date could be expected at a few specific localities. The formation is not an homogenous deposit in that a number of soil complexes evident, indicating significant climate changes over the period from about 6 Ma to 2.5 Ma. Grain-size, magnetic susceptibility (Ding *et al.*, in preparation) and isotope measurements (Chen MY verbal communication) suggest that climate change over this interval is dominated by a periodicity of roughly 400 ka, attributable to the Earth's eccentricity cycle. The cause of the dominance of this periodicity is not clear, but may be related to a complex internal response of the climate system to orbital forcing. Our initial results indicate that during this time cold periods are less severe than Quaternary glacials while warm intervals are warmer than most Quaternary interglacials. An apparent small-amplitude fluctuation pattern prior to 2.5 Ma may be attributable to a weaker winter monsoon system relative to after 2.5 Ma. The “Red Clay” formation reveals a series of climatic events which may be correlated to land and ocean-based proxies from the other parts of the world, such as: the desiccation of the Mediterranean [23]; Pliocene warmth, Mediterranean sapropels, and the so-called Gelasian stage [36].

LONG-TERM ARIDIFICATION HISTORY OVER THE LAST 2.5 MA

The 2.5 Ma age boundary, near the Matuyama/Gauss reversal and synchronous with intense growth of the polar ice-sheets, marks an important global climatic shift. In China, this boundary marks the limit between the “Red Clay” formation and the overlying loess-palaeosol sequence, and is conventionally considered the Pleistocene/Pliocene boundary [29]. Climate fluctuations after 2.5 Ma are dominated by large amplitude, high frequency glacial-interglacial changes, suggesting a much stronger winter monsoon system and generally greater continental aridity. This shift may be synchronous with a critical uplift stage of the Tibetan Plateau [30, 45].

The Loess Plateau in northern China, with an area of about 430,000 km^2 [29], is located within the zone of the east Asian monsoon [1]. The loess-palaeosol sequences, usually more than 100 m in thickness, have been studied by several research groups [29, 50] in using various methods [18, 19, 32, 44]. Loess-palaeosol sequences from different parts of the Loess Plateau are spatially correlative (Fig. 2), contain 37 major cycles [9] and at least 56 clear soil-forming intervals that represent 112 climatic stages [16]. At the beginning of 1980s research demonstrated that loess stratigraphy correlates very well with the marine $\delta^{18}O$ record [20]. Apparently, global ice-volume variations largely influence loess deposition, and by implication continental aridity and the strength of the winter monsoon. Both loess stratigraphy and the marine $\delta^{18}O$ record clearly display Milankovitch orbital periods [30, 8] (Fig. 3). The time interval from 0.9 to 0.55 Ma is critical and marks a major climatic shift in northern China in both amplitude and frequency domains. Before this time, climatic oscillations were characterized by

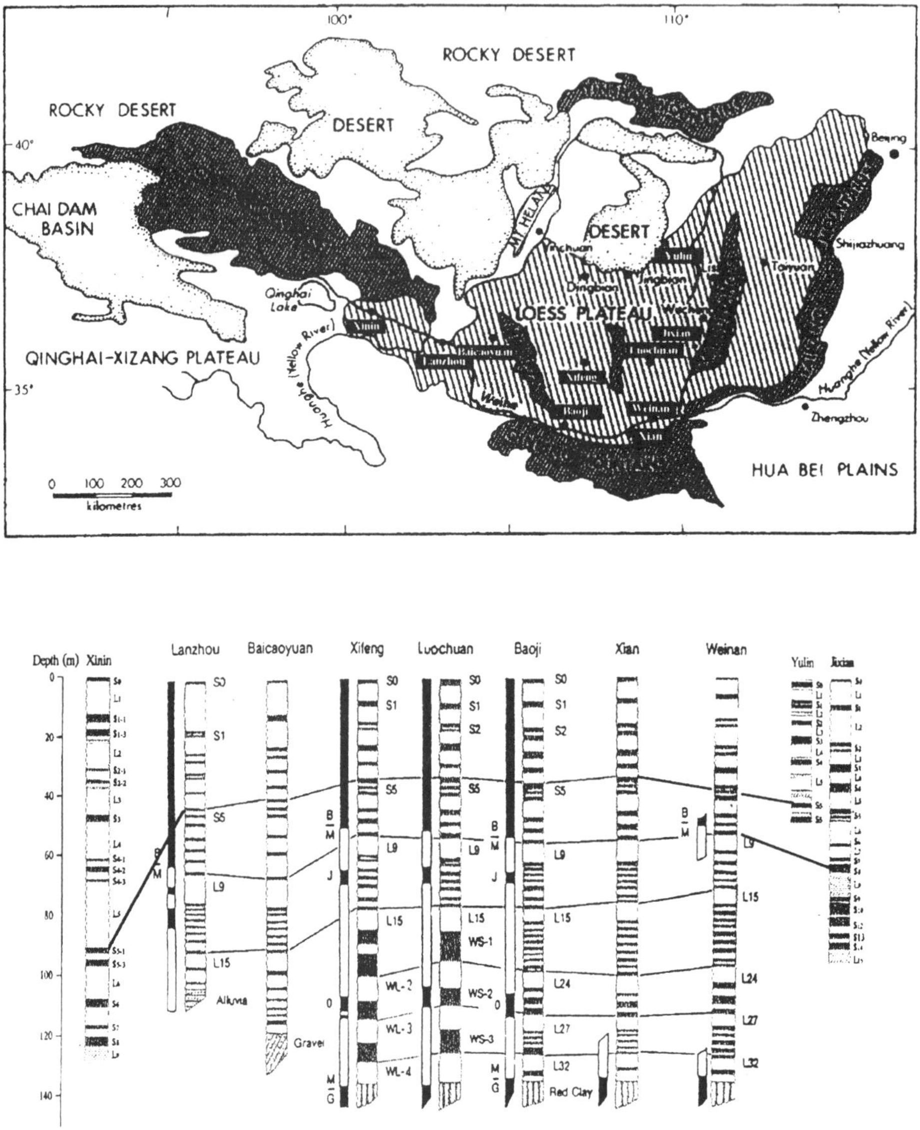

Figure 2. Distribution of loess in China and stratigraphic correlation of loess sections at different sites.

relatively weak glacial-interglacial contrasts [17] with a frequency dominated by the 40 ka obliquity cycle [8]. Oscillations after this interval are dominated by a 100 ka frequency [8] with greater glacial-interglacial contrasts, suggesting that the summer and winter monsoons were stronger [17]. The strengthened monsoon systems may also be associated with another intense uplift stage of the Tibetan Plateau [30]. The climatic shift after about 0.6 Ma is also synchronous with desert expansion in northern China. A comparison between the loess-paleosol sequences of China and climatic proxies in Australia suggests that the intensification of the Australian High may have lead to

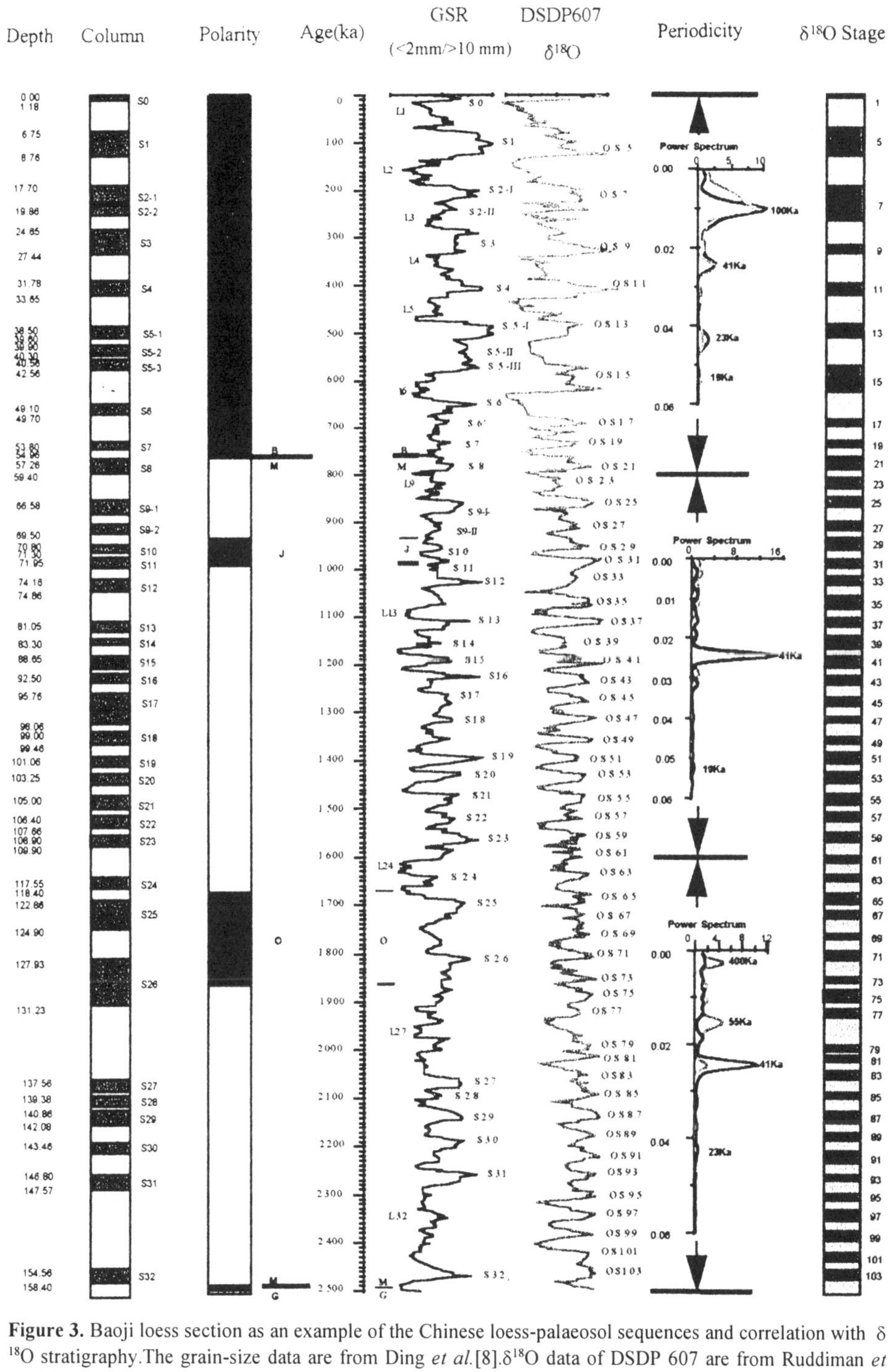

Figure 3. Baoji loess section as an example of the Chinese loess-palaeosol sequences and correlation with $\delta^{18}O$ stratigraphy. The grain-size data are from Ding *et al.*[8]. $\delta^{18}O$ data of DSDP 607 are from Ruddiman *et al.*[38].

aridification and desertification in Australia. Intensifying the Australian High may have played an important role in strengthening the Asian summer monsoon through cross-equatorial circulation. The interaction of the monsoon climates between the southern and northern hemispheres is more evident after 0.6 Ma than before this time [30]. Cross-equatorial transport of energy and moisture may also be an important factor at much shorter time scales, as shown in a comparative study of ice cores from Tibet and Peru (see a review in An *et al.*, in press).

Some important differences exist between the loess-palaeosol record and marine $\delta^{18}O$ data, the causes of which remain enigmatic. A comparison of the frequency domains suggests an ambiguous link for the period from 2.5 to 1.6 Ma. From 2.5 to 0.9 Ma four sandy loess layers, evidence of extremely cold-dry intervals with an approximate 400 ka periodicity, have no counterpart in the $\delta^{18}O$ record. From 0.9 to 0.13 Ma, the most striking irregularity between the two records is the strongly developed S4 and S5 soils, evidence of great warmth in China, which again have no analogues in the $\delta^{18}O$ record. The Brunhes/Matuyama boundary, identified in glacial loess unit L8 (equivalent to isotope stage 20), is located within interglacial stage 21 in the marine record. During the last climate cycle, stages 1 and 5e show comparable $\delta^{18}O$ values, while in our record the pedogenic signature of the Holocene soil (SO) greatly differs from the stage 5e soil (Luvic Phaeozem compared with Chromic Luvisol). In the last climatic cycle, the ~20 ka precessional cycle is significantly stronger in the loess record than throughout most of the $\delta^{18}O$ record. This supports the suggestion that loess deposition is primarily controlled by global ice-volume variations through influencing the strength of the Siberian high, and thus the winter monsoon. Soil formation appears more strongly coupled with low latitude solar insolation changes that influence summer monsoon intensity [30].

An interesting aspect of the geology-climate system in China involves the sea level changes and those in the size of the South China Sea in response to global glacial-interglacial cycles. The eastern coastline of China at the last glacial maximum (LGM) moved eastwards about 1000 km compared to the present-day position; consequently, a large portion of the continental shelf emerged. The South China Sea, the major inland moisture source, became a semi-closed basin [42]. This at least partly explains the apparent increase in continental aridity at the LGM. Moreover, the reduction in size of the South China Sea and its location within the modern west Pacific warm pool must have profoundly affected the role played by the global warm pool. A more pronounced seasonality of climate and greater glacial-interglacial contrasts in SST of the warm pools, compared with the open oceans, may explain an important tropical climate enigma: CLIMAP estimated a less than 2°C cooling of the tropical ocean at the LGM, while terrestrial evidence suggests a 6-8°C decrease. Enhanced seasonality of the marginal sea may explain this land-sea discrepancy [42]. Dramatic glacial-interglacial changes in vegetation, desert expansion and other geo-ecological systems in China must have closely related to these ocean changes [2].

RAPID AND HIGH FREQUENCY CHANGES IN THE LAST CLIMATIC

CYCLE AND IN THE HOLOCENE

High resolution records from Greenland and the North Atlantic have demonstrated that the circum-Atlantic region experienced rapid and large amplitudes climate changes (Heinrich events and Dansgaard-Oeschger cycles) during the last glaciation [7, 3], and probably during the last interglacial [14]. This suggests that the Earth's climate system has been highly unstable. The cause and global extent of these rapid, high amplitude changes remains unclear. Recent evidence from the Loess Plateau has shown that the winter monsoon experienced similar changes that are apparently coupled with iceberg discharge events (Heinrich events) in the North Atlantic [34] and with the Dansgaard-Oeschger cycles in the Greenland ice-cores [10] (Fig. 4). Measurements on palaeo-weathering intensity, primarily dependent on the amount of moisture brought by the summer monsoon, exhibit similar signals in the strength of the east Asian summer monsoon [16]. Fluctuating CH_4 concentrations during similar intervals in Greenland ice-cores [4] support the rapid climate changes in the low-latitudes. Data from Chinese loess also indicate that similar changes occurred during the penultimate glaciation. Climatic signals synchronous with the Heinrich events are also illustrated in several pollen records from the South China Sea [41]. The fact that most of the low weathering events occur at or near the stratigraphic boundary or transition between loess and soil units may be important toward understanding the mechanisms that drive the rapid events [16]. In addition, the instability of the last interglacial period (substage 5e) is contentious. Recent results from the western part of the Loess Plateau [11]support the notion of an unstable climate for this period, although more detailed time controls are needed to clarify this problem. All of these results clearly indicate that the Asian monsoon system is linked with the North Atlantic climate. A great number of ocean [25] and terrestrial records [13] have documented the synchroneity between two step deglaciation and the Younger Dryas event in the North Atlantic region. Moreover, a number of well-dated records have shown that climate fluctuations, from 14 to 10 ^{14}C ka, correlate with SST changes in the Norwegian Sea. Both geologic and modeling results indicate that the southwest and southeast summer monsoons intensified in response to insolation forcing at the beginning of the Holocene. A startling aspect of the Chinese Holocene record is a succession of marine transgressions generally leading to about 200 km of westward movement of the coastline in the Yangtze delta region [40], and a re-opening of the South China Sea [42]. Various geologic records (lake, ice-core, marine sediments and palaeo-deserts, *etc.*) tend to show that the Holocene climate in Asia has been highly unstable and is characterized by several century to millenary scale alternations of humid and arid episodes [28]. For example, a severe episode of aridification occurred near 4.0 - 3.6 ^{14}C-ka BP and seems to be synchronous with changes in the South China Sea SST records (Wang P.X., verbal communication). This episode has been reported in other parts of the world and is thought to have caused the collapse of several important ancient civilizations [43], such as Mesopotamia and Anatolia. The Chinese ancient civilization may also have been affected by this severe aridification event.

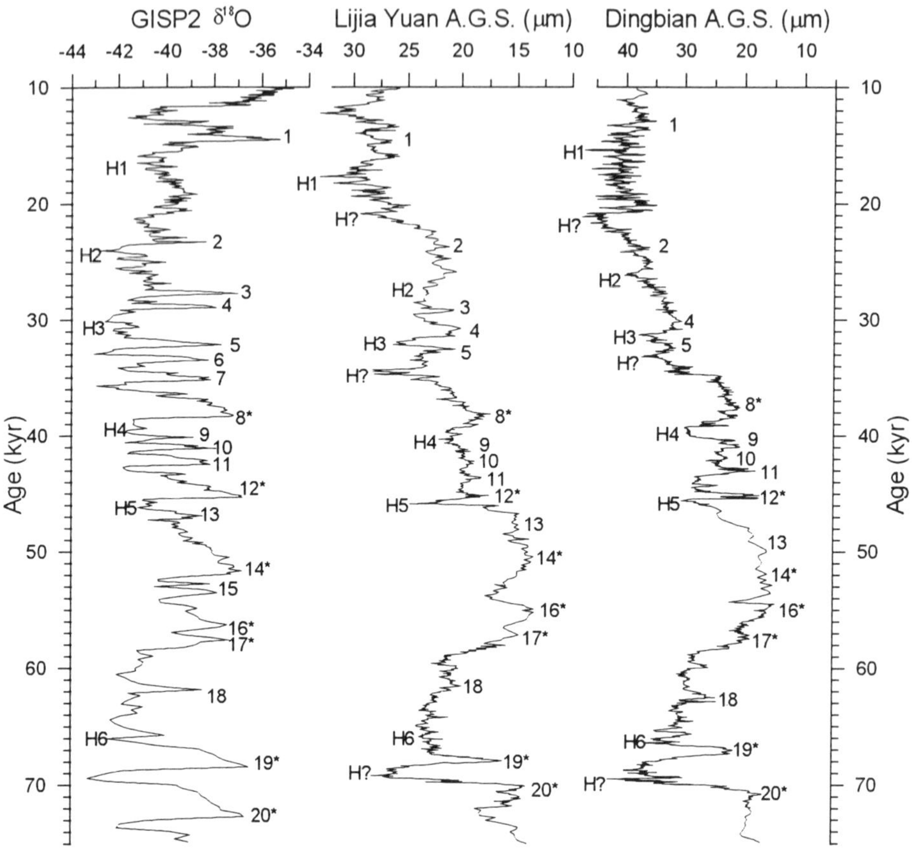

Figure 4. Correlation of Lijiayuan grain-size data and oxygen isotope of GISP 2 (from Ding *et al.*) [10].

The decade to century scale environmental changes during the last several thousand years in China can be reconstructed based on historical records, tree rings, speleothems [46], *etc.* Historical data indicate that the frequency of aeolian dust activity is closely related to alterations in atmospheric circulation patterns [48]. Our high resolution speleothem data indicate that temperature fluctuations in China are well correlated with the Greenland ice-core records, implying once again a strong link between the Asian monsoon climate and the circum-North Atlantic region.

A very startling and disturbing fact is that the Chinese deserts have expanded under the Late Holocene humid interglacial conditions, and that the extent of expansion is close to that during the last glacial maximum. This is an obvious consequence of human activity during historical time. The Saharan desert has responded similarly [33], as well many other sandy deserts on our globe. At question, is how these deserts will evolve in the future? The "green new great wall" (a planted forest belt) in northern China constructed in the last decades provides some encouraging insight for the answer of this question.

Acknowledgements

Acknowledgements are extended to Prof. Zheng M.P. for some of his unpublished materials, and to several colleagues from a number of research and educational institutions for helpful discussions.

REFERENCES

[1] Z. S. An, G. Kukla, S. C. Porter and J. L. Xiao. Magnetic susceptibility evidence of monsoon variation on the loess plateau of central China during the last 130,000 years. *Quaternary Research* **36**, 29-36 (1991)

[2] An Zhisheng, Wu Xihao, Lu Yanchou, Zhang De'er, Sun Xiangjun and Dong Guangrong. Variations of palaeo-environments in China since 20,000 years. In: *Loess, Quaternary Geology, Global Change*. T. S. Liu (Ed.). Science Press, Beijing, 1-26 (1990)

[3] G. Bond, W. S. Broecker, S. Johnson, J. McManus, L. Labeyrie, J. Jouzel and G. Bonanl. Correlations between climate records from North Atlantic sediments and Greenland ice. *Nature* **365**, 143-147 (1993)

[4] J. Chappellaz, T. Blunier, D. Raynaud, J. M. Barnola, J. Schwander and B. Stauffer. Synchronous changes in atmospheric CH4 and Greenland climate between 40 and 8 kyr BP. *Nature* **366**, 443-445 (1993)

[5] Chen Longxun. *The East Asian Monsoons*. China Meteorological Press, Beijing, 1-362 (1991)

[6] M. Y. Chen, W. G. Huo, Y. Y. Yao and M. R. Shao. Antarctic ice-sheet and aeolian deposits in China. *Quaternary Science* **3**, 261-271 (1990)

[7] W. Dansgaard, S. J. Johnsen, H. B. Clausen, D. Dahl-Jensen, N. S. Gundestrup, C. U. Hammer, C. S. Hvidberg, J. P. Steffensen, A. E. Svelnbjornsdottir, J. Jouzel and G. Bond. Evidence for general instability of past climate from a 250-kyr ice-core record. *Nature* **364**, 218-220 (1993)

[8] Z. L. Ding, Z. W. Yu, N. W. Rutter and T. S. Liu. Towards an orbital time scale for Chinese loess deposits. *Quaternary Science Review* **13**, 39-70 (1994)

[9] Z. L. Ding, D. S. Liu, X. M. Liu and M. Y. Chen. Thirty-seven climatic cycles in the last 2.5 Ma. *Chinese Science Bulletin* **35**, 667-671 (1990)

[10] Z. L. Ding, J. Z. Ren, T. S. Liu, J. M. Sun and X. Q. Zhou. Irregular changes in the monsoon-desert system on millennial time scales and the mechanisms. *Science in China (Series D)* **26**, 385-391 (1996)

[11] X. M. Fang, X. R. Dai, J. J. Li, J. X. Cao, D. H. Guan, Y. P. Hao, J. L. Wang and J. M. Wang. Instability in the evolution of the Asian monsoon. *Science in China (Series D)* **26**, 154-160 (1996)

[12] B. P. Flower, J. P. Kennett. Middle Miocene ocean-climate transition: high-resolution oxygen and carbon isotopic records from Deep Sea Drilling Project site 588A, southwest Pacific. *Palaeoceanography* **8**, 811-843 (1993)

[13] F. Gasse, M. Arnold, J. C. Fontes, M. Fort, E. Glbert, A. Huc, B. Y. Li, Y. F. Li, Q. Liu, F. Melieres, E. Van Campo, F. B. Wang, Q. S. Zhang. A 13,000-year climate record from western Tibet. *Nature* **353**, 742-745 (1991)

[14] GRIP Members. Climate instability during the last interglacial period recorded in the GRIP ice-core. *Nature* **364**, 203-207 (1993)

[15] Z. T. Guo. *Succession des palaeosols et des loess du centre-ouest de la Chine, approche micromorphologique* [unpublished thesis]. Univ. Paris VI, France, 1-266 (1990)

[16] Z. T. Guo, Z. L. Ding, T. S. Liu. Pedo-sedimentary events recorded in loess of China and Quaternary climatic events. *Chinese Science Bulletin* **41**, 56-59 (1996)

[17] Z. T. Guo, D. S. Liu, N. Fedoroff and Z. S. An. Shift of monsoon intensity on the Loess Plateau at *ca.* 0.85 Ma BP. *Chinese Science Bulletin* **38**, 586-591 (1993)

[18] J. M. Han, H. Y. Lu, N. Q. Wu, Z. T. Guo. The magnetic susceptibility of modern soils in China and its use for palaeoclimate reconstruction. *Studia Geophysica and Geodaetica* **40**, 262-275 (1996)

[19] J. M. Han, W. Y. Jiang, T. S. Liu, H. Y. Lu, Z. T. Guo and N. Q. Wu. Carbon isotopic records of climate changes in Chinese loess. *Science in China (series D)* **39**, 460-467 (1996)

[20] F. Heller and T. S. Liu. Magnetostratigraphical dating of loess deposits in China. *Nature* **300**, 431-433 (1982)

[21] D. A. Hodell and F. Woodruff. Variations in the strontium isotopic ratio of seawater during the Miocene: stratigraphic and Geochemical implications. *Palaeoceanography* **9**, 405-426 (1994)

[22] S. A. Hovan and D. K. Rea. Post-Eocene record of eolian deposition at sites 752, 754, and 756, eastern IndianOcean. *Proceedings of the Ocean Drilling Program, Scientific Results* **121**, 219-240 (1991)

[23] H. J. Hsu, W. B. F. Ryan and M. B. Cita. Late Miocene desiccation of the Mediterranean. *Nature* **242**, 240- 244 (1973)
[24] M. R. W. Johnson. Volume balance of erosional loss and sediment deposition related to Himalayan uplift. *Journal of Geological Society* **151**, 217-220 (1994)
[25] H. R. Kudrass, H. Erlenkeuser, R. Vollbrecht and W. Weiss. Global nature of Younger Dryas cooling event inferred from oxygen isotope data from Sulu Sea cores. *Nature* **349**, 406-409 (1991)
[26] R. M. Leckie and P. N. Webb. Late Oligocene-Early Miocene glacial record of the Ross Sea, Antarctica: evidence from DSDP site 270. *Geology* **11**, 578-582 (1983)
[27] J. J. Li, S. X. Wen, Q. S. Zhang, F. B. Wang, B. X. Zheng and B. Y. Li. Timing, amplitudes and behaviour of the Tibetan Plateau uplifting. *Science in China* **6**, 608-616 (1979)
[28] P. M. Liew and Q. Y. Huang. A 5000-year pollen record from Chitsai Lake, central Taiwan. *Terrestrial, Atmospherics and Ocean Sciences (TAO)* **5**, 411-419 (1994)
[29] T. S. Liu. *Loess and the Environment*. China Ocean Press, Beijing, 1-251 (1985)
[30] T. S. Liu, Z. L. Ding. Stepwise coupling of monsoon circulation to global ice volume variations during thelate Cenozoic. *Global and Planetary Change* **7**, 119-130 (1993)
[31] X. M. Liu, T. C. Xu and T. S. Liu. A study of anistropy of magnetic susceptibility of loess from Xifeng. *Geophys. J. R. Astr. Soc.* **92**, 349-353 (1988)
[32] H. Y. Lu, N. Q. Wu, T. S. Liu, J. M. Han, X. G. Qing, X. J. Sun and Y. J. Wang. Seasonal climatic variations recorded by phytolith assemblages from Baoji loess sequence in central China over the last 150,000 a. *Science in China (Series D)* **26**, 131-136 (1996)
[33] N. Petit-Maire. Will greenhouse green the Sahara? *Episodes* **13**, 103-107 (1990)
[34] S. Porter and Z. S. An. Correlation between climate events in the North Atlantic and China during the last glaciation. *Nature* **375**, 305-308 (1995)
[35] W. L. Prell and J. E. Kutzbach. Sensitivity of the Indian monsoon to forcing parameters and implications for its evolution. *Nature* **360**, 647-652 (1992)
[36] D. Rio, R. Sprovieri and E. D. Distefano. The Gelasian stage: a proposal of a new chronostratigraphic units of the Pliocene series. *Riv. It. Paleont. Strat* **100**, 103-124 (1994)
[37] W. F. Ruddiman, W. L. Prell and M. E. Raymo. Late Cenozoic uplift in southern Asia and the American West: rational for general circulation modeling experiments. *Journal of Geophysical Research* **94**, 18379- 18391(1989)
[38] W. F. Ruddiman, M. E. Raymo, D. G. Martinson, B. M. Clement and J. Backman. Pleistocene evolution: northern hemisphere ice sheets and North Atlantic Ocean. *Palaeoceanography* **4**, 353-412 (1989a)
[39] N. J. Shackleton and M. A. Hall. Stable isotope records in bulk sediments (Leg 138). *Proceedings of the Ocean Drilling Program, Scientific Results* **138**,797-805 (1995)
[40] Y. F. Shi. Mid-Holocene climates and environments in China. *Global and Planetary Change* **7**, 219-233 (1993)
[41] Sun Xiangjun. Pollen records of environmental evolution in the South China Sea. In: *The South China Sea over the last 150,000 years*. P. X. Wang (Ed.). pp. 65-74. Tongji University Press (1995)
[42] Wang Pinxian. *The South China Sea Over the last 150,000 years*. Tongji University Press, 1-184 (1995)
[43] H. Wess, M. A. Courty, W. Wetterstrom, F. Guichard, L. Senior, R. Meadow and A. Curnow. The genesis and collapse of Third Millennium North Mesopotamian civilization. *Science* **261**, 995-1004 (1993)
[44] N. Q. Wu, D. Rousseau, T. S. Liu. Land Mollusk records from the Luochuan loess sequence and their palaeoenvironmental significance. *Sciences in China (Series D)* **39**, 494-502 (1996)
[45] X. H. Wu and Z. S. An. Loess-palaeosol sequence on Loess Plateau and uplift of the Tibetan Plateau. *Science in China (Series D)* **39**, 121-133 (1996)
[46] D. X. Yuan, B. Li and Z. H. Liu. Karsts in China. *Episodes* **18**, 62-65 (1995)
[47] L. P. Yue. Palaeomagnetic polarity boundary were recorded in Chinese loess and Red Clay and geological significance. *Acta Geophisica Sinica* **38**, 311-320 (1995)
[48] D. E. Zhang. Synoptic-climatic studies of dust fall in China since the historic times. *Scientia Sinica* **27**, 825-836 (1984)
[49] Zhang Jiacheng and Lin Zhiguang. *Climate in China*. Meteorology Press, Beijing, 1-325 (1987)
[50] Z. H. Zhang and Y. Chen. New developments in the study of loess in China. *Episodes* **18**, 58-61(1995)
[51] H. B. Zheng, Z. S. An and J. Shaw. New contributions to Chinese Plio-Pleistocene magnetostratigraphy. *Physics of the Earth and Planetary Interiors* **70**, 146-153 (1992).

Proc. 30th Intern. Geol. Congr. Vol. 1, pp. 27-45
Wang *et al.* (Eds)

Sustaining Our Life Support System

PETER J. COOK
British Geological Survey, Keyworth, Nottingham, NG12 5GG, UK.

Abstract

Earth scientists have made a major contribution over the last century to extracting materials from the ground. Many of those same skills will be used to be applied to put materials into the ground in a more intelligent and more sustainable manner, whether that material is water, nuclear or toxic waste or perhaps even CO_2. We also need to refocus some of the available earth science expertise so that we can develop a better understanding of soil formation and erosion, surficial processes and landscape evolution. Universities, research institutions and geological surveys must become more concerned about the surface and near surface issues and perhaps less concerned with the core and lower mantle. Finally, the earth scientist must be active in Our life support system is the consequence of, and dependent on a complex interplay of physical chemical and biological processes. Those processes are being increasingly perturbed by human activity. A common perception is that the earth scientist is mainly concerned with "the problem" through involvement in resource extraction but in fact the earth sciences have a key role to play in "the solution". Contributing to solving problems of sustainability requires earth science involvement in determining the nature magnitude and rate of change, both pre-anthropogenic and anthropogenic. But more than this, earth scientists can contribute directly to the resolution of many identified problems whether in the urban environment, the coastal zone, the subsurface or the atmosphere. But this will require the application of existing skills in new ways. For example seeking a strengthened dialogue with other disciplines including planners, environmentalists, sociologists, economists, and many others if we are to sustain our life support system for the benefit of present and future generations.

Keywords: life support system, water, food, air, living space, minerals, energy

INTRODUCTION

The complex interplay between physical, chemical and biological processes, whether at the micro, macro or global scale, requires that we take a holistic approach and examine total systems rather than just one or other element of the system. Human activities are for the most part just normal physical, chemical and/or biological processes yet, at the same time, they can have some distinctive features compared to natural processes, in particular the speed or the scale at which they may occur. Additionally, there are some materials, for example certain chemicals, and increasingly certain genetically modified biological materials, which are entirely or largely the consequence of human intervention. Therefore it is not unreasonable to see human activity as a distinctive process which has some unique features and consequences. Together it is the interactions of these four processes, physical, chemical, biological and human, which not only define the form of processes but also the long-term viability or sustainability of processes and their consequences. The earth sciences are pivotal to understanding the interaction between these processes because they impact upon Earth materials and because Earth materials are themselves a consequence of those same interactions and processes. In some cases the intervention of human activity is of no consequence either because that activity is compatible with the natural world or because it is minuscule in scale compared to natural processes. However, in some cases the impact of human activity is so profound that it may call into question the sustainability of a natural process, an ecosystem or life itself.

In recent years the concept of sustainability has been spelt out by the World Commission on Environment and Development as meaning that state when humanity ensures ... it meets the needs of the present without compromising the ability of future generations to meet their own needs ...[1]. At its most simplistic this has been seen as meaning we must ensure we do not cheat on our great grandchildren! At the other end of the spectrum, the precise definition of the term has become the subject of a great deal of discussion. In particular, how is it possible to decide what is best for future generations? Is it best to leave a resource untouched for that future generation or is it better to exploit that resource, create wealth and use that wealth for an infrastructure from which future generations will benefit?

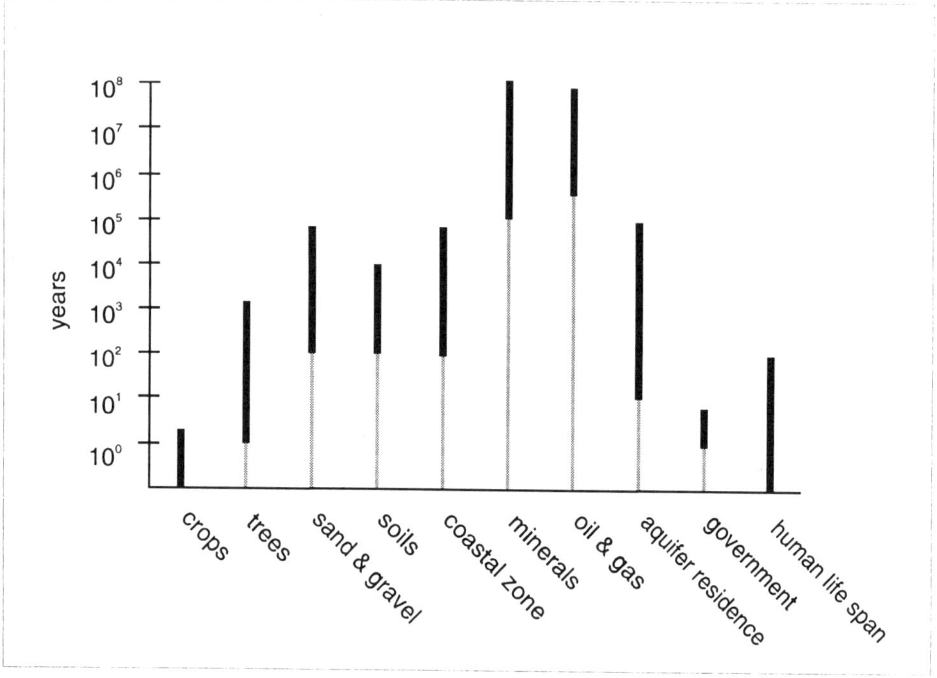

Figure 1. "Renewability"

The concept of sustainability is perhaps also flawed to some extent by the fact that there is no attempt to build a time scale into the concept other than the mention of future generations. If we consider renewability we could argue that, given enough time, all resources will be renewed and therefore, in theory at least, their exploitation is sustainable (Figure 1). But to try to set sustainability within geological time is disingenuous and we must surely see sustainability within a human time frame which could be seen as little as tens of years or as much as a few thousand years.

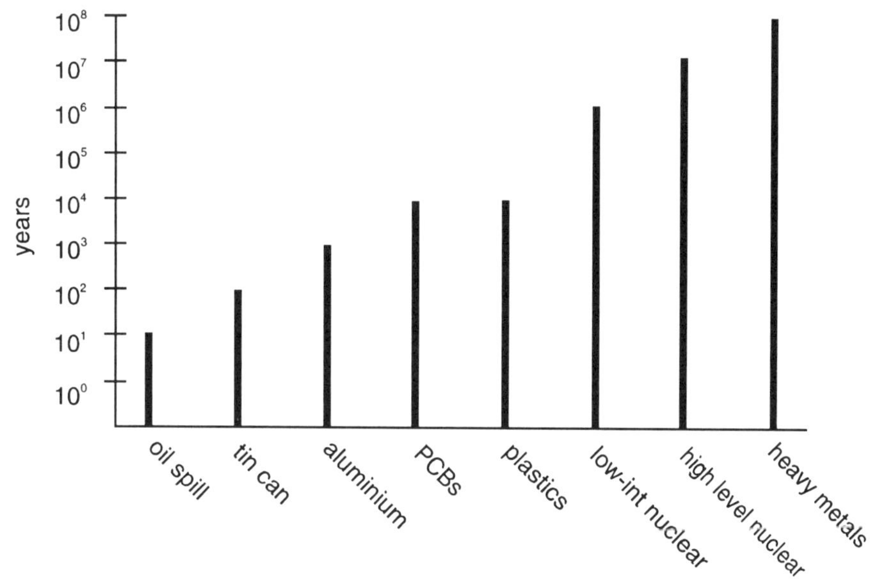

Figure 2. Lifespan of contaminants.

Set against this time frame the present generation becomes much more accountable for its actions. Consider the concept of pollution within this time frame (Figure 2). When set against the life span of a human, we are introducing a wide range of pollutants into the natural environment that will be there for time periods that are several orders of magnitude greater than the human life span - or the average term of a government charged with considering the problem. Therefore, again, given that the environments in which these pollutants are being deposited have only limited carrying capacities, it is likely that we are not disposing of many of our wastes in a sustainable manner at the present time. Indeed there are two global trends which suggest that unless we get smarter we will be much worse off in the future. The first of these trends is the inexorable growth in the world population. The present world population is about six billion; it will not stabilize until around 2100 and by that time will be about ten billion. Most of the extra 4 billion people born in the next 100 years will be in developing countries. The other trend that will have a profound effect on future sustainability issues is that of the rising expectation that people have; they want more and better food; a better education; better health; better roads and more material possessions, whether cars, refrigerators or television. The net result is likely to be a greater rate of usage of the world's resources and greater degradation of the environment. Clearly if this happens, rather than people enjoying an improved standard of living, or quality of life, there will be net degradation of the human life support system.

Let us therefore examine some of the elements of that system. More than two thousand years ago the Greeks saw the four primary elements as earth, air, fire and water, and in many ways they can still be regarded as the building blocks of the system, or the necessities for all life on earth to flourish. However, I would prefer to think of them in terms of water, food, air, living space, energy and minerals. These are what we need to sustain our life support system and these are the components of the system that I would like to consider briefly from an earth science perspective, examining likely future trends and the manner in which the earth sciences might impact upon them.

WATER

Water is our most vital earth resource; it is also the most ubiquitous. However, freshwater - water fit for human consumption and use - is much less common. It comprises only 2.5% of the world's water [5] and of that more than two-thirds is locked up in glaciers and permanent snow cover (Figure 3). The remainder comprises 30% fresh groundwater and only 1% or less of surface water (lakes, rivers, swamps, soil moisture).

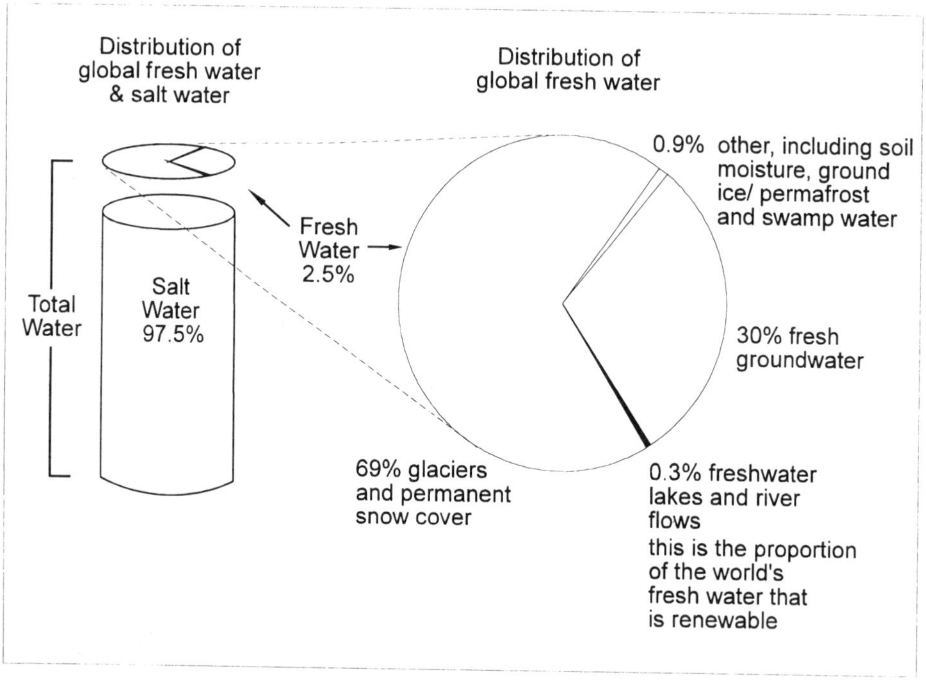

Figure 3. The World's water (modified after Gleick [5]).

In other words, of the potentially available freshwater more than 95% is groundwater and less than 5% is surface water, yet it is this surface water that we use as our primary source of water for human consumption. But whilst water availability is the key, it is also important that it is safe drinking water. The 1977 resolution resulting from the United Nations Conference held at Mar del Plata spelt out this need when it stated that "...all people have the right of access to drinking water in quantity and of a quality equal to their basic needs . ." The importance of this is highlighted if we look at the impact of access to safe drinking water and child mortality (Figure 4). There is a clear correlation; the more restricted the access to safe water the higher the proportion of children who die before reaching their fifth birthday [3]. What are future trends likely to be? Gleick [5] believes that after a recent improvement because of various United Nations initiatives, the situation will become progressively worse with, by the year 2000, more than 2 billion people not having access to safe drinking water,

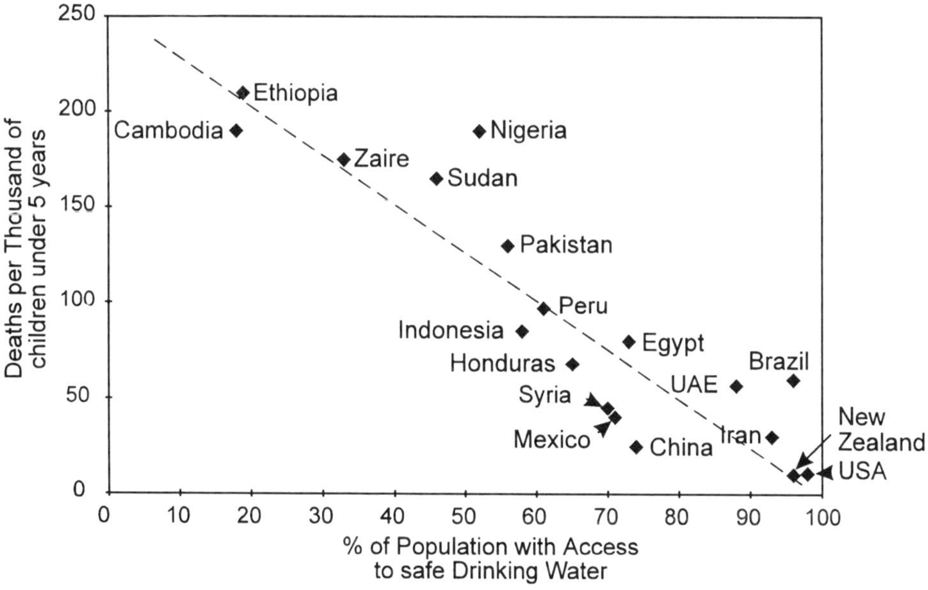

Figure 4. Water quality and child survival [3]

and almost 3 billion not having adequate sanitation (Figure 5). Similarly in a study by Population Action International [3] using the example of India, it is shown that even at the lowest potential birth rate (1.56 children per woman by the year 2025) there will be a significant level of water stress (defined as less than $150m^3$ of available fresh water per person per annum) by the year 2020 (Figure 6). With the low projection this will improve by about 2070. However, there is no improvement based on the medium projection and, for the high projection, India will enter a period of water scarcity (less than $1000m^3$ of water per person per annum) by 2050, by which time the population of India will be over one billion people. Against this background, it must be remembered that polluted water is one of the most common routes for the spread of infectious diseases and that nearly half the world's population suffers from water-borne diseases. Clearly then, there needs to be a strategy for tackling the issue of access to water, recognising that whilst that strategy may be global the action must be local. This is a highly complex issue that cannot be adequately dealt with here, but the estimate by Postal [8] gives us a starting point and graphically illustrates (Figure 7) that agriculture is the dominant user of water and that increased efficiency of 10% in that usage would provide a potential increase of 75% in the amount of water available for municipal/domestic supply. Needless to say it is not as simple as this in that the water used for agriculture may be far from where the municipal need exists. Also, it is no simple matter to increase agricultural efficiency by 10%, though undoubtedly pricing of water to reflect its true value could be an important instrument for doing this. As pointed out earlier, for many parts of the world, groundwater is the cheapest, most reliable and cleanest source. But in many areas that resource is being over-exploited. Obviously the earth scientist has a particular role in ensuring a more sensible approach to that exploitation. Let us consider just one example of how this might be done in both the developed and the developing world - artificial aquifer recharge. During the drier part of the year there is significant drawdown of groundwater; during the wet season there is some recharge but that recharge may be insufficient to fully restore the water table and consequently there is progressive drawdown. The earth scientists have for the most part concentrated on strategies to understand those processes of natural recharge and drawdown. However to date little effort has been put into establishing the feasibility and cost-effectiveness of artificially recharging the aquifer by pumping the excess wet season run-off back down into the aquifer thus enhancing the rate of recharge. There are potential problems in terms of biological and chemical activity

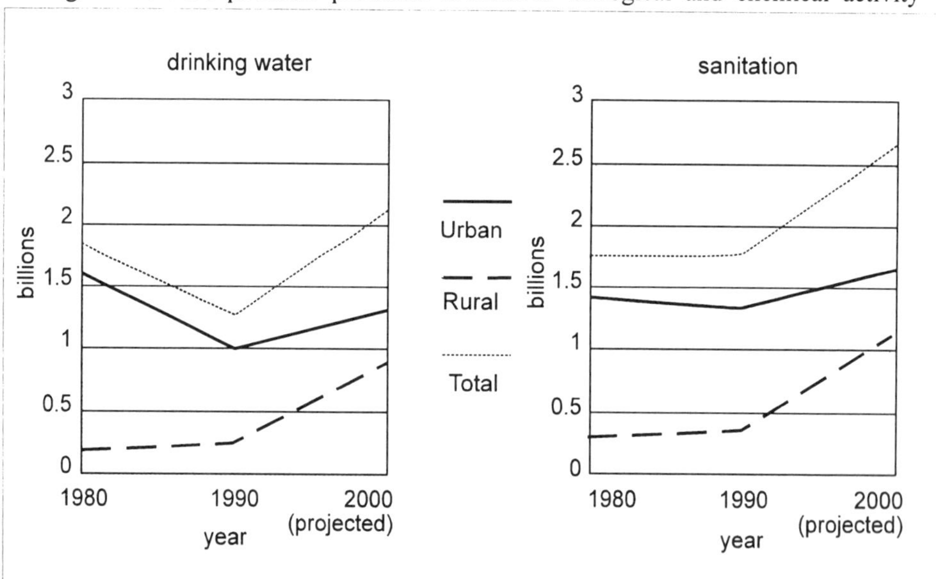

Figure 5. Population lacking access to safe drinking water and sanitation in developing countries, 1980-2000 (Modified after Gleick [5]).

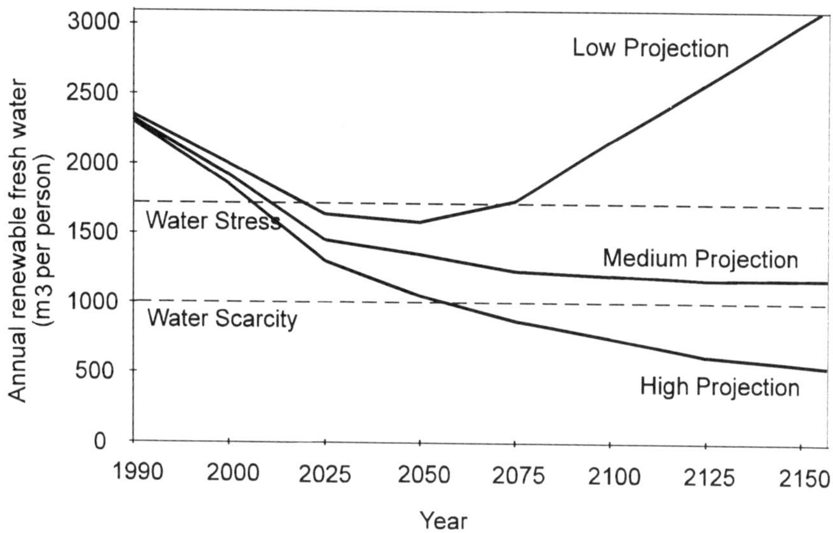

Figure 6. Annual renewable fresh water per person under UN population projections-India. (modified after Population Action International [3]).

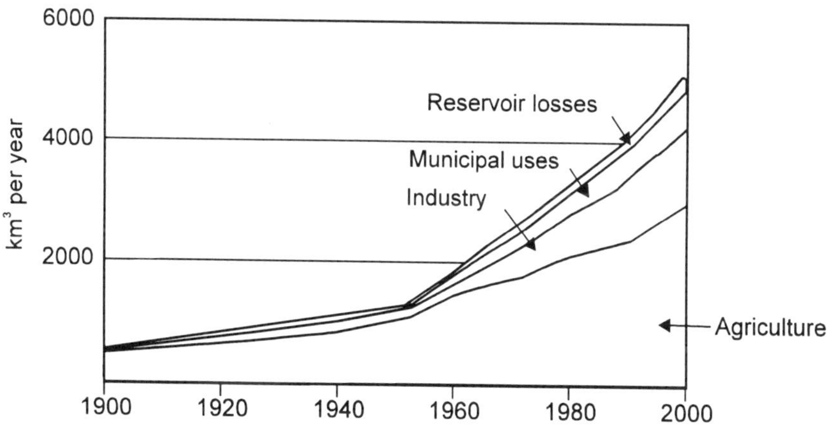

Figure 7. Estimated annual world water use (modified after Postel [8]).

within the artificially recharged aquifer system and it must be used selectively. To date earth scientists and particularly hydrogeologists have been concerned primarily with how to extract water more efficiently from the ground; we must now start to understand better how to put that water back into the ground in order to provide a more sustainable supply of clean water. The message must be conveyed to governments and water companies that not only is groundwater the cheapest, most reliable source of high quality water in many parts of the world but that most of the available storage capacity for freshwater is not as-yet-unflooded valleys but in surficial sediments and rocks. The earth scientist has a key role in getting that message across.

FOOD

It was Jonathan Swift who put the importance of this role into perspective when he said "...Whoever could make two ears of corn or two blades of grass to grow upon a spot where only one grew before would deserve better of mankind and do more essential service to his country than the whole race of politicians put together..."!

I suspect the common reaction of most earth scientists to the issue of ensuring security of food supplies for the world's growing population is to consider it to be somebody else's problem. But food security requires soil security: it commonly requires water for irrigation and usually requires food minerals (phosphate, potash, sulphur). Therefore geology has a fundamental role to play in many of the food issues that are affecting large parts of Africa, Asia and South America.

Swift also put the problem into perspective by emphasising the need to grow twice as much on the same spot, rather than merely growing more elsewhere, for only about 13% of the world's surface is arable land and man cannot continue to turn forests or deserts into arable land. Part of the response to the need to feed the world's growing population has been to increase the amount of irrigation (Figure 8). The other response has been through the so-called green revolution and the development of new strains of rice and other grains. This, in turn, has required increased usage of fertilisers. As a result the use of phosphatic fertilisers has increased by 500% over the last 50 years though it is evident from Figure 9, that there has been a significant decrease in usage over the past five years, perhaps a consequence of the removal of subsidies for fertiliser usage in the countries of the former Soviet Union.

But despite all this, world grain stocks have decreased in recent years and there are indications that global levels of productivity may no longer be increasing and may in fact be falling. One reason for this is the increase in salinization of land as a result of inappropriate large-scale irrigation schemes. In addition, land clearance and replacement of trees by shallow-rooted crops or use of cleared land for grazing are also responsible for salinization. Salinization as a result of intrusion of saline sea water in over-used aquifers in coastal regions or the clearance of mangrove swamps is also a feature of many coastal areas, Ghassemi [4] estimate that about 6% of the world's total cultivated land is now suffering from human-induced salinization. But whilst

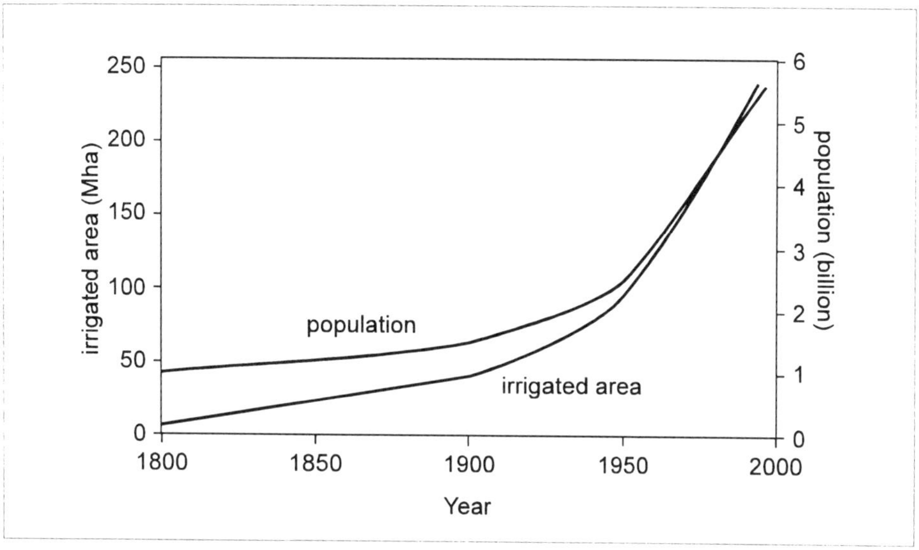

Figure 8. World population and irrigated area [4].

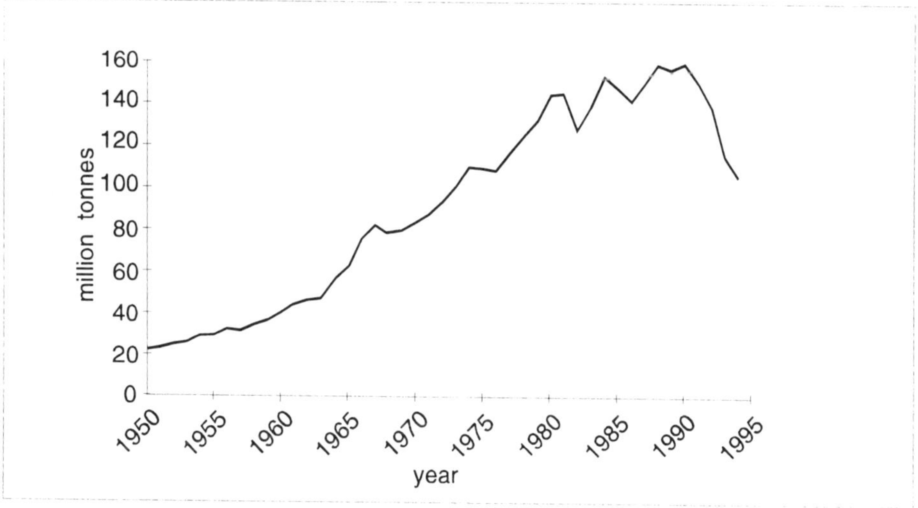

Figure 9. World phosphate rock production [12].

salinization is a very significant cause of soil degradation and loss of productivity it is certainly not the only one; soil erosion due to water or wind, chemical, weathering or physical degradation contribute to the total amount of human-induced soil degradation which may add up to as much as 15% of the total area of cultivated land. Therefore if we are not careful, attempts to grow even more food will lead to even more soil erosion which will ultimately lead to less food production rather than more. Thus, for example, in China the rate of soil erosion which is probably of the order of 0.1 to 2 tonnes of soil per hectare per annum may rise to as much as 200 tonnes per hectare in cultivated fields

[7]. Earth scientists have a role to play in assisting with the task of growing more food in existing agricultural areas by helping to develop strategies to minimize soil degradation. Improved agricultural practices are obviously vital here and these may be more dependant on financial incentives (or disincentives) than science. Nonetheless, the geologist along with the biologists, chemists, engineers and many others must work together to better understand the sensitive and complex ecosystems that make up soils; mineralogy, geochemistry, geotechnics, sedimentology etc can make a contribution to this. The documentation of present and past (pre-anthropogenic) rates of erosion must be documented and again the earth scientist has a particular role to play. However this may require a change of emphasis in the earth sciences: at the present time most geologists know more about the Earth's core and mantle than they know about soil and landscapes. Earth science departments in the universities spend more time teaching students about subduction zones than surficial sediments and weathering. If the earth sciences are to make a significant contribution then soils surfical processes and landscape development must be accorded a higher priority in the earth sciences generally and by the universities in particular.

AIR

Just as in the case of food, the reaction of most geologists when asked about issues relating to air quality or the greenhouse effect or the problems of CFCs is to feel that for the most part they are somebody else's problem - the atmospheric chemist or the atmospheric modeller perhaps? A geologist is often more inclined to associate with the problem such as the use of fossil fuels producing greenhouse gases rather than feel part of the solution. One of the important contributions that the geologist has made is to use the geological record to establish natural variability in the composition of the atmosphere. This is an important issue, but of course for most people living in the urban environment the most pressing problem is not greenhouse gases but air quality. Geology can help in a marginal way here by providing alternative energy sources, the most obvious one being the increased use of liquefied petroleum gas (LPG) as a fuel for motor vehicles. However, here the issue is more one of taxation and the attitude of the governments than the need to identify new resources of LPG. Within the urban environment, geology, and particularly geochemistry, have a valuable role to play in documenting the problem. For example, work by the British Geological Survey (BGS) and Imperial College, London, has shown that lead is concentrated not along roads but at the intersections of roads, where the cars are stationary but with their engines still producing exhaust fumes. Increasingly countries are insisting on the use of lead-free fuels and/or the use of catalytic converters, but this too may have its problems for the BGS-Imperial work in the London area suggests that we are now starting to see an increased concentration of finely particulate platinum.

Looking more creatively at the potential role in alleviating some of the problems, an increased use of tunnels for vehicles in urban areas might provide scope for large scale filtration and cleaning of exhaust fumes in a way that is not possible for individual vehicles. There are of course other benefits in using the sub-surface more, and I believe that the more intelligent use of the sub-surface, through tunnelling, is going to be one of the major developments of the future. Perhaps even more futuristically, the BGS and a number of partners have been looking into the feasibility of deep underground disposal of CO_2 from fossil-fuel fired power stations. The research to date has shown that it is technically feasible to dispose of CO_2 in this way. The CO_2 will dissolve in formation waters or displace the water and fill the existing pore space. It will then commonly react with alumino silicates precipitating calcite which then acts as a seal for any free CO_2 within the reservoir rock. Holloway et al, [6] have shown that all of the CO_2 from Western Europe's power stations could be disposed of in this way in the rocks underlying the North Sea for the foreseeable future. The problem is one of cost, and in particular the cost of separating out the CO_2 from the flue gases. Nonetheless, the method will shortly be utilized by Norway, with CO_2 from the Sleipner West gas field being pumped into the reservoir rocks of the Utsira Formation. It is the taxation system, and in particular the imposition of a greenhouse tax by Norway, which makes the whole process economically viable. Nonetheless, it may be an indication of a future strategy and serves to indicate that earth sciences may have an important role in the future in providing some of the solutions rather than just being part of the problem.

LIVING SPACE

As pointed out earlier the fundamental problem facing the world is the growth in population. This in turn impacts on issues such as supplies of food and water. However, there is also a physical problem of housing all these people; of providing them with adequate living room and a congenial environment. The most dramatic manifestation of population growth has been in the growth of the cities and particularly megacities (that is those with a population of 10 million or more). By the year 2000 it is estimated that there will be 23 megacities in the world, the largest being Mexico City with a population in excess of 25 million. Of these 23 cities more than half will be in the developing countries. This growth of urbanization brings with it a complex interplay of land-use issues relating to resources and building materials, changing water tables, pollution, waste disposal and over-exploitation of the coastal zone. Indeed the problem of urban growth cannot be separated from the problems of the coastal zone [2], for this is where urban development is preferentially taking place; this is where 17 out of the 23 megacities are located. Therefore to the issues must be added the problem of rising sea levels, though in many cases the problem is more the consequence of land subsidence and compaction in urban areas than of eustatic sea level rise associated with global warming. Eustatic rise of the order of one metre will potentially affect over 200 million people including more than 70 million people in Bangladesh and about the same number in China. Indeed a rise in sea level of one metre would result in the loss of 17.5% of the total area of Bangladesh [10].

But it is not only present-day activities in our cities or the future consequences of sea-level rise or climate change which impact upon our living space; it is also our past activities. For example, in Britain there are 40,000 hectares of derelict land at the moment, the largest single element in this being general industrial dereliction (about a quarter of the total), but mine dumps, excavations and quarries and areas of mining subsidence contribute a significant proportion of the total dereliction. Not all of this dereliction is documented; geological, geophysical and geochemical surveys in urban and adjacent areas are needed if we are to fully comprehend the scale of the urban problem and start to develop strategies for its resolution. But the earth sciences are only a small part of the solution. Our approach to issues in the cities and in the coastal zone has been fragmented. There must be more communication, more co-ordination and more integration not only between the natural sciences but between and within the social sciences. It must also be recognised by geologists that they should be less concerned with undertaking research in nice places and more with tackling the issues of urban areas.

ENERGY

Energy is almost as essential to humanity as food or water. Consequently, as the human population increases and as the aspirations of that population increases, so the use of energy rises inexorably. Figure 10 illustrates that rise schematically. Some developments, for example telecommunications, increased environmental awareness and perhaps the Internet have perhaps worked to slow that growth but such inhibitors have been overwhelmed by drivers such as population growth, rising living standards, global markets and travel.

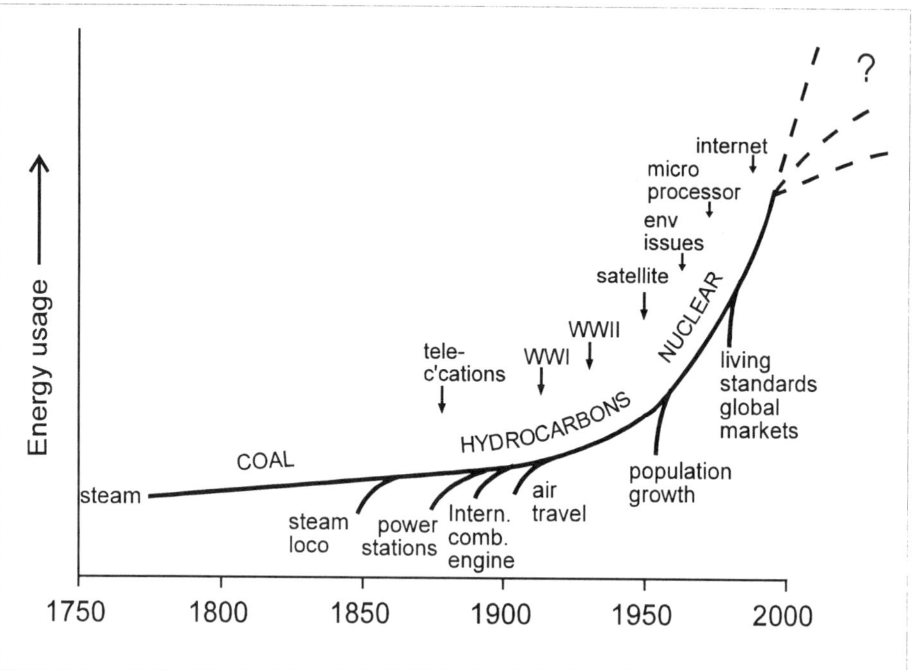

Figure 10. Energy usage, "modifiers" and "drivers".

To date the role of the earth sciences has been primarily in identifying new reserves of fossil fuels to meet that demand and there can be no question that as fossil fuels will continue to be the key to meeting the world's energy demands until well into the next century, the earth sciences will have an increasingly pivotal role. But this will move beyond the conventional role of exploring for oil and gas to improving secondary and tertiary oil and gas recovery. This will be achieved through a better understanding of reservoirs, through improved monitoring of reservoir processes during production and by improved management of oil reservoir depletion generally. Techniques such as applied seismic anisotropy will be particularly important. Obviously continuous improvement in exploration techniques will also be important. However, we may see increasing stress placed on natural gas reserves as these offer a cleaner fuel. Gas hydrates constitute an enormous potential resource of 10^{13} to 10^{14} gigatonnes of carbon. Nuclear power is likely

to continue to be important, perhaps more important in the future. Here the role of the earth scientist is crucial in determining how to dispose safely of the radioactive waste, with deep disposal the most appropriate strategy—here again we see the geologist more involved in intelligently putting material into the ground than with extracting it from the ground.

But perhaps all these geological issues will become irrelevant as fossil (and nuclear) fuels are replaced by renewable energy? Some of these, such as geothermal (or hot dry rock), in fact have an important geological element in them and there are geological problems still to be overcome. Hydroelectric, tidal wave and wind energy certainly have a contribution to make to energy supply, but they have a physical or visual impact which is not always welcome. Bio fuels may make a contribution, but as pointed out earlier, the area of arable land is limited and any attempt to grow more crops on it for energy will result in increased soil exploitation and degradation. Wood is still the prime energy source for 2 billion people; it is also probably the single most important reason for desertification, salinization and coastal erosion in Africa and Asia.

Therefore, perhaps solar power constitutes the main hope for the future? In hot climates the energy payback period may be as little as one to two years and the average life span of a solar panel may be as much as 20 years. There are also benefits such as there being no need to have expensive electricity distribution systems for rural areas. However, there is an important geological dimension to any major move to photovoltaic cells whether silicon, cadmium telluride or copper indium selenite and this is the source of the raw materials. If, for example, we were to attempt to use cadmium telluride thin film photo voltaic cells to meet the whole of the world's current energy needs (10 terrawatts) then we would use up all of the world's existing cadmium reserves, all of its identified tellurium reserves several times over and most of its known molybdenum reserves. Whilst this is something of a caricature in that it is inconceivable that we would shift to solar power to this extent, it does nonetheless serve to illustrate the point that whatever the source of energy in the future the earth sciences will continue to be the key part of meeting that demand.

MINERALS

Since human beings first used flint to make fire, clay to make pots for carrying water or salt to preserve food minerals have been one of the essential elements in our life support system. Despite some substitution by plastics the use of minerals continues to grow and a significant number of geologists are employed by the mineral exploration companies. Indeed if all people are to aspire to use the world's mineral resources at the same rate that they are used by the average American (Figure 11) then the growth in mineral exploration and usage will become even steeper, especially for industrial minerals (Figure 12). It is now accepted that the Club of Rome was wrong in its pessimistic projection for the future and there is no evidence of global shortages of any minerals though, of course, local shortages will develop if a country lacks the foreign exchange to buy a mineral commodity on the open market.

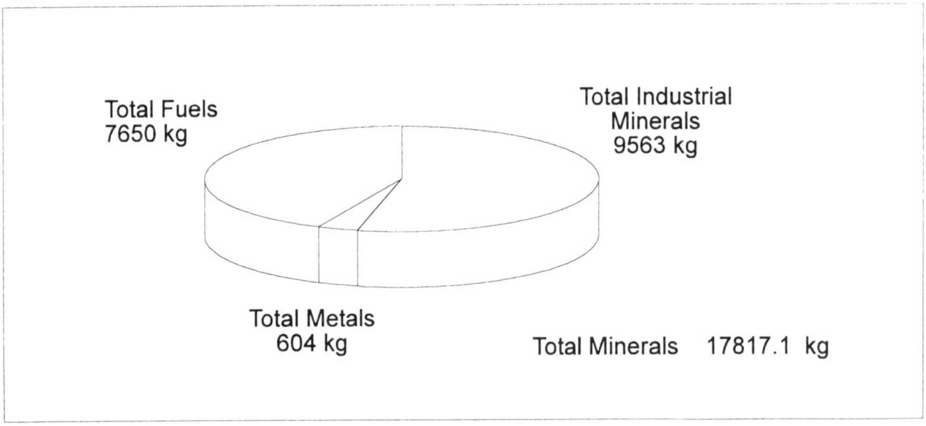

Figure11. Per capita consumption of mineral resources: USA 1989 [9].

It can be argued that whilst there are no shortages of any minerals and new technologies are allowing us to exploit lower grade ores, eventually we will get to the stage where commodity prices will have to rise and we will start to invest an increasing and perhaps unacceptably high proportion of our GDP in order to obtain minerals from progressively lower grade ores. But if we look at metal prices over this century (Figure 13) we see no indication of this. In fact the trend is the opposite with a very marked decrease in the metal price index. With the conclusion of the 'Cold War', the concept of strategic minerals is now questioned, the philosophy being that in a global market you can buy anything you want. Whilst this is true for the moment, it is a questionable assumption for the long term, for post cold-war geopolitics are presently characterized by instability and there are a number of major world commodities which have limited geographic (and national) distribution. Most notable amongst the commodities with such a limited distribution (taken as being when the majority of production is derived from four producers or less), (Figure 14), we see that manganese, cobalt, phosphate. chromium, tungsten and vanadium all fall into this category at the present time. In the future this may be a cause for concern.

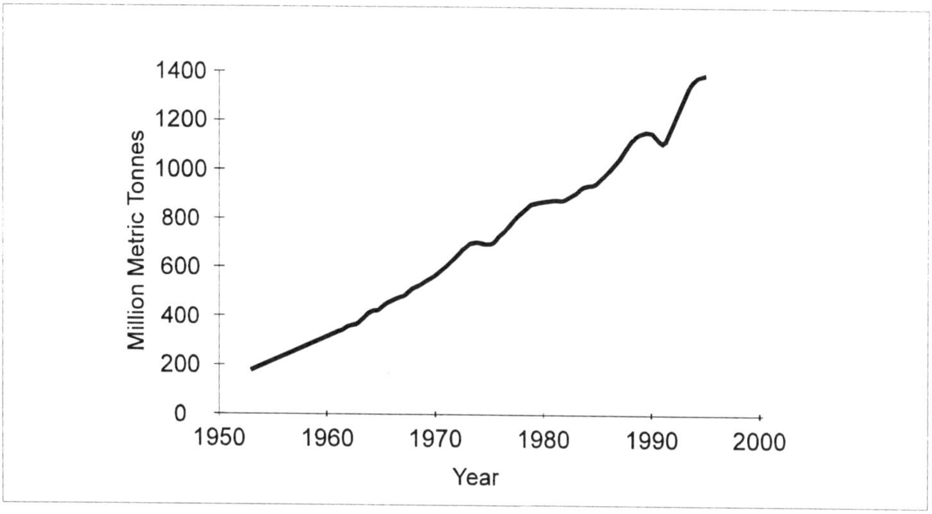

Figure 12. World cement production [12].

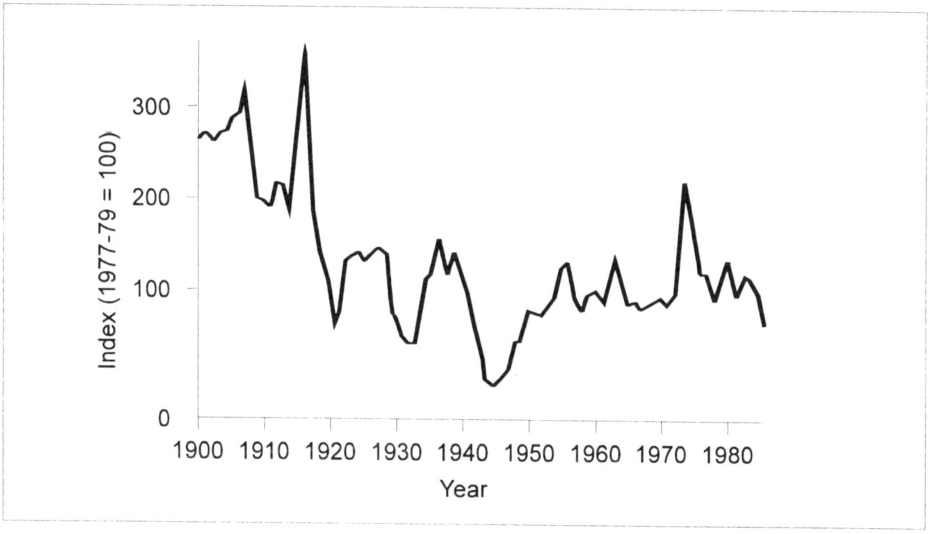

Figure 13. World metal prices 1900 - 1988 [11].

However, at the present time it is not political instability which is likely to be the overwhelming commodity issue for developed countries, but the impact of socio-economic issues on mineral deposits. The concept of mineral reserves and resources is well known. As resources (which may or may not be recoverable at present day prices) geologically become better known and more amenable to economic recovery so they move from the 'resource' category into the 'reserve' classification. However, as commodity prices drop so 'reserves' can move into the 'resource' (*i.e.* uneconomic) category. This philosophy is encapsulated in what is known as the McKelvy Box

(Figure 15). However, increasingly it is becoming necessary to use a cube to represent the situation (Figure 16), for there is now a third axis characterized by what I have termed here 'accessibility'. Accessibility may be limited by the legislative framework of a state or country, by political instability, by environmental concerns or by issues related to indigenous people. In other words a deposit can be economic, and it can be geologically proven but it may still fall into the unusable resource category because it is inaccessible for socio-political rather than economic reasons. This, more than anything else will limit our access to minerals in the future and we must learn to better factor this issue into exploration and exploitation philosophies. In turn this will require geologists to communicate more effectively with other disciplines, including planners, environmentalists, anthropologists, and with the community at large, if exploration successes are to translate into mines. If we do not do this then mineral shortages could develop in the future.

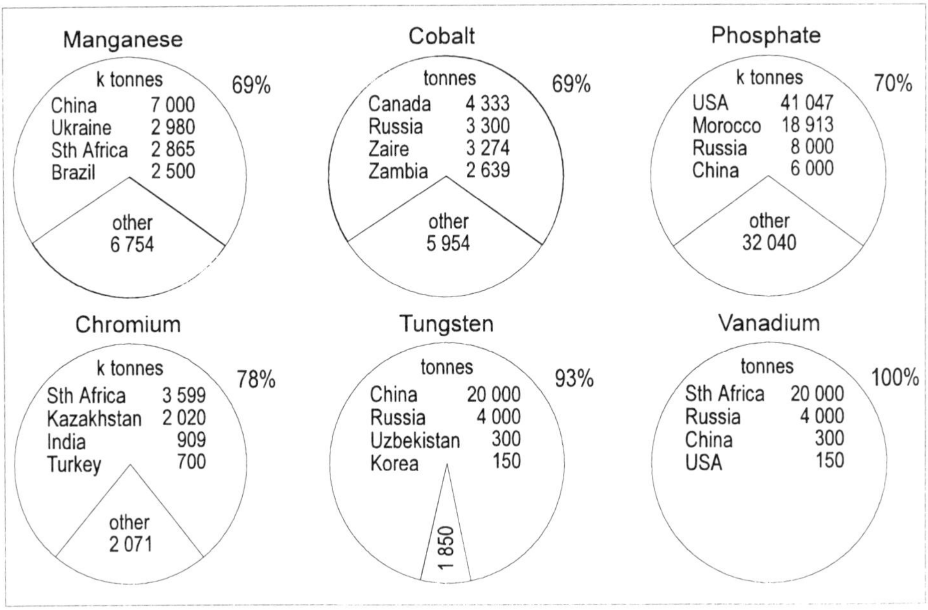

Figure 14. Minerals with a limited geographic production. Shows top four producers and other producers [12].

CONCLUSIONS

1. We must plan on global growth but work to minimize its impact.

2. The sustainability of our life support system depends on an extraordinarily complex interplay of factors affecting the availability and use of water, air, soils, land, energy and minerals. Geoscience is a key to this understanding and this will require better documenting, monitoring, modelling and understanding of the system as a whole.

3. For most people the important sustainability issues are not availability of metals or oil but access to clean water, good soils, adequate building materials and a secure place to live. Geological education and values must reflect this and educators must be prepared to move well outside the traditional approach to geological education.

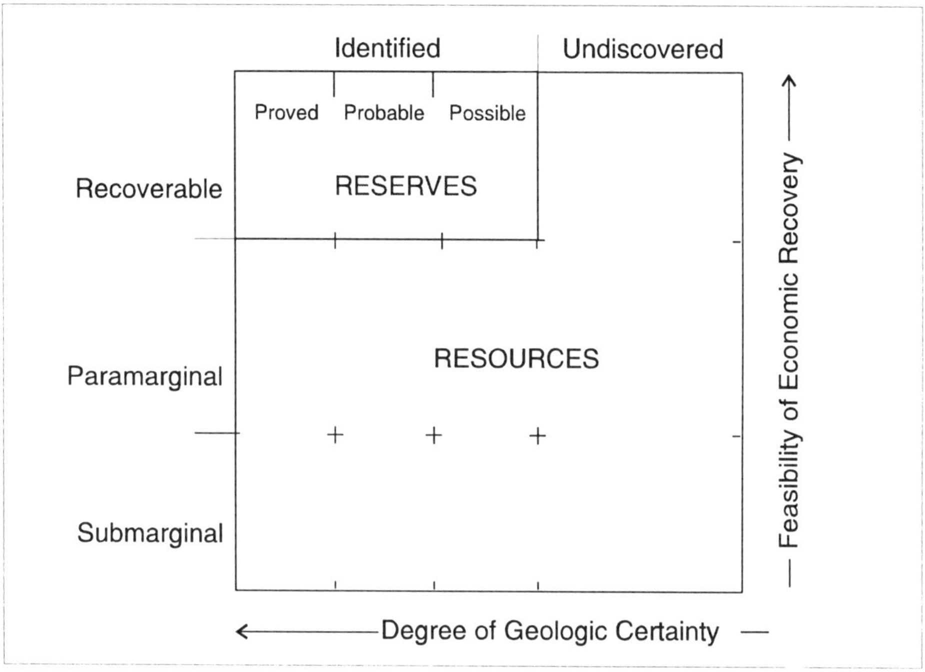

Figure 15. The "Mckelvey Box" to illustrate the reserve-resource concept.

4. Interdisciplinarity and effective networking are essential for this future, particularly as some of the major resource companies 'downsize' and 'outsource' their science, thus placing greater reliance on external expertize.

5. A narrow short term market-driven approach underlies many sustainability problems; a broad scale long term market driven approach may provide answers, but this must take into account principles such as intra/inter generational equity, polluter-pays, proper natural resources accounting, appropriate returns to the public and the rights of indigenous people. Earth scientists must be prepared to proselytise, communicate and speak out on such issues.

6. Politics is short-term, whereas resource and environment issues are long-term. Consequently we must ensure that robust institutions exist, notably national geological surveys that are able to undertake long term strategic geoscience mapping, surveying, monitoring, databasing and underpinning research and development in association with academia and industry.

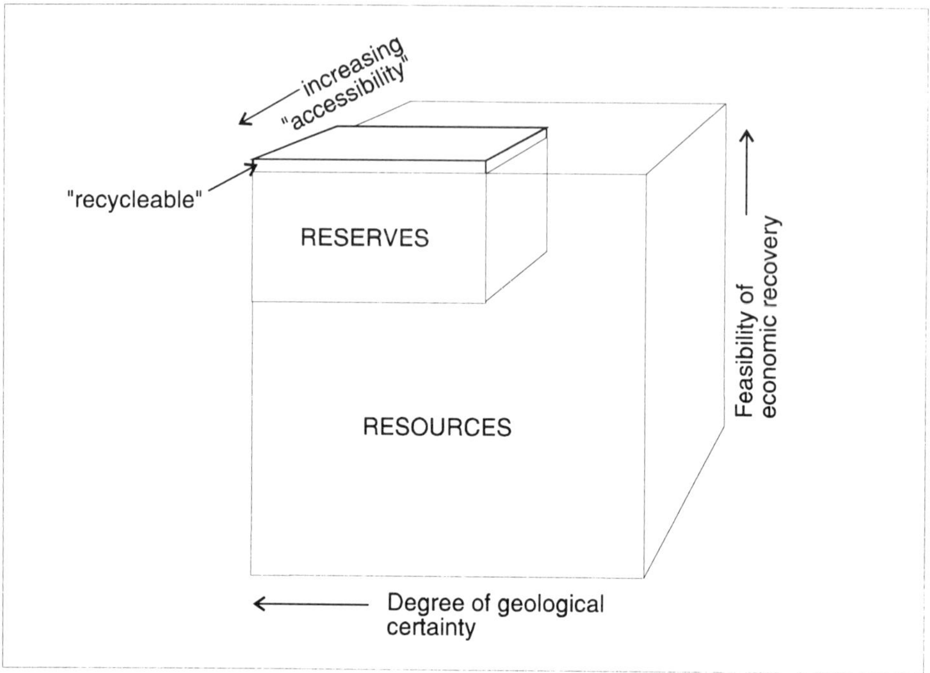

Figure 16. The impact of "accessibility" issues on reserves and resources

7. The important geoscience challenges of the future are likely to be found in inner city areas, waste dumps, low cost bulk commodity quarries, over-exploited farmlands, polluted rivers and coastal swamps rather than in the beautiful mountain settings that lured many of us into geology in the first place.

Acknowledgements

One of the privileges of being Director of a major geological survey is the opportunity it offers to be exposed to an enormous range of activities. discoveries and ideas from around the world. The thing that is generally lacking is the time to pull all these ideas together into a coherent view. This paper is an attempt to do this. I am grateful to my family for being understanding about the constraints on my time and my colleagues in the BGS for being generous with their time. Thanks are due to the organizers of the Geological Congress for not only did they do an outstanding job of organizing the IGC but it was their invitation to deliver one of the Opening Addresses for the Congress which gave me the impetus to prepare this paper in the first place. I am most grateful to Dan Sheath for providing many of the compilations used in this paper and for drafting the figures. The text was typed by Sarah McIlfatrick and Wanda Crosby.

REFERENCES

[1] G. H. Bruntland. (Chair). Our Common Future. In: *Report of World commission on Environment and Development*. Oxford University Press, Oxford (1987).

[2] P. J. Cook. *Societal trends and their impact on the coastal zone and adjacent seas.* British Geological Survey Technical Report WQ/96/3 (1996).
[3] R. Engelman and P. LeRoy. Sustaining Water: *Population and the Future of Renewable Water Supplies.* Population Action International, Washington, DC (1993).
[4] F. Ghassemi, A. J. Jakeman and H. A. Nix. *Salinisation of land and water resources.* UNSW Press, Kensington NSW (1995).
[5] P. Gleick. *Water in Crisis: A Guide to the World's Fresh Water Resources.* Oxford University Press (1993).
[6] S. Holloway. *The Underground disposal of carbon dioxide.* Final report of Joule II project No. JOU2 CT92-0031. British Geological Survey, Keyworth (1996).
[7] R. P. C. Morgan. *Soil Erosion and Conservation.* Longman (1995).
[8] S. Postel. *The last oasis: Facing water scarcity.* Worldwatch Institute, Earthscan Publications Ltd (1992).
[9] B. J. Skinner. Resources in the 21st Century, can supply meet demand? *Episodes* **12:3**, 267-278 (1989).
[10] R. K. Turner, S. E. Subak and W. N. Adger. *Pressures, trends and impacts in the coastal zone: Interactions between socio-economic and natural systems.* CSERGE Working Paper GEC 95-09 (1995).
[11] *The World Bank Economic Review* **2:1**, 8-20 (1988).
[12] *World Mineral Statistics.* British Geological Survey, Keyworth (1996).

Proc. 30th Intern. Geol. Congr., Vol. 1, pp. 47-60
Wang *et al.* (Eds)

Tectonic Evolution and Uplift of the Qinghai-Tibet Pateau

XIAO XUCHANG and LI TINGDONG
Chinese Academy of Geological Sciences, Beijiang, 100037, P. R. China

Abstract

Recent studies on palaeobiogeography, sedimentology and palaeomagnetism show that the Indian plate and the terranes of the Qinghai-Tibet plateau accreted to Eurasia during the Late Palaeozoic to Mesozoic did not drift northward across a vast ocean. At that time there was only an epeiric sea or a "limited ocean basin" rather than a "vast eastward opening Tethys". With the acquisition of the data of geodetic leveling, GPS and fission track dating, the uplift rates of the Qinghai-Tibet plateau (QTP) have been inferred. The uplifting of the QTP can be subdivided into three major stages, the slow uplifting stage in Late Cretaceous-Oligocene, the intermediate uplifting stage in Miocene-Pliocene and the rapid uplifting stage in Pleistocene-Holocene. The uplift rates increased with time. There were probably two abrupt epochs: one was around Late Miocene and the other around Late Pliocene-Pleistocene. Based on the recent geological and geophysical data, we suggest that the causes of the rise of the QTP are complex and involve many factors. Our preliminary study indicates that the compression from the India terrane and the resistive force (or hysteresis effect) from the rigid blocks(the Tarim and Yangtze terranes), the thermal effects and expansion, and the isostatic adjustment in the recent period are the three major factors leading to the crustal shortening, thickening and uplifting of the plateau.

Keywords: Tibet, tectonics, uplift of plateau

INTRODUCTION

Since the founding of the People's Republic of China, Chinese geoscientists have long paid attention to the study of the Qinghai-Tibet plateau, and the geological, geophysical and geochemical surveying have been organized and carried out by on the plateau. Particularly in the past two decades more extensive geoscientific investigations and projects have been performed by Chinese geoscience expeditions as well as co-operative geoscience research teams with foreign geologists. During this period much information and data have been obtained, and the formation of the high Himalayan mountains, the tectonic evolution and uplift of the Qinghai-Tibet plateau have been studied. Some main mineral-deposits, such as chromites, poly-metallic and auriferous deposits, have been explored, and great scientific achievements have been acquired. Therefore geoscientific investigation on the plateau has entered into a new stage.

Based on part of the results obtained in the recent decades, we intend to give a brief review of the tectonic evolution and uplift of the Qinghai-Tibet plateau.

TECTONIC EVOLUTION

The Qinghai-Tibet plateau is the highest plateau in the world, an active earthquake zone, a strongly uplifted mountainous region aprobably of unique morphogenesis. It forms an area of utmost interest and subject of lively discussion among geoscientists worldwide. The tectonic evolution of the Qinghai-Tibet plateau is closely related to the splitting of Gondwanaland and to the evolution of Tethys in Mesozoic and Cenozoic.

Since the emergence of the plate tectonics theory in the late 1960s, many geoscientists have believed that the Indian plate,the Himalaya terrane and the southern terranes of the Qinghai-Tibet plateau were successively accreted to Eurasia after their separation from northern Gondwanaland, and drifted northward across a vast deep ocean about 7000 km wide, the so-called Palaeo-Tethys, around 200 Ma ago [8, 9].

Recent studies on palaeobiogeography, sedimentary facies and palaeomagnetism indicate that during the Late Palaeozoic-Early Mesozoic (Triassic), most of the Tethys was an epeiric sea rather than a vast deep ocean, except perhaps for the belts along the Jinsha River and the north of the Bayan Har Mts. where rifting might have formed narrow deep troughs or small ocean basins as represented by the dismembered ophiolites. A quantitative palaeobiogeographical analysis of the Carboniferous-Triassic faunas and floras of Qinghai-Tibet has been presented by Smith *et al.* [27], which indicates that there was no suture line consistently marking the position of a faunal or floral break during the Late Palaeozoic-Late Triassic and there appears to have been no physical barrier, such as a vast ocean, to biotic dispersal between terranes of the Qinghai-Tibet plateau at that time. According to recent mapping and traverses of the Karakoram Himalaya, Muhammad N. Chaudhry, Munir Ghazanfar and David Spencer [21] suggest that "*there is a distinctive stratigraphical correlation between the units of the Karakoram Himalaya (Eurasian plate) and the Himalaya (India plate) ... that these similarities are so strong that the likelihood should be considered that the Karakoram Himalaya (Eurasian plate) and the Indian plate were, until the Early Mesozoic, attached as part of the same block or plate.*"

Table 1. Similarity of faunas and floras of the Gondwana facies between India and Qinghai-Tibet plateau

<table>
<tr><td colspan="2">Area</td><td>India</td><td colspan="2">Qinghai-Tibet plateau</td><td rowspan="2">Australia</td><td rowspan="2">Antarctica</td></tr>
<tr><td colspan="2">Flora and Fauna</td><td>Subcontinent</td><td>S.Tibet</td><td>N.Tibet</td></tr>
<tr><td rowspan="6">Flora</td><td>Glossopteris</td><td>+</td><td>+</td><td>+</td><td>+</td><td>+</td></tr>
<tr><td>Gangamopteris</td><td>+</td><td>−</td><td>−</td><td rowspan="6">lack of data</td><td rowspan="10">lack of data</td></tr>
<tr><td>Schizoneura</td><td>+</td><td>−</td><td>+</td></tr>
<tr><td>Sphenophyllum</td><td>+</td><td>+</td><td>+</td></tr>
<tr><td>Cordaites</td><td>+</td><td>−</td><td>+</td></tr>
<tr><td>Tizygia</td><td>+</td><td>−</td><td>−</td></tr>
<tr><td rowspan="4">Fauna</td><td>Eurydesma</td><td>+</td><td>+</td><td>+</td></tr>
<tr><td>Stepanoviellia</td><td>+</td><td>+</td><td>+</td><td>+</td></tr>
<tr><td>Monodiexodina</td><td>+</td><td>−</td><td>+</td><td rowspan="3">lack of data</td></tr>
<tr><td>Lytvolasma</td><td>+</td><td>+</td><td>+</td></tr>
<tr><td>Index</td><td></td><td>100</td><td>50</td><td>80</td></tr>
</table>

(Data of India subcontinent after Ahmad [1], data of Qinghai-Tibet plateau after Smith and Xu Juntao [27], Yang Zuryi *et al.*, 1990, Li Xingxue, 1984)

Evidence from the palaeobiometrics indicates that the similarity index of the floras and faunas of Late Carboniferous to Permian (Gondwana facies) for the Qinghai-Tibet plateau is high, reaching 80% (Table 1) . The terrestrial tetrapods and ammonites of Triassic to Early Jurassic age for China and north Tibet is also high, reaching 71-80% (Table 2, 3), the ammonites of Late Jurassic to Early Cretaceous age for north Tibet is highest, reaching 87% (Table 4).

Table 2. Similarity of the tetrapods (T-J_1) between India and Qinghai-Tibet plateau

Indian Genera and Families	N. America	Europe	Africa	China	East Asia	S. America	Australia	Antarctica
Metoposaurus Metoposauridae	+	+	+	+(T_3)	+	–	lack of data	lack of data
Paradapedon Rhynchosauridae	+	+	+	+(J)	–	+		
Parasuchus Paraschidae	+	+	+	–	+	–		
Typothorax Stagonolepididae	+	+	–	–	+	+		
Exaeretodon Traversodontidae	+	+	–	+($T_{2\text{-}3}$)	–	+		
Malerisaurus Protorosauridae	+	–	–	+(T)	–	–		
Walkeria Podokesauridae	+	+	+	+(J_1)	–	–		
Index	100	86	57	71	43	43		

(Data after Chatterjee [5] and Sun Ailing *et al.* [30])

Table 3. Similarity of the Triassic ammonites for India–Qinghai-Tibet plateau

Time	Area	Nepal and Himalaya	Lhasa area	N. Tibet-Qinghai plateau	
				Qiangtang	other areas
T1	*Owenites*	+	+	+	+
	Flmingites	+	+	+	+
	Proptychites	+	?	+	?
	Anotoceras	+	?	?	+
T2	*Gymnites*	+	+	+	+
	Hollandites	+	+	+	+
	Leiophyllites	+	+	+	+
	Japonites	+	+	+	+
	Ptychites	+	+	+	+
	Paraceratites	+	?	?	?
	Balatonites	+	+	+	+
	Procladiscites	+	?	+	+
	Sturia	+	+	?	?
T3	*Protrachyceras*	+	?	+	+
	Trachyceras	+	?	+	+
	Index	100	60	80	80

Table 4. Similarity of ammonites (J_3-K_1) between India and Qinghai-Tibet plateau

Area	India Subcontinent			Nepal	Qinghai-Tibet		Australia	Antarctic
	Kachchh	Jaisalmer	Spiti		S. Tibet	N. Tibet		
Virgato-sphinctes	+	+	+	+	+	+	lack of data	+
Aspidoceras	+	+	-	+	-	+		lack of data
Berriasella sp.	+	+	+	+	+	+		
Himalayites	+	-	+	+	+	+		
Spiticeras	+	-	+	+	+	+		
Neocomites	+	-	-	+	+	+		
Lytocera-tinae	+	-	-	+	-	-		
Acanthoho-plites sp.	+	-	-	+	-	+		
Index	100	37	50	100	63	87		

(Data of India subcontinent after Jai Krishna, 1981; data of Qinghai-Tibet plateau after Liu Guifang, 1990)

A palaeomagnetic study also indicates that in the Late Palaeozoic, there was no 7000km wide ocean between Eurasia and Gondwana, across which a number of solitary blocks, including South China, Indochina, Afghanistan, Iran, Kunlun, Qiangtang and Karakoram *etc*., had accreted to form the Cathaysian composite continent [16].

So it is believed that the Indian plate and Asia remained an integrated geographic unit during the Late Triassic and even Early Jurassic. In the Jurassic to Early Cretaceous, although the Yarlung-Tsangpo suture zone seemed to have resulted from the closing of an ocean basin , the palaeobiocoenosis, palaeogeography and sedimentary facies both to the north and the south of the Yarlung-Tsangpo suture zone were characterized by an epibenthic environment [37]. The above-mentioned palaeobiometrics of the Late Jurassic to Early Cretaceous (Table 2), the high similarity index of the ammonites, and particularly the recent discovery of the Indian endemic fauna *Indocephalites*, *Spiticeras* and *Himalayites* in northern Tibet (Liu Guifang, manuscript) indicate a close relation of the areas. Moreover, *Ptilophyllum* of the so-called upper Gondwanian flora was also found in the Lhasa area [37]. Therefore, the ophiolites along the Yarlung-Tsangpo suture zone probably represents relics of oceanic crust-upper mantle rocks of a limited ocean basin or a narrow marine trough of Jurassic to Early Cretaceous age.

The REE, the bulk composition, stable elements and isotopic values of the ophiolites can be used to deduce indirectly the spreading rate parameters. Such data from the Yarlung-Tsangpo ophiolites suggest a slow spreading rate (less than 2 cm/a, averaging about 1.5 cm/a) [20]. The duration of the ocean basin represented by the Yarlung-Tsangpo ophiolites is from about 190 Ma (Jurassic) to 110 Ma (Early Cretaceous). The width of the basin is roughly estimated at 2400 km [(190-110) Ma×1.5 cm/a)×2].

All the above evidences indicate that the Indian terrane was near the Gangdise (Lhasa) terrane and never far from Asia. From the Jurassic to Cretaceous, most of the area between the Indian terrane and the QTP was covered by an epeiric sea. The Yarlung-Tsangpo suture zone resulted only from the subduction and consumption of a " limited (small) ocean basin ". During that time, Antarctica and Australia might have probably split from India and drifted to the south.

It is appropriate to recall the words of Alfred Wegener [33], written 60 years ago, on the evolution of the Himalayas and Tibetan high land: *The case of India is somewhat different: it was originally connected by a long continental tract, mostly, it is sure, covered by shallow sea, to the Asiatic continent...this long connecting portion was more and more and constitutes today the mightiest mountain folds of the earth, the Himalayas and the numerous folded ranges of high lands of Asia folded together through the continuous gradual approach of India to Asia.*

CRUSTAL THICKENING, SHORTENING AND UPLIFTING

From studies of palaeomagnetism, sedimentology, tectonic history, stratigraphic palaeontology, magmatism, highpressure and low temperature metamorphic belts and ophiolites, most geoscientists believe that the "Neo-Tethys ocean basin" closed and the Indian terrane collided with Eurasia, in the period of Cretaceous to Eocene. Since then (about 65-50Ma), there have occurred crustal thickening, shortening and uplifting of the QTP.

Until now, estimates of the northward displacement of the Indian terrane since the Tertiary have been varied, and not generally accepted. But some tentative estimate of the rates is possible after preliminary analysis.

The northward drift rate of the Indian terrane after its convergence with Eurasia in the Eocene was 16-17cm/a at 65-50Ma, about 10cm/a in the Oligocene-Miocene, and 2-10cm/a in the Late Pliocene [10, 14, 19, 22, 41]. The generally accepted rate of the northward motion of the Indian terrane since the Pliocene is 50-60 mm/a. Among the total value, according to recent studies by Chinese geoscientists [28], 5 mm is absorbed by compressive faults or thrusts in the Himalaya terrane, 15 mm by strike-slip faults in the Gangdise terrane, 12 mm by strike-slip faults in the Qiangtang terrane, 8 mm by strike-slip faults and the Red River right-lateral strike-slip fault in the Yunnan block, about 11 mm by the Altun, Kunlun and Qilian fault systems, and lastly about 3-5 mm by the Tianshan fault system.

Table 5 is based on palaeomagnetic studies of recent years, which show that from the Himalaya terrane to Qaidam terrane, the displacement successively decreased from 2664 to 110km. Based on the movement distance of the above terranes, the crustal shortening from the Gangdise terrane to the Qiangtang terrane can be roughly estimated at about 990 km. This is more than the shortening between the Himalaya and the Gangdise terrane or that between the Qiangtang and the Qaidam terrane (Table 5). These data are consistent with geophysical surveying, which indicate that the Gangdise terrane and the area north of it have a “thick crust”, indicating that the Qaidam terrane in the north of QPT was an obstruction to the compressive force from the south.

Table 5. Tertiary palaeomagnetism and crustal shortening of the Qinghai-Tibet plateau.

Terrane	Age	Palaeo latitude (D°)	Modern latitude (D_1°)	Northward displacement and latitude (D_1°-D°)×10	Distance between terranes	Shortening of crust (km)
Himalaya	Tertiary	4.5°N (2.5°-5°, 4.6°N)	28.5°N	24° 2664 km		
Gangdise	Tertiary	12°N (10-13.8)	30°N	18° 1980 km	130 km	2664–(1980+130)=554
Qiangtang	Tertiary	27.5°N 25.4-29.5	32.5°N	5° 550 km	440 km	1980–(550+440)=990
Qaidam	Tertiary	37°N (34°-40°)	38°N+	1° 110 km	480 km	550–(110+480)=–40

(Data of Himalaya after Sino-French joint investigation in Himalaya region, 1984, Dong Xuebing *et al.*, 1990; data of Gangdise after Michele Westphal, Zhou Yaoxin *et al.*, 1983, Dong Xuebing, 1990; data of Qiangtang after Zhou Yaoxin, Courtillot *et al.*, 1984; Dong Xuebing, 1990; data of Qaidam after Dong Xuebing, 1992; li Yongan, 1992)

The rapid uplift of the QTP since the Late Tertiary is fully demonstrated by data of neotectonics, geomorphology, palaeontology and sedimentology [15,23,42]. The uplift rates of the QTP obtained from various methods in the last few years are shown in Table 6.

The rate of uplift has increased with time: 0.01-0.07 mm/a in the Eocene to Oligocene. 0.87 mm/a (the average uplift rate) in the Late Miocene to Pliocene (10-3 Ma), 4.0 mm/a in the Late Pliocene to Pleistocene (2.0-0.5 Ma), and since the Holocene, the uplift rate has increased rapidly from north to south. There were probably two abrupt periods, one is around Late Miocene and other around Late Pliocene-Pleistocene (Fig. 1).

The uplift process of QTP can be subdivided into three major stages:

I. Slow uplift stage in the Late Cretaceous-Oligocene, mainly resulted from the collision between the Indian terrane and the southern terranes of the QTP .

II. Intermediate uplift stage in the Miocene

III. Rapid uplift stage from the Pleistocene to the Holocene

Concerning the theories and models for explaining the crustal thickening, shortening and uplifting of the Qinghai-Tibet plateau, there have been lively discussions during the past two decades. The space limitation here prevents all of the diverse models to be reviewed. With the acquisition of the recent geological and geophysical data, especially those from deep geophysical studies, we regard that the following factors should be taken into consideration:

1. Tectonic deformation is the main factor. Geophysical data obtained in the last ten years reveal that the QTP may be subdivided into three structural layers. The first layer, considered as the upper crust, extends to the depth of 10-30 km, in which brittle and ductile deformation are dominant. A low-velocity/low-resistivity layer at the base of the first layer [36] acts as a detachment level along which listric thrusting caused crustal thickening, shortening, and uplifting of the plateau. The second structural layer, regarded as the lower crust, is between the low-velocity/low-resistivity layer and the depth of 50-60 km, where crustal thickening and shortening were induced mainly by ductile and plastic deformation. The third structural layer is at the depths of 50-60 to 70-80km and is characterized by anomalous mantle or the crust-mantle mixtures (transitional zone) due to the upwelling of mantle materials and their interaction with the crust. This layer is dominated by plastic deformation which caused crustal thickening and shortening.

2. Thermal effect and partial melting are the important factors which cannot be neglected for the formation of the QTP (Fig. 2-1, 2-2).

(1) Heat flow of about 71 mW/m^2 has been measured to the south of the QTP in Katmandu, Nepal, and about 70 mW/m^2 to the north, the Qaidam basin. In thc plateau proper, particularly in the south central part of the plateau, the heat flow increases, however, to 92~146 mW/m^2 [26]

(2)Evidence from the seismic tomography using the method of SKS and PKS wave splitting by Sino-French co-operative investigation [40] indicates that there exists the lowest velocity from the upper crust to lithospheric mantle in the south central part of Tibet, which is consistent with the data of heat flow. Results from another seismic tomography [3] also show a prominent low velocity in the crust-upper mantle in central Tibet.

(3) The recent MT investigations by Drs.Chen Leshou, Wei Wenbo, and Booker L.R., *et al*, 1996 (Persional communication) reveal that there is lower resistivity in the south-central part of the Tibet plateau.

(4) Very low Q values prevail in the QTP. Q_β is 65~100 at a depth of 13~33 km in the upper crust and 18~110 below 55 km in the lower crust. These values are consistent with the upper and lower low-velocity layers respectively [32].

(5) Long-period seismic records show that the Lg phase is weak or missing in the QTP.

The geophysical data mentioned above indicate that there are a low viscosity and highly molten layer with intensive geothermal activity in the deeper part of the QTP. The thickening, shortening and uplifting of the QTP were most probably brought about by these thermal effect and consequent heat expansion.

Table 6. Uplift rate of the Qinghai-Tibet plateau since Eocene to Holocene

Method	Region (from S. To N.)	Duration (Ma)	Uplift Rate (mm/a)	Source of data
Geodetic leveling, GPS + SLR	1.Qomolangma (Everest) 2. Area near Qomolangma 3.Lhasa(Gangdise) terrane 4.Karakoram-Kunlun Mtn. 5.N. Altun Mtn	last 30 years last 30 years last 30 years last 30 years last 30 years	37 4- 15 9-10 6-9 5.1	①
Fission Track	N. W. Himalaya	55-20 23-11 10-3 2.0-0.5 0.5-Holocene	0.07-0.22 0.13-0.39 0.30-1.10 1.6-4.5 >4.5	②
	Middle Himalaya (Nyulam area)	10-3 2.0-0.5 0.6-Holocene	0.07-1.05 3.50-4.67 12	③
	M. E. Himalaya	11-8	0.45-0.49	④
	E. Himalaya (mainly Namjagbarwa)	25± 11.25-3.82 3.82 0.5-Holocene	0.31 0.62-1.14 1.14 14.19	⑤
	Extreme east of Gangdise terrane	25-11 10-3 2-0.5	0.31-0.62 0.66-2.05 1.82-5.35	⑥
	Lhasa (Gangdise) terrane	60-55 25-27	0.01-0.064 0.25	⑦
Neotectonics and geomorphology	Karakoram (N.W. Tibet region)	Holocene	4-5	⑧

(①J.Y. Chen *et al.* (National Bureau of Surveying and Mapping, P.R. China); Chang Qingsong, 1991; Roger Bilham, 1990 ;State seismological Bureau [28];②P. K. Zeitler *et al.*, 1985-1991;③Xiao Xuchang, Wang Jun and Wang Yanbin, 1995; Kagunori Arita *et al.*, 1995;④Liu Shunsheng *et al.*, 1987;⑤-⑥Ding Lin, Zhong Dalai *et al.*, 1995;⑦Wang Jun and Xiao Xuchang *et al.*, 1995;Liu Shunsheng *et al.*, 1987;⑧Liu Qing (doctoral thesis, 7th University of Paris, 1993)

3. Isostatic adjustment is one of the major factor in the last stage of the uplift of the QTP. Since the Tertiary, lithosphere downwarping on the plateau occurred inevitably due to the static load pressure resulting from the thickening of the crust. So the QTP was in a state of isostatic compensation. Later, in the Pleistocene to Holocene, the compressive force from the Indian terrane was decreased or relieved [22,39], and induced an upwarp of the previously downwarped area.

Rapid uplift and high positive gravity anomalies ($60{\sim}80\times10^{-5}m/s^2$), high resistivity and high velocity in the extreme south of the Himalaya region present since the Tertiary [18] are due to intensive compressive stress and strong deformation events such as the thrusting, over-faulting and overlapping, which affected both the lithosphere and upper mantle and induced the upper mantle material to be squeezed upwards.

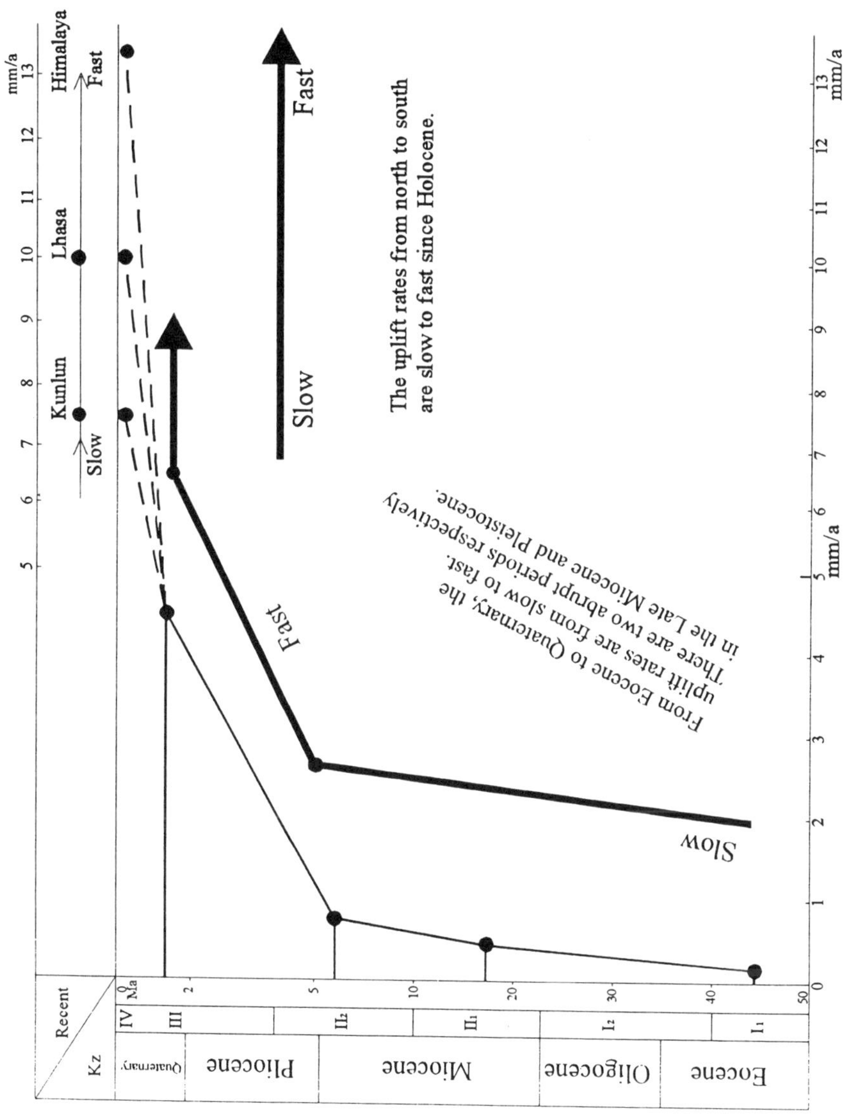

Figure 1. Simplified figure of the uplift rates of the Qinghai-Tibet plateau since Cenozoic.

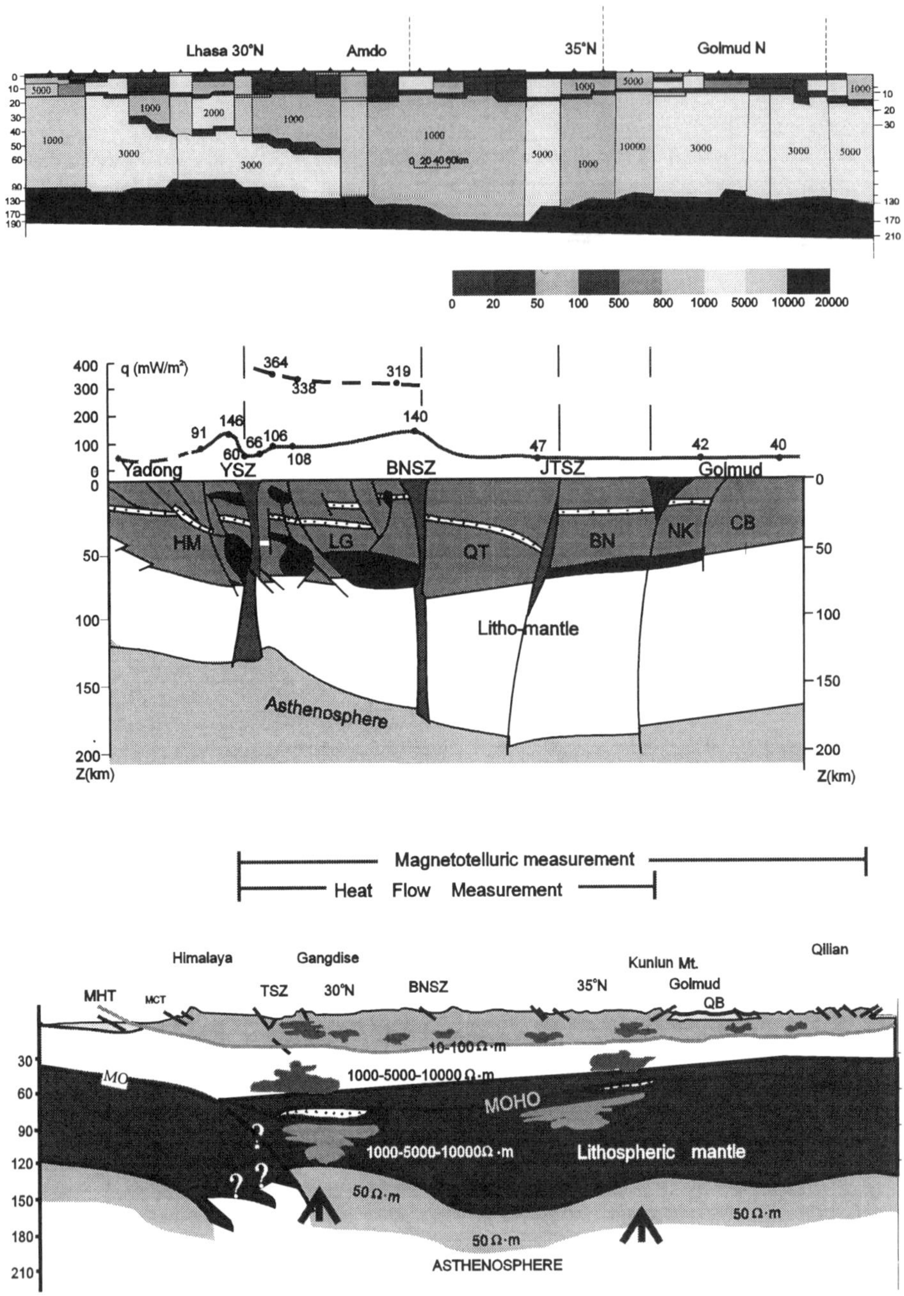

Figure 2-1. Magnetotelluric and heat flow measurements in the Qinghai-Tibet plateau and model of India-Qinghai-Tibet collision.

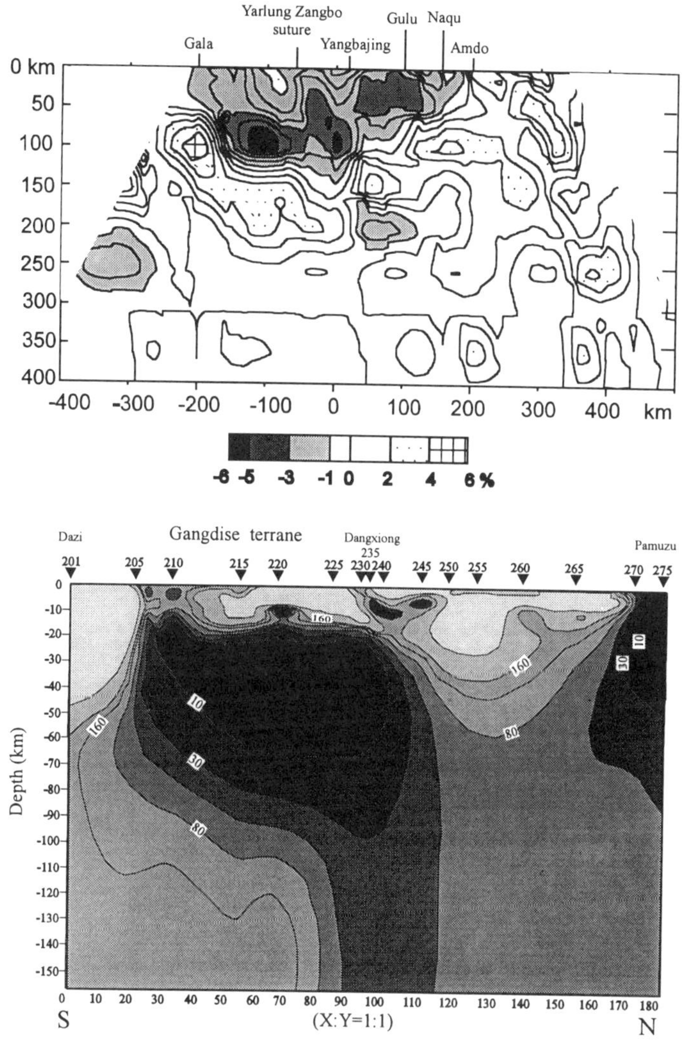

Figure 2-2. Seismic tomography section from Tangula pass to Gala and MT inversion in south-central part of the Qinghai-Tibet plateau.

CONCLUSION

Multidisciplinary studies indicate that during the Late Palaeozoic to Early Mesozoic, most of the Tethys was an epeiric sea , not a vast deep ocean.The presence of ophiolites along the Yarlung-Tsangpo (YTSZ), Lancang-Bangong (LBSZ) and Jinshajiang (JSZ) suture zones are fully demonstrated to be narrow deep troughs or small ocean basins formed by rifting.

Based on the geological and geophysical data above mentioned, we speculate that

Indian lithospheric mantle (mantle lid) might have been, with moderate north-dipping, subducted just beyond the Tsangpo suture zone, and that the Gangdise calc-alkalic magmatism was possibly generated in such a northward subduction zone environment. To the north, we infer, by integrating geophysical studies [32] that an intracontinental subduction, with high-angle dipping to the south, occurred between the Tarim terrane and Kunlun Mountains, which might have induced the Cenozoic calc-alkalic--alkalic volcanics developed to the south of the Kunlun Mountains.

The causes for the uplift of the Qinghai-Tibet plateau are complicated and heterogeneous both in time and space. However, the *compression* from the Indian terrane and resistive force (or hysteresis) from its arounding rigid blocks (the Tarim and Yangtze terrane), the *thermal effect* and partial melting, and the *isostatic adjustment* in the latest period, are the three major factors inducing the uplift of the Qinghai-Tibet plateau

Acknowledgements

The authors are thankful to Dr. Wang Shichen for providing fission track data.

REFERENCES

[1] F. Ahmad. *Gondwanaland, the concept that failed.* Birbal Sahni, Institute of Palaeobotany, Lucknow, (1978).

[2] C. Burchfiel *et al.*. The south Tibetan detachment system, Himalayan Orogen. *Geological Society of American, Special Paper* **269**, 41 (1992).

[3] L. Bourjot and B. Romanowicz . Crust and upper mantle Tomography in Tibet using surface waves. *GRL*. **19**:**9**, 881-884 (1992).

[4] S.W. Carey. *The Expanding Earth.* Elservier, Amsterdam (1976).

[5] S. Chatterjee. A new tetrapod dinosaur from India with remarks on the Gondwana -Eaurasia connection in the Late Traissic. In: *Proceedings of Conference Gondwana 6: Stratigraphy, Sedimentology and Palaeontology*. pp. 183-190 (1987).

[6] M. N. Chaudhry, M. Ghazanfar and D.A. Spencer. Evidence for pre-Mesozoic affinity of the Karakoram Himalaya to the Indian plate, Northern Pakistan. *Abstractvoe 11th Himalaya-Karakoram-Tibet workshop*. pp. 33 (1996).

[7] J. A. Y. Chen *et al.* Crustal movement and gravity field of the Mt. Qomolangma (Everest) and its adjacent area. *Chinese Science Bulletin* **39:13**, 1204-1207 (1994).

[8] J. F. Dewey, R. M. Shackleton, Chang Chengfa and Sun Yiyin. The tectonic evolution of the Tibet plateau. In: *the Geological Evolution of Tibet*. pp. 379-408 The Royal Society, London (1990).

[9] R. S. Dietz and J. C. Holden. The breaking up of Pangaea. *Scientific American* **223:4**, 30-41 (1970).

[10] Dong Xuebin *et al.*. New palaeomagnetic results from Yadong-Golmud geoscience transect and a preliminary study on the model of terranes evolution in Qinghai-Tibet plateau. *Bulletin of the CAGS* **21**, 139-148 (1990).

[11] T. M. Harrison *et al.*. Raising Tibet. *Science* **225**, 1663-1669 (1992).

[12] K. L. Kaila *et al.*. Structure of Kashmir-Himalayan from deep seismic soundings. *Jour. Geol. Soc. India* **19**, 1-20 (1978).

[13] W. S. F. Kidd and P. Molnar. Quaternary and active faulting observed on the 1985 Academic Sinica-Royal Society of Geotraverse of Tibet. In: *Report of the Royal Society-Academic Sinica Geolotraverse of Qinghai-Tibet Plateau*. pp. 348-371 (1990).

[14] C. T. Klootwijk *et al.*. An early India Asia contact, Palaeomagnetic constraints from Ninetyeast Ridge, ODP Leg 121. *Geology* **20**, 395-398 (1992).

[15] Li Tingdong. Deep structure and uplift mechanism of the Qinghai-Tibet plateau. *Geological Memoir, Ministry of Geology and Mineral Resources, China* **5:7**, 174-185 (1988).

[16] Lin Jinlu and D. R Watts. Palaeomagnetic results from the Tibetan plateau. In: *Report of the Royal Society- Academic Sinica Geotraverse of the Qinghai-Tibet plateau*. pp. 242-281 (1990).

[17] Liu Xun *et al.*. The Stratigraphy, Palaeobiography and sedimentary tectonic evolution of Qinghai-Tibet plateau. *Geological Memoir, Ministry of Geology and Mineral Resources, China* **2:15**, 169 (1992).

[18] Meng Lingshun. Gravity survey and the lithosphere structure in Qinghai-Tibet plateau. *Geological Memoir, Ministry of Geology and Mineral Resources, China* **5:13**, 113 (1992).

[19] P. Molnar *et al.*. Geomorphic evidence for active faulting in the Altyn Tagh and N. Tibet and qualitative estimates of its contribution to the convergence of India and Eurasia. *Geology* **15**, 249-253 (1987).

[20] Mo Xuanxue. New facts and concepts on the volcanism and volcanic rocks in the orogenic belts in W. China. *Contributions to Petrology* **11**, 47-55 (1992).

[21] N. C. Muhammad, M. Ghazanfar and D. A. Spencer. Evidence for pre-Mesozoic affinity of the Karakoram-Himalaya to the Indian plate,N. Pakistan. *11 KHT workshop, Abst.* 33-35 (1996).

[22] P. Patriat *et al.*. India-Eurasia collision chronology has implications for crustal shortening and driving mechanism of plates. *Nature* **311**, 615-621 (1984).

[23] Pan Guitang *et al.*, Cenozoic tectonic evolution of Qinghai-Tibet Plateau. *Geological Memoir, Ministry of Geology and Mineral Resources, China* **5:9**, 141-152 (1990).

[24] M. N. Qureshy. Vertical tectonics in middle Himalayas, an appraisal from regravity data. *Bulletin, Geological Society of America* **85**, 201-230 (1974).

[25] Research Group of the CIT Harvard University and MIT. Science and citizen. *Scientific American* **256**:**2**, 42-44 (1987).

[26] Shen Xianjie *et al.*. Heat flow and tectonic thermal evolution of terranes of the Qinghai-Tibet plateau. *Geological Memoir, Ministry of Geology and Mineral Resources, China* **5:14**, 94 (1992).

[27] A. B. Smith and Xu Juntao. Palaeontology of the 1985 Tibet Geotransverse, Lhasa to Golmud. In: *Report of the Royal Society Academic Sinica Geotraverse of Qinghai-Tibet plateau*. pp. 49-106 (1990).

[28] State Seismological Burreau (SSB). *The active Altyn faul* (1992).

[29] J. Stocklin. Himalayan orogeny and Earth expansion. In: *Expanding Earth Symposium, Sydney*. pp. 119-130 (1983).

[30] Sun Ailling *et al.*. *The Chinese fossil reptiles and their kins*. Science press, Beijing (1992).

[31] P. Tapponnier *et al.*. On the mechanics of the collision between India and Asia. In: *Collision tectonics*. pp. 115-157. Geological Society of London Special Publication 19 (1986).

[32] Teng Jiwen. An introduction to geophysical study on the Tibetan plateau area. *Acta.Geophysica Sinica* **28**:**1**, 1-5 (1985).

[33] A. Wegener. *The origin of continents and oceans*. Methuen and Co. Ltd., London (1924).

[34] Wen Shixuan. New palaeontological evidence for continental drift in the Qinghai-Tibet plateau. In: *Proceeding of the 1st Symposium on the Tibet plateau*. pp. 308-315 (1992).

[35] S. D. Willett *et al.*. Subduction of Asian mantle beneath Tibet inferred from models of continental collision. *Nature* **369:23**, 642-645 (1994).

[36] Wu Gongjian, Cui Junwen and Gao Rui *et al.*. An introduction to the "Yaodong-Golmud Geo-Transect " in Qinghai-Tibet plateau. *Bulletin of the CAGS* **21**, 1-9 (1990).

[37] Xia Daixiang *et al.*. Regional geology of Tibet (Xizang). *Geological Memoir, Ministry of Geology and Mineral Resources, China* **1:31**, 707 (1993).

[38] Xiao Xuchang *et al.*. Ophiolites of the Tethys-Himalayan of China and their tectonic significance. In: *26th Intern. Geol. Congress, Colloque C5*. pp. 149-151 (1980).

[39] Xiao Xuchang *et al.*, Tectonic evolution of the lithosphere of the Himalayan (General Principles). *Geological Memoir, Ministry of Geology and Mineral Resources, China* **5:7**, 236 (1988).

[40] Xu Zhiqin *et al.*, Tectonophysical process at depth for the uplift of the N. part of the Qinghai-Tibet plateau. *Acta Geologica Sinica* **70:3**, 196-206 (1996).

[41] Zhou Yaoxin *et al.*. The palaeomagnetic study on the Tibet plateau and preliminary discussion on its tectonic evolution. *Geological Memoir, Ministry of Geology and Mineral Resources, China* **7:6**, 186 (1990).

[42] Zhang Qingrong *et al.*. On the present uplift speed of Qinghai-Tibet plateau. *Chinese Science Bulletin* **36:21**, 285-288 (1992).

ORIGIN AND HISTORY OF THE EARTH

Proc. 30th Intern. Geol. Congr. Vol. 1, pp. 61-78
Wang *et al.* (Eds)

Mantle Degassing and Origin of the Atmosphere

YOUXUE ZHANG
Department of Geological Sciences, The University of Michigan, Ann Arbor, MI 48109-1063, USA

Abstract

This paper first reviews noble gas isotopic data and mantle degassing models. The different models for mantle degassing and origin of the atmosphere are then critically evaluated. All noble gas constraints are presented in a model-independent form so that different models can be examined. Common features and differences of most degassing models are discussed. He-Ne systematics are shown to be more consistent with the notion that Ne degassed more rapidly relative to He. That is, the systematics are more consistent with a solubility-controlled mantle degassing model. However, due to relatively large uncertainties in the isotopic and concentration ratios of various reservoirs, the systematics do not rule out other degassing models at present. Whether the atmosphere originated entirely from mantle degassing is more complicated. Radiogenic nuclides of noble gases (^{40}Ar and $^{21}Ne^*$) almost certainly originated from mantle degassing. Nonradiogenic nuclides of noble gases in the atmosphere may be entirely from mantle degassing if Loihi mantle is 71% degassed. They cannot be entirely from mantle degassing if Loihi mantle is truly undegassed. Fractionation of nonradiogenic Xe isotopes between air/mantle Xe and U-Xe most likely occurred before Xe accreted on earth.

Keywords: air, atmosphere origin, earth evolution, mantle degassing, volatiles

INTRODUCTION

Noble gas isotopic systems have provided powerful constraints on understanding mantle degassing and origin/evolution of the atmosphere. Many models have been proposed. Most of these publications only discuss degassing using a set of model assumptions without a critical evaluation of other models. Previous authors often treat mantle degassing and origin of the atmosphere as one topic in which the atmosphere originates entirely from mantle degassing [*e.g.*, 2, 3, 33, 45, 46, 26]. Ozima and Zahnle [26] discussed in detail some of the major noble gas constraints. Marty [19] proposed that the atmosphere did not entirely originate from mantle degassing. To allow the possibility that only part of the atmosphere originated from mantle degassing and the rest from other sources (such as accretion degassing, cometary injection, etc. [5]), mantle degassing and origin of the atmosphere will be discussed separately in this paper. I first review various noble gas constraints on mantle degassing and examine ways to test models for mantle degassing and atmospheric evolution. The constraints from actual data are presented, instead of model calculations as in many previous publications. I then evaluate different degassing models incorporating recent development in noble gas geochemistry. The origin of the atmosphere is discussed by evaluating whether production of radiogenic and nucleogenic nuclides in the solid earth is enough to provide all the radiogenic nuclide in the atmosphere and whether the amount of primordial nonradiogenic noble gas nuclides in the solid earth is enough to account for their abundance in the atmosphere.

All noble gas elements have isotopes that receive a radiogenic or nucleogenic (for simplicity, hereafter "radiogenic" will be used to refer to both radiogenic and nucleogenic) contribution. Of the two stable helium isotopes (^{3}He and ^{4}He), ^{3}He receives only a minor nucleogenic contribution that is often ignored, whereas most ^{4}He in the earth is produced by the a-decay of ^{232}Th, ^{235}U and ^{238}U. Of the three stable isotopes of Ne (^{20}Ne, ^{21}Ne and ^{22}Ne), on a global scale, only ^{21}Ne receives a significant nucleogenic contribution through reactions ^{18}O(a,*n*)^{21}Ne and ^{24}Mg(*n*,a)^{21}Ne [44]. Of the three stable Ar isotopes (^{36}Ar, ^{38}Ar, and ^{40}Ar), only ^{40}Ar is radiogenic. Practically all ^{40}Ar in the earth is due to the decay of ^{40}K. Kr isotopic system is rarely discussed in depth because it mimics the Xe isotopic system but shows smaller effects that are more difficult to quantify. Xe isotopic system is complex with nine stable isotopes (^{124}Xe, ^{126}Xe, ^{128}Xe, ^{129}Xe, ^{130}Xe, ^{131}Xe, ^{132}Xe, ^{134}Xe and ^{136}Xe). Among these, ^{129}Xe, ^{131}Xe, ^{132}Xe, ^{134}Xe and ^{136}Xe receive a fission-genic contribution due to fission of ^{235}U, ^{238}U, ^{232}Th, and ^{244}Pu (an extinct nuclide with a half life of 80 million years). The fission-genic component contributes mostly to ^{136}Xe and ^{134}Xe but only negligibly to ^{129}Xe. ^{129}Xe receives a significant radiogenic contribution from ^{129}I, an extinct nuclide with a half life of 15.7 million years.

All the parental nuclides that decay to produce the radiogenic noble gas nuclides are less volatile than the noble gases. Therefore, as a part of the mantle becomes degassed (see [45, 26] for degassing mechanisms), the concentrations of the noble gas elements decrease and the ratios of the parent to the nonradiogenic noble gas nuclide (such as ^{238}U/^{3}He, ^{238}U/^{22}Ne, ^{40}K/^{36}Ar, ^{129}I/^{130}Xe, and ^{244}Pu/^{130}Xe) increase. With the passage of time, the radiogenic nuclides of the noble gas elements are produced by the parental nuclides, leading to a higher ratio of radiogenic isotope to a nonradiogenic isotope (such as ^{4}He/^{3}He, ^{21}Ne/^{22}Ne, ^{40}Ar/^{36}Ar, ^{129}Xe/^{130}Xe, ^{136}Xe/^{130}Xe; collectively referred to as the R/N ratio). The more degassed the part of the mantle, the greater the parent/daughter ratio, and the more rapid increase of the R/N isotopic ratios with time. Therefore, isotopic ratios of noble gas elements in a mantle-derived rock indicate the relative degree of degassing of the mantle source when contamination is negligible or can be corrected.

True (uncontaminated and unfractionated) noble gas isotopic ratios and concentration patterns in different mantle reservoirs are essential in interpreting noble gas isotopic systematics. Because isotopic ratio variations in noble gases are large, isotopic fractionations due to melting, crystallization, and degassing are ignored in discussing noble gases in mantle-derived rocks. However, because noble gas concentrations in mantle-derived rocks are low compared to those in air (except for He), measurement of noble gas isotopic ratios is difficult and is often plagued by air contamination, either in nature before the rock cooled down or in the lab. Easiness of contamination is related to the concentration ratio of the nuclide in the atmosphere to that in the rock. Because He in the atmosphere escapes to outer space, atmospheric He content is low compared to He in rocks. Hence, air contamination to measured ^{4}He/^{3}He ratios is often negligible. Therefore, mantle-derived rocks with a smaller ^{4}He/^{3}He ratio indicates a lower degree of degassing of the mantel source compared to those with greater ^{4}He/^{3}He. As shown later, even though Ne isotopic ratios are almost always contaminated by air Ne, it is possible now to correct for air contamination. He isotopic ratios and the contamination-corrected Ne isotopic ratios may provide powerful constraints in understanding mantle degassing.

NOBLE GAS ISOTOPIC RATIOS IN DIFFERENT RESERVOIRS

In using noble gas isotopic systems to model mantle degassing, the mantle is often divided into two reservoirs [2, 3, 33, 45, 46, 26] based on information from He and other noble gases: one is MORB (mid-ocean ridge basalt) mantle (the degassed mantle) and the other is Loihi mantle (the less degassed mantle). The MORB mantle has also been referred to as the upper mantle. However, among others, Zhang and Zindler showed that ^{40}Ar in the air requires that at least 43% and probably 70% of the whole mantle is degassed [45]; that is, the volume of degassed mantle is greater than that of the upper mantle. The Loihi mantle has also been referred to as the *undegassed* mantle. The justification for the undegassed mantle claim is not clear since noble gas isotopic ratios can only show that it is less degassed than MORB mantle. In this paper, the Loihi mantle is not a priori assumed to be *undegassed* [26]; it is less degassed. The atmosphere and continental crust (including oceans) are often treated as another single reservoir in terms of volatile components. Therefore, there are three volatile reservoirs for the earth, referred to as AC (atmosphere plus continental crust), MM (MORB mantle), and LM (Loihi mantle or less degassed mantle). The following is a review of noble gas isotopic data.

He

$^4He/^3He$ ratio in air is 7.15×10^5 [25]. The initial $^4He/^3He$ ratio in the mantle (that is, the ratio at the time when the earth formed) is not well known. The present-day solar ratio is 2190 [4]. The ratio measured in meteorites (the "planetary" component) is ~7000 [25]. Since initial Ne in the mantle is not "planetary" (see below), the planetary $^4He/^3He$ ratio may not be applicable to the primordial mantle. However, the present-day solar ratio may not represent the primordial ratio either because it may have been affected by nuclear reactions in the sun [25]. Recently, $^4He/^3He$ ratio in Jupiter's atmosphere was determined to be ~9000 [22], which may represent the initial solar ratio. Therefore, the initial $^4He/^3He$ ratio in the mantle may be between 2190 and 9000. The $^4He/^3He$ ratio in MORB glasses is relatively uniform, 8.5×10^4 [15, 18], which is taken to be the ratio in MM. This ratio is much less than the ratio in the atmosphere. The high $^4He/^3He$ ratio in the atmosphere is due to the preferential escape of the lighter ^3He from the atmosphere to the outer space compared to the heavier ^4He. $^4He/^3He$ ratio in OIB glasses are variable and can be as low as 2.2×10^4 [3, 10, 11, 36]. The lowest ratio is regarded to be representative of LM because contamination by MM and by air increases the ratio. The less radiogenic $^4He/^3He$ ratio in LM suggests that LM is less degassed. The variability in $^4He/^3He$ ratio in OIB glasses has been attributed to mixing with MM and/or ingrowth of ^4He in the magma chamber [47].

Ne

The $^{20}Ne/^{22}Ne$ and $^{21}Ne/^{22}Ne$ ratios in air are 9.80 and 0.0290 [25]. The initial $^{20}Ne/^{22}Ne$ ratio in the mantle is not completely constrained. The solar wind $^{20}Ne/^{22}Ne$ and $^{21}Ne/^{22}Ne$ ratios are 13.8±0.1 and 0.0328±0.0005 [4]. The meteoritic Ne contains two major components, one (component B, a solar Ne component) with $^{20}Ne/^{22}Ne$ and $^{21}Ne/^{22}Ne$ ratios of 12.5±0.2 and 0.0335±0.0015, and the other (component A, the "planetary component") with $^{20}Ne/^{22}Ne$ and $^{21}Ne/^{22}Ne$ ratios of 8.2 and 0.024 [25]. Note that the $^{21}Ne/^{22}Ne$ ratio in the solar wind is the same within error as that in component B despite a difference in $^{20}Ne/^{22}Ne$. Because the accuracy of the solar wind $^{21}Ne/^{22}Ne$ ratio is greater,

0.0328 will be used for the $^{21}Ne/^{22}Ne$ ratio in both component B and solar wind Ne. To determine the Ne isotopic ratios in the initial mantle requires an examination of the present-day mantle. $^{20}Ne/^{22}Ne$ and $^{21}Ne/^{22}Ne$ ratios are variable in MORB [6, 9, 19, 34]. In a $^{20}Ne/^{22}Ne$ vs. $^{21}Ne/^{22}Ne$ plot (three isotope plot), most MORB glass samples lie along a straight line, trending from air ratios toward an endmember with high $^{20}Ne/^{22}Ne$ and $^{21}Ne/^{22}Ne$ [34]. The straight line is explained by mixing of mantle Ne (with high $^{20}Ne/^{22}Ne$ and $^{21}Ne/^{22}Ne$ ratios) with air Ne [10, 32, 31]. That is, air contamination to Ne isotopic ratios is ubiquitous and the endmember with high $^{20}Ne/^{22}Ne$ and $^{21}Ne/^{22}Ne$ ratios is taken to represent MM. Most $^{20}Ne/^{22}Ne$ ratios in MORB are below or at 12.5 and a few approaching 13.8 but with large uncertainties [34], suggesting that the $^{20}Ne/^{22}Ne$ ratio in the initial and MM is at least 12.5 [6] and may be as high as 13.8. Hence, primordial mantle Ne is solar (instead of "planetary"), with a $^{20}Ne/^{22}Ne$ ratio of 12.5 or 13.8, and a $^{21}Ne/^{22}Ne$ ratio of 0.0328. The low $^{20}Ne/^{22}Ne$ and $^{21}Ne/^{22}Ne$ ratios in the atmosphere can be explained by the preferential loss of the lighter ^{20}Ne and ^{21}Ne during hydrodynamic escape of Ne from the atmosphere [26]. The $^{20}Ne/^{22}Ne$ ratio in the present-day MM is assumed to be the same as the initial ratio because nucleogenic production of ^{20}Ne and ^{22}Ne is negligible and isotopic fractionation due to degassing is assumed to be negligible. The $^{21}Ne/^{22}Ne$ ratios in the present-day MM are greater than the primordial ratio due to nucleogenic growth. $^{20}Ne/^{22}Ne$ and $^{21}Ne/^{22}Ne$ are also variable in OIB, forming a trend from air Ne to an end member with high $^{20}Ne/^{22}Ne$ ratio [21]. At the same $^{20}Ne/^{22}Ne$ ratio, OIB samples have lower $^{21}Ne/^{22}Ne$ ratio compared to MORB glasses. That is, LM is less nucleogenic than MM in terms of ^{21}Ne. The result implies lower degree of degassing for LM, consistent with He isotopic systematics. Contamination of OIB samples by MM causes some Loihi and other OIB samples to have greater $^{21}Ne/^{22}Ne$ ratios.

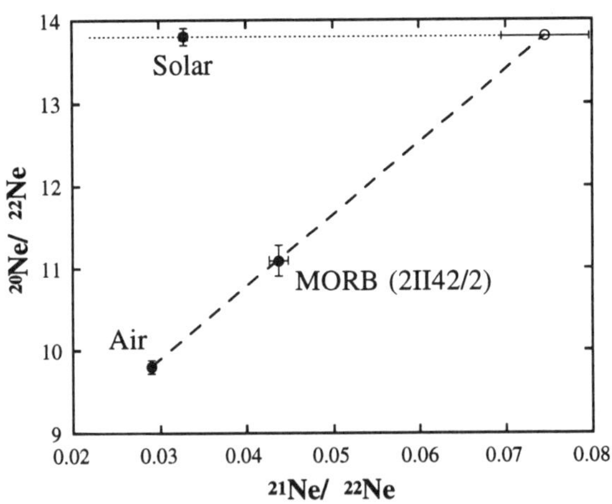

Figure 1. Calculation of $(^{21}Ne/^{22}Ne)_{13.8}$, defined as the $^{21}Ne/^{22}Ne$ ratio of a samples at $^{20}Ne/^{22}Ne = 13.8$ assuming the measured Ne isotopic ratio represents a mixture between the uncontaminated sample and air. The data for the sample is from Sarda *et al.* [34]. The "decontamination" can be graphically achieved by drawing a straight line from air Ne ratios to samples Ne ratios (labeled) and extending the straight line to $^{20}Ne/^{22}Ne = 13.8$ (open circle with error bar). The error bar of the calculated $(^{21}Ne/^{22}Ne)_{13.8}$ is obtained by statistically processing errors in $^{20}Ne/^{22}Ne$ and $^{21}Ne/^{22}Ne$ ratios in air and the sample. The calculated $(^{21}Ne/^{22}Ne)_{13.8}$ ratio is assumed to represent the true $^{21}Ne/^{22}Ne$ in the mantle before any air contamination.

On the basis of the above understanding, it is possible to correct for air contamination to Ne isotopic ratios by examining the three isotopes of Ne. True $^{20}Ne/^{22}Ne$ in uncontaminated MORB and OIB is 12.5 or 13.8. Air-contamination-free $^{21}Ne/^{22}Ne$ ratio in each MORB and Loihi sample can be calculated by subtracting the air component so that $^{20}Ne/^{22}Ne$ ratio is 12.5 or 13.8 (see Figure 1). The corrected ratios will be referred to as either $(^{21}Ne/^{22}Ne)_{12.5}$ or $(^{21}Ne/^{22}Ne)_{13.8}$.

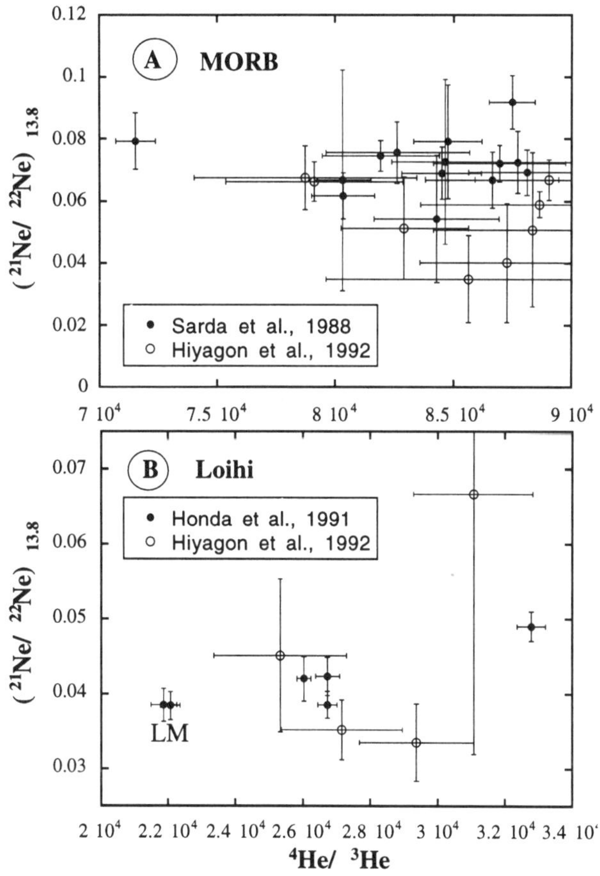

Figure 2. $(^{21}Ne/^{22}Ne)_{13.8}$ versus $^{4}He/^{3}He$ in MORB and Loihi glass samples. $(^{21}Ne/^{22}Ne)_{13.8}$ is the $^{21}Ne/^{22}Ne$ ratio calculated by removing the air component so that $^{20}Ne/^{22}Ne = 13.8$. It is calculated as $^{21}Ne/^{22}Ne + (13.8 - {}^{20}Ne/^{22}Ne)(^{21}Ne/^{22}Ne - 0.029)/(^{20}Ne/^{22}Ne-9.8)$ where $^{21}Ne/^{22}Ne$ and $^{20}Ne/^{22}Ne$ are the measured ratios reported in literature (see below). Shown are also 1s errors calculated from errors in original data. Data points with large error bars extending through the entire diagram are not included.

Figure 2 shows $(^{21}Ne/^{22}Ne)_{13.8}$ versus $^{4}He/^{3}He$ for MORB and Loihi samples. For MORB samples, $(^{21}Ne/^{22}Ne)_{13.8}$ ratios by Hiyagon *et al.* [9] are consistently smaller than those by Sarda *et al.* [34]. Lower $(^{21}Ne/^{22}Ne)_{13.8}$ ratios in MORB (approaching the ratio in Loihi mantle) are probably due to some measurement difficulties or the involvement of plume material (though the high $^{4}He/^{3}He$ ratio is not entirely consistent with this contention). For the Loihi samples, the data by Hiyagon *et al.* [9] do not define a reasonable trend. Hence, the data by Hiyagon *et al.* are not used in deriving the $^{21}Ne/^{22}Ne$ and $^{4}He/^{3}He$ for the Loihi and MORB mantle. The $(^{21}Ne/^{22}Ne)_{13.8}$ ratio in MORB is obtained by simply averaging the data by Sarda *et al.* [34] in Figure 2. The true $^{21}Ne/^{22}Ne$ in LM is chosen to be the lowest $(^{21}Ne/^{22}Ne)_{12.5}$ or $(^{21}Ne/^{22}Ne)_{13.8}$ in Loihi glasses, similar to the case for $^{4}He/^{3}He$. The data by Honda *et al.* [10] show a rough correlation between $(^{21}Ne/^{22}Ne)_{13.8}$ versus $^{4}He/^{3}He$, presumably due to the mixing of MORB mantle component with the Loihi mantle component. The two samples with lowest $^{4}He/^{3}He$ and $(^{21}Ne/^{22}Ne)_{13.8}$ (labeled LM in Figure 2b) are assumed to represent $^{21}Ne/^{22}Ne$ and $^{4}He/^{3}He$ for the least degassed component in LM. The choice of $^{4}He/^{3}He$ and $(^{21}Ne/^{22}Ne)_{13.8}$ from the same samples to represent LM assures internal consistency. This internally consistent choice differs from that of Porcelli and Wasserburg [31] who used the lowest $^{4}He/^{3}He$ ratio in Loihi to represent He in LM but the average $(^{21}Ne/^{22}Ne)_{12.5}$ ratio in Loihi to represent Ne in LM. Although the diagrams show how $(^{21}Ne/^{22}Ne)_{13.8}$ ratios are obtained, $(^{21}Ne/^{22}Ne)_{12.5}$ in MM and LM can be obtained in a similar fashion.

On the basis of the above discussion, the best estimates of $^{4}He/^{3}He$ and $^{21}Ne/^{22}Ne$ ratios (air-contamination corrected) in the MORB mantle (MM) and Loihi mantle (LM) are shown in Table 1. The $^{3}He/^{22}Ne$ in LM can be estimated from $(^{4}He^{*}/^{21}Ne^{*})[(^{21}Ne/^{22}Ne)_{LM} - (^{21}Ne/^{22}Ne)_{0}]/[(^{4}He/^{3}He)_{LM} - (^{4}He/^{3}He)_{0}]$, where the superscript * signifies radiogenic. Using a $^{4}He^{*}/^{21}Ne^{*} = 10^{7}$ [17], the $^{3}He/^{22}Ne$ ratio in LM is 1.6 if $(^{21}Ne/^{22}Ne)_{12.5}$ is used, and is 3.5 if $(^{21}Ne/^{22}Ne)_{12.5}$ is used, similar to the results of Honda *et al.* [10].

Table 1. He and Ne isotopic ratios in LM and MM

Isotopic ratio	Initial	LM	MM
$^{4}He/^{3}He$	$2.2x10^{3}$, $9x10^{3}$	$2.2x10^{4}$	$8.5x10^{4}$
$(^{21}Ne/^{22}Ne)_{13.8}$	0.0328	0.0384	0.075
$(^{21}Ne/^{22}Ne)_{12.5}$	0.0328	0.0354	0.060

Ar

No significant variation in $^{38}Ar/^{36}Ar$ ratio has been found in the earth, implying negligible fractionation [26]. $^{40}Ar/^{36}Ar$ ratio in the air is 296 [25]. The initial $^{40}Ar/^{36}Ar$ ratio is ~0.0003 [25], meaning essentially all ^{40}Ar in the earth is radiogenic. $^{40}Ar/^{36}Ar$ ratio in MORB samples is much greater than the air ratio and highly variable. Highest $^{40}Ar/^{36}Ar$ ratio in MORB is $3x10^{4}$ [33, 2, 38], ~100 times greater than that in air. The highest ratio is interpreted to approach the ratio in MM, and the lower ratios due to air contamination. This interpretation is supported by the inferred ubiquitous air contamination to Ne isotopes in MORB. $^{40}Ar/^{36}Ar$ ratio in Loihi is also variable but is not much greater than the air ratio [3, 11, 33]. Because MORB contamination increases $^{40}Ar/^{36}Ar$ ratio and air contamination reduces it, there is no consensus on a true estimate of $^{40}Ar/^{36}Ar$ ratio in LM [27, 28, 37]. The $^{40}Ar/^{36}Ar$ ratio in LM can be as low as 390 and as high as 10^{4} [37, 6, 26].

Kr and Xe

Variations in Kr isotopic ratios are usually not discussed. Variations in Xe isotopic ratios are complex. Xe isotopic ratios in the atmosphere are: $^{124}Xe/^{130}Xe = 0.02337$, $^{126}Xe/^{130}Xe$

= 0.02180, $^{128}Xe/^{130}Xe$ = 0.4715, $^{129}Xe/^{130}Xe$ = 6.496, $^{131}Xe/^{130}Xe$ = 5.213, $^{132}Xe/^{130}Xe$ = 6.607, $^{134}Xe/^{130}Xe$ = 2.563, and $^{136}Xe/^{130}Xe$ = 2.176 [25]. The initial Xe isotopic ratios in solar Xe are very different from those in air and the difference can be attributed to mass fractionation [25]. Because there are some variations and irregularities in measured solar Xe pattern, the initial solar Xe isotopic ratios are estimated from fractionated air Xe and other considerations [29]. The estimated initial solar pattern is referred to as U-Xe with $^{124}Xe/^{130}Xe$ = 0.02947, $^{126}Xe/^{130}Xe$ = 0.02541, $^{128}Xe/^{130}Xe$ = 0.5087, $^{129}Xe/^{130}Xe$ = 6.287, $^{131}Xe/^{130}Xe$ = 4.996, $^{132}Xe/^{130}Xe$ =6.048, $^{134}Xe/^{130}Xe$ =2.129, and $^{136}Xe/^{130}Xe$ = 1.663 [29]. In terms of radiogenic Xe, compared to the air, Xe in MORB is enriched in ^{129}Xe (recall that ^{129}Xe is radiogenic from ^{129}I) and fission-genic Xe, especially ^{136}Xe. The $^{129}Xe/^{130}Xe$ and $^{136}Xe/^{130}Xe$ ratios in MORB can be as high as 7.5 and 2.68 [35, 38], compared to the ratios of 6.5 and 2.18 in air. In a $^{129}Xe/^{130}Xe$ vs. $^{136}Xe/^{130}Xe$ plot, most MORB glasses lie along a straight line, trending from air ratios to higher ratios. The highest ratios are believed to approach the true $^{129}Xe/^{130}Xe$ and $^{136}Xe/^{130}Xe$ ratios in the MM, and the lower ratios due to air contamination. For OIB samples, $^{129}Xe/^{130}Xe$ and $^{136}Xe/^{130}Xe$ ratios may be just barely greater than those in air. The true ratios in LM are difficult to obtain because contamination by MM increases the ratios and contamination by air reduces the ratios.

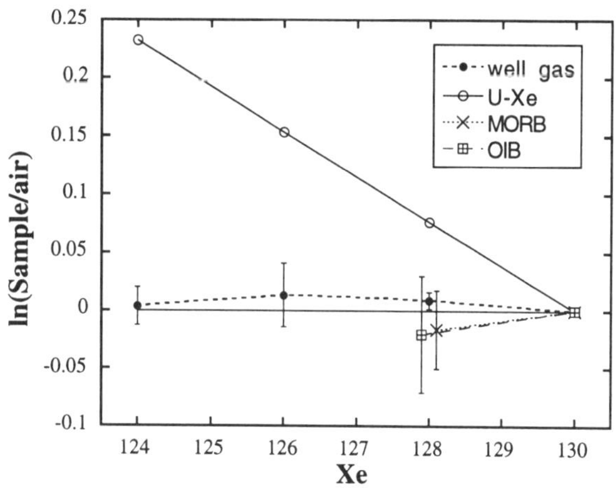

Figure 3. Nonradiogenic Xe isotopes normalized to air Xe. Values of $\ln[(^{i}Xe/^{130}Xe)_{sample}/(^{i}Xe/^{130}Xe)_{air}]$ are plotted on the vertical axis, where *i* can be 124, 126 and 128. Isotopic ratios of U-Xe are from Pepin [29]. Data for Xe in CO_2 well gas are from Phinney *et al.* [30]. Data published in the abstract form are not used because there is not enough explanation of the data. Data for Xe in MORB and OIB are averages using data of Allegre *et al.* [3, 9]. $(^{i}Xe/^{130}Xe)_{air}$ ratios from Ozima and Podosek [25] are used for normalization except for well-gas Xe, for which Phinney *et al.*'s air measurement [30] is used for self-consistency. Linear regression assuming Xe in CO2 well gas is a mixture of U-Xe and air Xe shows that the fraction of U-Xe is 0.05 ± 0.06 (2σ). That is, the fraction of U-Xe is indistinguishable from zero, but can also be as high as 11%. For clarity, datum on ^{128}Xe is plotted at "mass number" of 128.1 for MORB and at 127.9 for OIB.

Besides data from air, MORB and Loihi (and other OIB, ocean island basalts), important noble gas isotopic data are also obtained from CO_2-rich well gases [30]. Because large quantities of the gas can be collected, contamination in the lab is negligible. Hence, the noble gas isotopic ratios can be and have been measured to a much better accuracy than those of MORB and OIB glasses [*e.g.*, 30]. The high quality of the well gas Xe data allows two important applications which are not possible with MORB and OIB data due to the low accuracy. One is to resolve whether radiogenic $^{131-136}$Xe in well gases is plutogenic or uranogenic. The radiogenic $^{131-136}$Xe isotopes in well gases are mostly uranogenic [30] but those in air are mostly plutogenic [29]. Radiogenic $^{131-136}$Xe isotopes in the mantle are expected to be mostly plutogenic [31]. Hence Xe in well gases may not be of entirely mantle origin, suggesting caution in using the data. The second is to examine possible variations in non-radiogenic Xe isotopic ratios, such as ^{124}Xe/^{130}Xe, ^{126}Xe/^{130}Xe, and ^{128}Xe/^{130}Xe, so as to constrain whether Xe experienced hydrodynamic escape in the early history of the earth [26]. If these Xe isotopic ratios in the mantle are indeed different from those in the atmosphere, accurate measurement of the ratios in rocks allows correction for air contamination, similar to the case of Ne isotopic system. Figure 3 compares nonradiogenic Xe isotopes in the atmosphere, in CO_2-rich well gases, in MORB and OIB, and in U-Xe. Figure 3 shows Xe isotopic ratios in the mantle (including well gas) are similar to those in the air but different from those in U-Xe.

MANTLE DEGASSING MODELS

Early degassing models are mostly qualitative or considers isotopes of only one noble gas element (*e.g.*, [16, 43], see [25] for a review). With the availability of data on isotopic ratios of several noble gas elements, comprehensive quantitative models on mantle degassing and evolution of the atmosphere have been developed to account for all the isotopic data. The big pictures emerging from all these models are the same: (i) The mantle is not homogeneous in terms of noble gas distribution and there are a degassed mantle reservoir and a less degassed mantle reservoir; (ii) There are still primordial gases (such as primordial ^{3}He) coming out of the mantle; (iii) The degassed mantle is almost completely degassed; and (iv) Mantle degassing was an early event (most of the noble gases were outgassed during the first 100 million years of the earth's history) and the atmosphere is very old. However, in terms of quantitative details and in terms of degassing mechanisms, there are differences in the degassing models. These models can be classified into three kinds: bulk degassing (BD), solubility-controlled degassing (SCD) and steady-state degassing (SSD) models.

In the context of the BD model, all the gases degassed at the same relative rate and have the same time constant for degassing. The degassing rate is often assumed to be an exponential function of time, such as $e^{-t/t}$ where t characterizes the overall degassing time and is a constant to be solved. The BD model was developed by Sarda *et al.* [33] and Allegre *et al.* [2] but encountered the following difficulties in quantitative aspects: He isotopic ratios in different reservoirs demand slower degassing compared to Ar, which also seems to have degassed slower than Xe. Some ad hoc assumptions must be invoked about the difference between Ar and Xe and between He and Ar. To reconcile Ar and Xe isotopic ratios, Sarda *et al.* [33] proposed a two-stage (instead of one-stage) degassing model. To reconcile He and Ar isotopic ratios, Allegre *et al.* [2] proposed a He flux from LM to MM without fluxes of other noble gases. The SSD model [14, 24, 31] can be viewed as a refinement of their suggested flux from LM to MM.

In the context of the SCD model, mantle degassing is accomplished by first melting the mantle and then degassing the melt through bubbling [13, 45, 26]. The degassing can occur during magma ocean stage and at the present-day mid-ocean ridges. The degree of degassing of each gas is controlled by its solubility. For a noble gas with greater solubility (such as He), the melt retains a greater fraction of the gas and hence the degree of degassing is lower than a noble gas with a lower solubility (such as Xe). In this model, different gases had different degassing history, depending on its solubility in silicate melts. Because noble gas solubilities in melts have been experimentally determined (*e.g.*, [13]), solubilities are not free parameters in quantifying the model to reconcile isotopic data. Hart *et al.* [7] first explained the isotopic systematics of He and Ar in MORB and Loihi glasses using the difference of solubility of He and Ar in silicate melts. Motivated by the inability of the BD model to account for all the noble gas isotopic data without ad hoc assumptions, Zhang and Zindler [45] quantified the solubility argument and constructed a comprehensive model to explain the isotopic systematics of He, Ar and Xe. Although the model reconciled all the He, Ar and Xe isotopic data available at the time, a re-examination of the model is necessary because (i) some of the data [2, 3, 33, 35, 36] used in the model were later interpreted by some authors to be due to atmospheric contamination [27, 28], (ii) new advances in Ne isotopic geochemistry [9-11, 19, 21, 32, 34, 38] have not been examined in the context of the model, and (iii) other new models have been developed.

The SSD model assumes that the upper mantle itself has been nearly completely degassed a long time ago and present-day degassing at the mid-ocean ridges is at steady state with respect to the noble gas nuclides; that is, the degassing is due to flux from the lower mantle and the crust into the MORB mantle plus ingrowth of radiogenic nuclides [23, 14, 24, 19]. Earlier SSD models treated isotopic ratios of only one noble gas element (*e.g.*, [23, 14]). Recently, the SSD model has been developed into a comprehensive model to account for isotopic systematics of all the noble gases [31]. Since the SSD model considers only the more recent degassing history, whereas the SCD and BD models are more sensitive to the earliest degassing history (pre-4.4 Ga), the SSD model and other degassing models are not exclusive of each other. However, it is of interest to examine whether the present-day mantle degassing and noble gas isotopic systematics can be interpreted using the SSD model.

PREDICTIONS OF DEGASSING MODELS

All degassing models predict that MM is almost completely degassed. Hence, noble gas concentrations in MM are predicted to be much lower than that in LM. The atmosphere would have the same nonradiogenic noble gas pattern as the LM if the atmosphere is entirely due to mantle degassing. That is, $^{36}Ar/^{82}Kr$, $^{36}Ar/^{130}Xe$ ratios in LM would be the same as those in air. (Due to escape of ^{3}He to outer space and possible escape of ^{20}Ne, $^{3}He/^{36}Ar$ and $^{20}Ne/^{36}Ar$ ratios in air are not necessarily the same as those in LM.) The SSD model is similar to the BD model in many predictions except in cases involving extinct nuclides. The BD and SCD models treat subduction of trapped gas in crustal material as a kind of contamination whereas the SSD model treats subducted gas as part of the MM signature.

Concentration patterns

The BD and SSD models predict that MM and LM have the same nonradiogenic noble gas pattern (*i.e.*, $^{3}He/^{22}Ne$, $^{3}He/^{36}Ar$, $^{3}He/^{82}Kr$, and $^{3}He/^{130}Xe$ ratios in MM are the same as those in LM), whereas the SCD model predicts that MM has higher $^{3}He/^{22}Ne$,$^{3}He/^{36}Ar$, $^{3}He/^{82}Kr$, and $^{3}He/^{130}Xe$ ratios than the LM. However, both air contamination and degassing affects ^{22}Ne, ^{36}Ar, and ^{130}Xe concentrations in a major way. Hence, measured concentration patterns in MORB and Loihi glasses are unreliable and requires critical evaluation. Fortunately, air contamination to He is negligible. For Ne system, air contamination can be corrected (Figure 1) and hence He/Ne ratios will be used to test the models.

Isotopic ratios

$^{4}He/^{3}He$. All models predict that the $^{4}He/^{3}He$ ratio in MM is greater than that in LM (subducted material does not contain much He).

$^{21}Ne/^{22}Ne$. All models predict that the contamination-free $^{21}Ne/^{22}Ne$ ratio in MM is greater than that in LM. The degree of contamination can be estimated from $^{20}Ne/^{22}Ne$ (Figure 1).
$^{40}Ar/^{36}Ar$ and $^{136}Xe/^{130}Xe$. The BD and SCD models predict that the $^{40}Ar/^{36}Ar$ and $^{136}Xe/^{130}Xe$ ratios in MM are greater than those in LM, whereas in the context of the SSD model $^{40}Ar/^{36}Ar$ and $^{136}Xe/^{130}Xe$ in MM can be either greater or less than those in LM (since the ratios in MM can be increased through ingrowth but decreased through subduction of the crust).

$^{129}Xe/^{130}Xe$. The SCD and BD models predict a greater $^{129}Xe/^{130}Xe$ in MM than in LM. However, the SSD model predicts that $^{129}Xe/^{130}Xe$ in the MM is less than or equal to that in LM because (i) ^{129}Xe is no longer produced in the MM (hence $^{129}Xe/^{130}Xe$ in MM cannot be greater than that in LM) and (ii) the SSD model treats the subducted crust as part of MM and the $^{129}Xe/^{130}Xe$ ratio in the subducted crust is low. The SCD model and BD models predict that the LM would have $^{129}Xe/^{130}Xe$ and $^{136}Xe/^{130}Xe$ ratios only slightly greater than that in air if air originated from mantle degassing. As summarized above, $^{129}Xe/^{130}Xe$ ratio in MORB glasses can be very high but that in Loihi and other OIB glasses is never significantly different from the air ratio. Therefore, evidence from $^{129}Xe/^{130}Xe$ does not seem to be consistent with the SSD model even though air contamination of Xe isotopic ratios may be invoked to explain the air-like ratio in Loihi.

β values. Another difference between the BD, SSD, and SCD models is in terms of the relative degree of radiogenic contribution to MM and LM for different noble gases. Following Zhang and Zindler [45], this relative degree of radiogenic contribution to MM and LM can be measured using a parameter β (refeered as parameter *S* in Zhang and Zindler [45]) defined as (using $^{21}Ne/^{22}Ne$ as an example):

$$\beta_{Ne21/22} = \frac{\left(^{21}Ne/^{22}Ne\right)^{T}{}_{MM} - \left(^{21}Ne/^{22}Ne\right)^{0}{}_{MM}}{\left(^{21}Ne/^{22}Ne\right)^{T}{}_{LM} - \left(^{21}Ne/^{22}Ne\right)^{0}{}_{LM}}.$$

where the superscripts "T" and "0" means t = T (at present) and t = 0 (at the beginning of the earth), and the subscripts "MM" and "LM" means MORB mantle and Loihi mantle. $\beta_{He4/3}$, $\beta_{Ar40/36}$, $\beta_{Xe129/130}$, and $\beta_{Xe136/130}$ can be defined similarly. The β value is a ratio of the radiogenic component in the MORB mantle divided by that in the Loihi mantle, and depends on the difference in the time-integrated parent/daughter ratio and hence the degassing history of the two mantle reservoirs. Because nucleogenic $^{21}Ne^*$ production is related to $^4He^*$ production with a constant $^{21}Ne^*/^4He^*$ production ratio of 10^{-7} [17], $\beta_{Ne21/22}$ and $\beta_{He4/3}$ are related. In the context of BD and SSD models, $\beta_{Ne21/22} = \beta_{He4/3}$, and $\beta_{He4/3}$ is 0.74 to 2.7 times $\beta_{Ar40/36}$ [45]. In the context of SCD model, $\beta_{Ne21/22} \approx 2\beta_{He4/3}$, and $\beta_{Ar40/36} \approx 10\beta_{He4/3}$ where the coefficients 2 and 10 are roughly the solubility ratio (the solubility of He is roughly 2 times that of Ne and 10 times that of Ar, [13]).

Summary of major differences

The SSD model predicts a $^{129}Xe/^{130}Xe$ ratio in LM to be greater than or equal to that in MM, contrary to the prediction of the SCD and the BD model. The SSD model and the BD model predicts identical $^3He/^{22}Ne$ and $^3He/^{36}Ar$ in MM and LM and that $\beta_{Ne21/22} = \beta_{He4/3}$, whereas the SCD model predicts that $^3He/^{22}Ne$ and $^3He/^{36}Ar$ in MM are greater than those in LM and that $\beta_{Ne21/22} \approx 2\beta_{He4/3}$. These different predictions between the models are used below to test them.

AN EVALUATION OF DEGASSING MODELS

The following discussion tests the BD, SCD and SSD models using He-Ne systematics by comparing (i) $\beta_{He4/3}$ and $\beta_{Ne21/22}$, and (ii) $^{22}Ne/^3He$ in MORB and Loihi glasses. Other constraints are also briefly discussed.

Comparison of $\beta_{He4/3}$ and $\beta_{Ne21/22}$

From Table 1, $\beta_{Ne21/22}$ is 10.5 for correction to $^{20}Ne/^{22}Ne = 12.5$, and is 7.5 for correction to $^{20}Ne/^{22}Ne = 13.8$. The value of $\beta_{He4/3}$ is 4.2 if an initial $^4He/^3He$ of 2.2×10^3 is used, and is 5.8 if an initial ratio of 9×10^3 is used. Even though $\beta_{Ne21/22}$ is roughly two times $\beta_{He4/3}$, the large errors in the isotopic ratios and uncertainties in the initial $^4He/^3He$ and $^{20}Ne/^{22}Ne$ ratios also allow $\beta_{Ne21/22}$ and $\beta_{He4/3}$ to be roughly the same. Therefore, this test shows that the SCD model is more consistent with He-Ne isotopes, but does not rule out the BD and SSD models.

Comparison of $^{22}Ne/^3He$ ratio in MORB and Loihi glasses

The BD and SSD models predict identical $^{22}Ne/^3He$ ratio in MM and LM. However, in the context of the SCD model, $^{22}Ne/^3He$ ratio is fractionated by degassing and MM is predicted to have a lower $^{22}Ne/^3He$ ratio. The predicted $^{22}Ne/^3He$ ratio in MM is 0.43 times that in LM. To obtain $^{22}Ne/^3He$ ratio in MM and LM, contamination by air and fractionation due to degassing (due to different solubility) must be corrected. Though it is difficult to correct for fractionation due to recent degassing, contamination can be undone by correction to $^{20}Ne/^{22}Ne = 13.8$ (correction to $^{20}Ne/^{22}Ne = 12.5$ does not alter the conclusion). Corrected ^{22}Ne concentration will be referred to as $^{22}Ne_{13.8}$. With the correction, $^{22}Ne_{13.8}$ concentrations in MORB and Loihi glasses are similar. Because ^3He concentration in MORB glass is greater than that in Loihi glass, the $^{22}Ne_{13.8}/^3He$ ratio is greater in MORB glass than in Loihi glasses. For example, using the data of Sarda *et al.* [34], Honda *et al.* [10], and Hiyagon *et al.* [9], the $^{22}Ne_{13.8}/^{3}He$ ratio in MORB ranges from 0.006 to 0.4 (with one exception that has a strange $^{4}He/^{3}He$ ratio > 100,000) with a geometric average of 0.06, and the ratio in Loihi glass ranges from 0.1 to 0.7 with a geometric average of 0.25. Hence $^{22}Ne_{13.8}/^{3}He$ ratio in MORB is on average ~0.24 times than that in Loihi. The results are more consistent with the SCD model and.less consistent with the BD and SSD models although there are still uncertainties related to fractionation during degassing. Excluding the data of Hiyagon *et al.* [9] does not change the above results significantly.

Other constraints

Constraints involving Ar concentrations and isotopic ratios are difficult to apply because (i) there is no consensus on the $^{40}Ar/^{36}Ar$ ratio in Loihi mantle [10, 16, 18, 24-26], and (ii) $^{40}Ar/^{36}Ar$ ratio in MM may be close to the highest measured ratio (3×10^4) if the uncontaminated $^{20}Ne/^{22}Ne$ ratio in MM is 12.5 (because highest measured $^{20}Ne/^{22}Ne$ is close to 12.5), but it may also be much greater than 3×10^4 if the uncontaminated $^{20}Ne/^{22}Ne$ ratio in MM is 13.8 [5, 27]. Therefore, $\beta_{Ar40/36}$ is not estimated in this work. Similarly, $\beta_{Xe129/130}$ and $\beta_{Xe136/130}$ are not estimated. Estimation of ^{36}Ar and ^{130}Xe concentrations in mantle reservoirs is difficult due to air contamination and degassing. Hence, it is impossible to evaluate whether the $^{3}He/^{36}Ar$, and $^{36}Ar/^{130}Xe$ ratios in mantle reservoirs are consistent with the SCD model [45, 26]. Since it has not been proven that the entire atmosphere originated from mantle degassing, $^{36}Ar/^{130}Xe$ ratios (or other ratios) in MM and in air should not be used to evaluate the validity of the models

Summary

In summary, the combined weight of the above discussion shows that the He-Ne systematics are more consistent with the SCD model than the BD and SSD models, although the BD and SSD models cannot be definitively ruled out. I argue that the solubility-controlled degassing played a dominant role in controlling the He-Ne systematics. This conclusion is independent of the origin of noble gases in air (whether there was initial air before mantle degassing, whether hydrodynamic escape is important, or whether there is continuous injection of noble gases from comets/meteorites) because only two mantle reservoirs are compared to reach the conclusion. Furthermore, the conclusion is independent of whether Loihi mantle is undegassed or partially degassed as long as it is less degassed than MM. Even though the SCD model seems to be more consistent with He-Ne systematics, no comprehensive remodeling is done using the SCD model due to difficulties in estimating the relevant β values.

ORIGIN OF THE EARTH'S ATMOSPHERE

Does the atmosphere originate entirely from mantle degassing?

Ozima and Podosek [25] reviewed arguments against the idea that the earth's atmosphere was formed by gravitational capture of ambient solar nebula gas. The volatile composition of the earth is simply different from such a captured nebula gas. For example, $N_2/^{36}Ar$ ratio in the earth's atmosphere is 2.4×10^4, whereas the ratio in the nebula gas is 16. Hydrodynamic escape cannot account for the high $N_2/^{36}Ar$ ratio because N_2 would have escaped from the earth preferentially. Similarly, one may infer that the earth did not acquire the gases through magma ocean equilibrium with nebula gas because the solubilities of N_2 and Ar are not very different (see [46] for a review). Two facts support that the atmosphere (at least part of the atmosphere) is due to mantle degassing: (i) the earth is still degassing at mid-ocean ridges and (ii) the only plausible source for nuclides such as ^{40}Ar is by degassing the interior of the earth. Therefore, many previous authors assumed closed system degassing and no initial atmosphere; that is, they assumed that the atmosphere completely originated from mantle degassing [*e.g.*, 3, 7, 33, 2, 45, 46], although comets and meteorites may also contribute to the volatile inventories of the earth [*e.g.*, 5, 19]. The origin of noble gases in the atmosphere can be constrained from whether the radiogenic and nonradiogenic noble gases in the mantle are enough to provide noble gases in air as reviewed below.

There is a general consensus on the U, Th and K concentration in the primitive mantle [12, 40, 41, 8, 20, 1]. Using the concentrations of K (260 ppm), U (21 ppb), and Th (86 ppb) in the primitive mantle recommended in the most recent three publications [8, 20, 1], production of radiogenic ^{4}He* and ^{40}Ar* over the entire history of the earth is 7.0×10^{18} mol and 3.8×10^{18} mol. Total ^{40}Ar in the atmosphere (1.65×10^{18} mol, [25]) can be derived by degassing 43% of the total mantle. Considering (i) the amount of ^{40}Ar in the continental crust and (ii) the amount of ^{40}Ar in the degassed mantle (though amount of ^{36}Ar is negligible), Zhang and Zindler [45] estimated that ~70% of the mantle has been degassed to provide ^{40}Ar in the atmosphere. The amount of ^{4}He in the atmosphere (9.3×10^{14} mol, [25]) is minuscule compared to the total amount of radiogenic ^{4}He*, consistent with our understanding of its escape to outer space.

The nucleogenic ^{21}Ne* to radiogenic ^{4}He* ratio in the mantle has been estimated to be 10^{-7} [17]. Using this ratio, production of ^{21}Ne* over the entire history of the earth is 7.0×10^{11} mol. ^{21}Ne* in air can be estimated as follows. If initially air had a ^{20}Ne/^{22}Ne ratio of 13.8 and ^{21}Ne/^{22}Ne ratio of 0.0328, fractionation to ^{20}Ne/^{22}Ne = 9.8 would leave a ^{21}Ne/^{22}Ne ratio of 0.0277. This would imply 3.9×10^{11} mol nucleogenic ^{21}Ne in air, ~56% of total ^{21}Ne* production in the mantle for a ^{21}Ne*/^{4}He* production ratio of 10^{-7}. If initial air had a ^{20}Ne/^{22}Ne ratio of 12.5 and ^{21}Ne/^{22}Ne ratio of 0.0328, fractionation to ^{20}Ne/^{22}Ne = 9.8 would leave a ^{21}Ne/^{22}Ne ratio of 0.029. There would be no significant nucleogenic ^{21}Ne* in air. Either way, ^{21}Ne* in air can be supplied by mantle degassing, in agreement with conclusions in [29, 31].

Because the production ratio of ^{136}Xe*/^{4}He* is ~2.3×10^{-9} [25] and ^{4}He* in the bulk silicate earth (BSE) is 7.0×10^{18} mol (above), total nucleogenic ^{136}Xe* in the BSE due to U is 1.6×10^{10} mol. Even if this amount is completely degassed, it accounts for only a quarter of the total ^{136}Xe* in air (6.3×10^{10} mol, [29]). Therefore, most of the nucleogenic ^{136}Xe* must come from ^{244}Pu [29]. Using estimated ^{244}Pu and ^{129}I concentrations in the initial earth, Porcelli and Wasserburg [31] showed that they can provide more than the inferred amount of ^{136}Xe* and ^{129}Xe* in the atmosphere.

The above argument shows that radiogenic nuclides in earth's atmosphere can be derived from mantle degassing. It is also important to estimate whether mantle degassing can supply all the nonradiogenic nuclides in the air. ^{3}He in the present-day air is minuscule and can of course be supplied by mantle degassing but most ^{3}He in air has escaped. For heavier noble gases, this is best done through the Ne system because air contamination can be corrected. To estimate total ^{22}Ne in the initial mantle, ^{21}Ne*/^{22}Ne ratio in the undegassed mantle is needed. If LM is undegassed, then the undegassed mantle would have a ^{21}Ne*/^{22}Ne ratio of 0.0384 - 0.0328 = 0.0056. Primordial ^{22}Ne in the BSE would be $7.0\times10^{11}/0.0056 = 1.25\times10^{14}$ mol, corresponding to a primordial ^{22}Ne concentration of 1.8×10^{10} atoms/g in the BSE [31]. The total amount of ^{22}Ne in the primordial BSE would be less than ^{22}Ne in air (2.97×10^{14} mol). That would mean that degassing of MM (70% of the whole mantle) would only provide 29% of ^{22}Ne in air [31]. (If a significant amount of ^{22}Ne escaped from air, the discrepancy is even greater.) However, Loihi mantle might have been degassed and the accuracy and applicability of the calculated ^{21}Ne*/^{4}He* production ratio are not known. If Loihi mantle does not represent undegassed mantle but was degassed by 71% early in Earth's history (*i.e.*, degassing of ^{21}Ne* from Loihi mantle was negligible), degassing of 70% of the mantle can supply all ^{22}Ne in air. In such a scenario, ~29% of the initial ^{22}Ne still remains in the Loihi mantle whereas only ~1% of the initial ^{22}Ne [45] still remains in the MORB mantle. Alternatively, if the ^{21}Ne*/^{4}He* production ratio were 3.4×10^{-7}, degassing of 70% of the mantle could supply all ^{22}Ne in air. A combination of some degree of degassing for Loihi mantle and a higher than 10^{-7} production ratio can also account for ^{22}Ne in air. If escape of ^{22}Ne from air (during the hydrodynamic escape that fractionated solar Ne into air Ne) is significant, the degassing of Loihi mantle must be more extensive or the ^{21}Ne*/^{4}He* production ratio in the mantle would have to be more extensive so as to derive ^{22}Ne in air by mantle degassing. Because

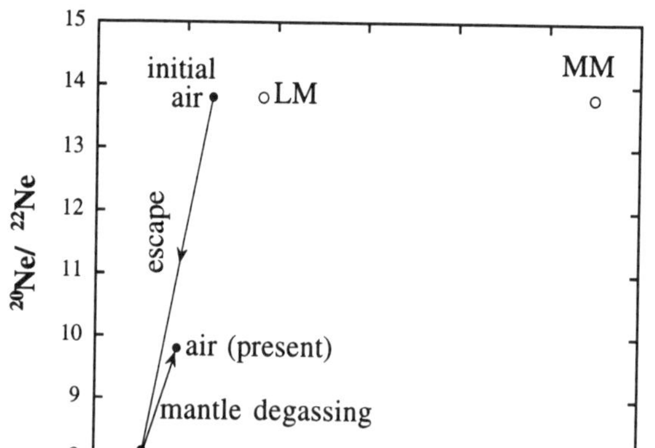

Figure 4. A possible scenario for isotopic evolution path of Ne in air (solid dots and lines with arrow). Initial Ne in air is assumed to be solar. Ne in present-day MM and LM are also shown (open circles). Hydrodynamic escape fractionated Ne isotopes to ^{20}Ne/^{22}Ne » 8.16 and ^{21}Ne/^{22}Ne » 0.0253. Mantle degassing added a component of Ne assumed to have an average ^{21}Ne/^{22}Ne » 0.038, slightly less nucleogenic than Loihi mantle Ne. In this scenario, the contribution of mantle degassing to air ^{22}Ne after the hydrodynamic escape is ~29%.

$^{38}Ar/^{36}Ar$ ratios in mantle-derived samples are indistinguishable from those in air, and because there does not seem to be a major fractionation in $^{15}N/^{14}N$ between mantle-derived samples and air, the loss of light gases during the early episode of hydrodynamic escape probably did not affect ^{36}Ar (mass 36) and N_2 (mass 28). Therefore, hydrodynamic escape probably did not extend much above mass 22. The Ne isotopes in air can be accounted for by starting from solar Ne, followed by hydrodynamic escape, and then followed by addition of Ne from degassing of the mantle (Figure 4).

Because the $^{40}Ar/^{36}Ar$ ratio in LM is not well constrained and because LM may not be the undegassed mantle, it is not possible to estimate the percentage of atmospheric ^{36}Ar that is from mantle degassing. If $^{40}Ar/^{36}Ar$ £ 480 in the undegassed mantle [2, 45, 26], then initial amount of ^{36}Ar in BSE was ³ 7.9×10^{15} mol and degassing of MM (70% of the whole mantle) can supply all atmospheric ^{36}Ar. If $^{40}Ar/^{36}Ar > 480$ in the undegassed mantle, then part of atmospheric ^{36}Ar does not originate from mantle degassing.

The above discussion suggests that an unambiguous determination of $^{40}Ar/^{36}Ar$ and $^{21}Ne^*/^{22}Ne$ in the undegassed mantle will constrain whether ^{36}Ar and ^{22}Ne in air are entirely from mantle degassing.

Hydrodynamic loss of Xe from air?

An important issue regarding the origin and early evolution of Earth's atmosphere is possible hydrodynamic loss and fractionation of Xe isotopes. The severe fractionation between atmospheric Xe and U-Xe prompted discussions of hydrodynamic escape of Xe from Earth. For example, Tolstikhin and O'Nions [24] proposed that (i) primordial mantle Xe was U-Xe, (ii) Xe in air originated from mantle degassing, and (iii) the difference in nonradiogenic Xe isotopic ratios in air and in U-Xe is due to hydrodynamic escape. This hypothesis is best tested by comparing mantle Xe with air Xe and U-Xe. Figure 3 shows the best published data for nonradiogenic Xe isotopes in CO_2 well gas [30], and $^{128}Xe/^{130}Xe$ in average MORB and OIB [3, 9] (other Xe isotopic ratios are not well constrained in MORB and OIB). Xe in well gas is treated to be mantle origin even though the radiogenic Xe isotopes are uranogenic [30]. (If the uranogenic nature of $^{131\text{-}136}Xe^*$ means that Xe in well gas does not represent mantle Xe, the following discussion of well gas data would not apply and there would be no constraint on the nonradiogenic Xe isotopes in the mantle.) Some authors claim that the data support the notion that mantle Xe is U-Xe (*e.g.*, [24]). However, in my opinion, the data show that nonradiogenic Xe isotopic ratios in CO_2 well gas are indistinguishable from those in air and are clearly different from those in U-Xe. The high $^{20}Ne/^{22}Ne$ and especially $^{21}Ne/^{22}Ne$ ratios, the high $^{40}Ar/^{36}Ar$ ratio, and the high $^{129}Xe/^{130}Xe$ and $^{136}Xe/^{130}Xe$ ratios in the CO_2 well gas suggest that air contamination to the well gas is not major. Therefore, if nonradiogenic Xe isotopic ratios in MM are similar to those in CO_2 well gas, then nonradiogenic Xe isotopic ratios in MM are similar to ratios in air within analytical error. That is, primordial nonradiogenic Xe is likely the same as nonradiogenic atmospheric Xe, or a mixture of predominantly air Xe with a minor U-Xe component. The similarity between nonradiogenic Xe in the mantle and in air is in clear contrast with the large difference in $(^{20}Ne/^{22}Ne)_{MM}$ and $(^{20}Ne/^{22}Ne)_{air}$. The large difference can be ascribed to hydrodynamic escape of Ne, whereas the similarity implies no hydrodynamic escape of Xe from Earth. The depletion of light Xe isotopes is likely a whole earth phenomena, not just an air phenomena. Some authors may invoke subduction of Xe to account for the similarity between mantle Xe and air Xe. However, such massive subduction to bring mantle Xe from U-Xe to the present-day Xe in CO_2 well gases would have wiped out the isotopic difference in $^{129}Xe/^{130}Xe$ and $^{136}Xe/^{130}Xe$ between mantle Xe and air Xe and between mantle Ne and air Ne [26]. In conclusion, the fractionation of Earth's Xe isotopes most likely occurred before or during the accretion of the earth and is not due to hydrodynamic escape of Xe from the atmosphere of the earth.

CONCLUSIONS AND FUTURE DIRECTIONS

Noble gas systems provide powerful constraints on the degassing history of the mantle and the origin of the atmosphere. Most mantle degassing models agree on the big picture of the mantle degassing process, though they disagree on details. Using model-independent constraints, the following conclusions can be reached:

(i) Isotopic and concentration ratios in MORB and Loihi mantle are more consistent with a solubility-controlled degassing model, but other models cannot be definitively ruled out.

(ii) Radiogenic nuclides in the atmosphere can be accounted for by mantle degassing. If Loihi mantle indeed represents undegassed mantle, mass balance calculations indicate that only part of the atmosphere originated from mantle degassing. However, if Loihi mantle has been partly degassed prior to the Loihi eruption, it is possible that all the atmosphere originated from mantle degassing (closed-system degassing assumption).

(iii) Hydrodynamic escape during the early evolution of the atmosphere probably did not significantly affect nuclides with mass numbers greater than 22. The fractionation of atmospheric Xe isotopes most likely occurred before the accretion of the earth.

The most important noble gas constraints in the future will probably come from major improvement in measuring Ne and Xe isotopic ratios because Ne and Xe isotopic systematics have the potential to resolve air contamination and mantle components (if nonradiogenic Xe isotopic ratios in the mantle are indeed somewhat different from those in the atmosphere). Determination of the $^{129}Xe/^{130}Xe$ ratio in LM (corrected for air contamination) will provide an important constraint to distinguish different degassing models.

Acknowledgements

This work was supported by the US NSF (EAR-9458368). I thank C.J. Ballentine, A.N. Halliday, D. Lee, M. Ozima, and L. Wang for discussions, and two anonymous reviewers for comments on an earlier version of this paper.

REFERENCES

[1] C. J. Allegre, J. P. Poirier, E. Humler and A. W. Hofmann. The chemical composition of the earth, *Earth Planet. Sci. Lett.* **134**, 515-526 (1995).

[2] C. J. Allegre, T. Staudacher and P. Sarda. Rare gas systematics: formation of the atmosphere, evolution and structure of the Earth's mantle, *Earth and Plan. Sci. Lett.* **81**, 127-150 (1986/87).

[3] C. J. Allegre, T. Staudacher, P. Sarda and M. Kurz. Constraints on evolution of Earth's mantle from rare gas systematics, *Nature* **303**, 762-766 (1983).

[4] J. P. Benkert, H. Baur, P. Signer and R. Wieler. He, Ne, and Ar from the solar wind and solar energetic particles in lunar ilmenites and pyroxenes, *J. Geophys. Res.* **98**, 13147-13162 (1993).

[5] C. F. Chyba. Impact delivery and erosion of planetary oceans in the early inner solar system, *Nature* **343**, 129-133 (1990).

[6] K. A. Farley and R. J. Poreda. Mantle neon and atmospheric contamination, *Earth planet. Sci. Lett.* **114**, 325-339 (1993).

[7] R. Hart, L. Hogan and J. Dymond. The closed-system approximation for evolution of argon and helium in the mantle, crust and atmosphere, *Chem. Geol.* **52**, 45-73 (1985).

[8] S. R. Hart and A. Zindler. In search of a bulk-earth composition, *Chem. Geol.* **57**, 247-267 (1986).

[9] H. Hiyagon, M. Ozima, B. Marty, S. Zashu and H. Sakai. Noble gases in submarine glasses from mid-oceanic ridges and Loihi seamount: Constraints on the early history of the Earth, *Geochim. Cosmochim. Acta* **56**, 1301-1316 (1992).

[10] M. Honda, I. McDougall, A. Doulgeris and D. A. Clague. Terrestrial primordial neon, *Nature* **352**, 388 (1991).

[11] M. Honda, I. McDougall, D. B. Patterson, A. Doulgeris and D. A. Clague. Noble gases in submarine pillow glasses from Loihi and Kilauea, Hawaii: A solar component in the Earth, *Geochim. Cosmochim. Acta* **57**, 859-874 (1993).

[12] E. Jagoutz, H. Palme, H. baddenhausen, K. Blum, M. Cendales, G. Dreibus, B. Spettel, V. Lorenz and H. Wanke. The abundances of major, minor and trace elements in the earth's mantle as derived from primitive ultramafic nodules, *Proc. Lunar Planet. Conf.* **10**, 2031-2050 (1979).

[13] A. Jambon, H. Weber and O. Braun. Solubility of He, Ne, Ar, Kr, and Xe in a basalt melt in the range of 1250-1600°C. Geochemical implications, *Geochim. Cosmochim. Acta* **50**, 401-408 (1986).

[14] L. H. Kellogg and G. J. Wasserburg. The role of plumes in mantle helium fluxes, *Earth Planet. Sci. Lett.* **99**, 276-289 (1990).

[15] M. D. Kurz and W. J. Jenkins. The distribution of helium in oceanic basalt glasses, *Earth Planet. Sci. Lett.* **53**, 41-54 (1981).

[16] M. D. Kurz, W. J. Jenkins and S. R. Hart. Helium isotopic systematics of oceanic islands and mantle heterogeneity, *Nature* **297**, 43-47 (1982).

[17] T. K. Kyser and W. Rison. Systematics of rare gas isotopes in basic lavas and ultramafic xenoliths, *J. Geophys. Res.* **87**, 5611-5630 (1982).

[18] J. E. Lupton. Terrestrial Inert gases: Isotope tracer studies and clues to primordial components in the mantle, *Ann. Rev. Earth. Planet. Sci.* **11**, 371-414 (1983).

[19] B. Marty. Neon and xenon isotopes in MORB: Implications for the earth-atmosphere evolution, *Earth Planet. Sci. Lett.* **94**, 45-56 (1989).

[20] W. F. McDonough and S. S. Sun. The composition of the Earth, *Chem. Geol.* **120**, 223-253 (1995).

[21] M. Moreira, T. Staudacher, P. Sarda, J. G. Schilling and C. J. Allegre. A primitive plume neon component in MORB: The Shona ridge-anomaly, South Atlantic (51-52°S), *Earth Planet. Sci. Lett.* **133**, 367-377 (1995).

[22] H. B. Niemann, S. K. Atreya, G. R. Carignan, T. M. Donahue, J. A. Haberman, D. N. Harpold, R. E. Hartle, D. M. Hunten, W. T. Kasprzak, P. R. Mahaffy, T. C. Owen, N. W. Spencer and S. H. Way. The Galileo probe mass spectrometer: Composition of Jupiter's atmosphere, *Science* **272**, 846-849 (1996).

[23] R. K. O'Nions and E. R. Oxburgh. Heat and helium in the Earth, *Nature* **306**, 429-431 (1983).

[24] R. K. O'Nions and I. N. Tolstikhin. Behavior and residence times of lithosphile and rare gas tracers in the upper mantle, *Earth Planet. Sci. Lett.* **124**, 131-138 (1994).

[25] M. Ozima and F. A. Podosek. *Noble Gas Geochemistry*. 367 pp., Cambridge U. Press, Cambridge, UK (1983).

[26] M. Ozima and K. Zahnle. Mantle degassing and atmospheric evolution: Noble gas view, *Geochem. J.* **27**, 185-200 (1993).

[27] D. Patterson, M. Honda and I. McDougall. Contamination of Loihi magmas with atmosphere derived noble gases: A reply to comments by T. Staudacher, P. Sarda and Allegre, *Geophys. Res. Lett.* **18**, 749-752 (1991).

[28] D. B. Patterson, M. Honda and I. McDougall. Atmospheric contamination: A possible source for heavy noble gases in basalts from Loihi Seamount, Hawaii, *Geophys. Res. Lett.* **17**, 705-708 (1990).

[29] R. O. Pepin. On the origin and early evolution of terrestrial planet atmospheres and meteoritic volatiles, *Icarus* **92**, 2-79 (1991).

[30] D. Phinney, J. Tennyson and U. Frick. Xenon in CO_2 well gas revisited, *J. Geophys. Res.* **83**, 2313-2319 (1978).

[31] D. Porcelli and G. J. Wasserburg. Mass transfer of helium, neon, argon, and xenon through a steady-state upper mantle, *Geochim. Cosmochim. Acta* **59**, 4921-4937 (1995).

[32] R. J. Poreda and K. A. Farley. Rare gases in Samoan xenoliths, *Earth Planet. Sci. Lett.* **113**, 129-144 (1992).

[33] P. Sarda, T. Staudacher and C. J. Allegre. $^{40}Ar/^{36}Ar$ in MORB glasses: Constraints on atmosphere and mantle evolution, *Earth Planet. Sci. Lett.* **72**, 357-375 (1985).

[34] P. Sarda, T. Staudacher and C. J. Allegre. Neon isotopes in submarine basalts, *Earth Planet. Sci. Lett.* **91**, 73-88 (1988).

[35] T. Staudacher and C. J. Allegre. Terrestrial xenology, *Earth Planet. Sci. Lett.* **60**, 389-406 (1982).

[36] T. Staudacher, M. D. Kurz and C. J. Allegre. New noble-gas data on glass samples from Loihi Seamount and Hualalai and on dunite samples from Loihi and Reunion Island, *Chem. Geol.* **56**, 193-205 (1986).

[37] T. Staudacher, P. Sarda and C. J. Allegre. Comment on "Atmospheric contamination: A possible source for heavy noble gases in basalts from Loihi Seamount, Hawaii" by D.B. Patterson, M. Honda, and I. McDougall, *Geophys. Res. Lett.* **18**, 745-748 (1991).

[38] T. Staudacher, P. Sarda, S. H. Richardson, C. J. Allegre, I. Sagna and L. V. Dmitriev. Noble gases in basalt glasses from a Mid-Atlantic Ridge topographic high at 14°N: geodynamic consequences, *Earth Planet. Sci. Lett.* **96**, 119-133 (1989).

[39] M. Stein and A.W. Hoffmann. Mantle plumes and episodic crustal growth, *Nature* **372**, 63 (1994).

[40] S. S. Sun. Chemical composition and origin of the earth's primitive mantle, *Geochim. Cosmochim. Acta* **46**, 179-192 (1982).

[41] S. R. Taylor and S. M. McLennan. *The Continental Crust: Its Composition and Evolution.* 311 pp., Blackwell Scientific, Oxford, UK (1985).

[42] I. N. Tolstikhin and R. K. O'Nions. The Earth's missing xenon: A combination of early degassing and of rare gas loss from the atmosphere, *Chem. Geol.* **115**, 1-6 (1994).

[43] K. K. Turekian. The terrestrial economy of helium and argon, *Geochim. Cosmochim. Acta* **17**, 37-43 (1959).

[44] G. W. Wetherill. Variations in the isotopic abundances of neon and argon extracted from radioactive minerals, *Phys. Rev.* **96**, 679-683 (1954).

[45] Y. Zhang and A. Zindler. Noble gas constraints on the evolution of Earth's atmosphere, *J. Geophys. Res.* **94**, 13719-13737 (1989).

[46] Y. Zhang and A. Zindler. Distribution and evolution of carbon and nitrogen in Earth, *Earth Planet. Sci. Lett.* **117**, 331-345 (1993).

[47] A. Zindler and S. Hart. Helium: Problematic primordial signals, *Earth Planet. Sci. Lett.* **79**, 1-8 (1986).

Proc. 30th Intern. Geol. Congr. Vol. 1, pp. 79-95
Wang *et al.* (Eds)

Pulsation Model of Mantle Differentiation: Evolution, Geochronological, Geochemical, Petrologic and Geodynamic Implications

YURI A. BALASHOV
Geological Institute of Kola Science Centre RAS, 14 Fersman St., 184200 Apatity, Russia

Abstract

Variations of εNd(T) in mantle and crustal rocks (over 4,000 analyses) have been systematized and plotted on a εNd(T) – T diagram. The εNd(T) value shows a synchronous variation in geological time, reflecting a periodical change-over of epochs of maximum (L-megacycles) and minimum (S-megacycles) differentiation of magma-generating systems in the upper mantle. By the degree of variations of the εNd(T) parameter, duration of L- and S-megacycles has been determined. It corresponds to the intervals 0.85-0.70-0.55-0.40-0.35-0.15 Ga for L(1-6)-megacycles, and 0.28-0.35-0.3-0.3-0.25-0.15 Ga for S(1-6)-megacycles. Distinguished are four epochs during which depleted mantle magma-generating reservoirs with the following parameters appeared into existence: DM-1 εNd(T)=38.6 (± 0.9) – 9.01 T , f=0.36; DM-2 εNd(T)=29.7 (±0.9) – 7.89 T , f=0.32; DM-3 εNd(T)=13.7(±0.5) – 3.81 T, f=0.16; DM-4 εNd(T)= 8.6(±0.5) – 2.01 T, f=0.08. The formation of DM mantle domains is confined to the following time intervals, respectively: 4280±60; 3400-3500; 3000-2800 and 4280 ? Ma. All the DM mantle domains formed in the Archaean time, which is therefore fundamentally different in the evolutionary nature from all younger periods of the planet life. By the degree of variations of the εNd(T) parameter, the Archaean can be subdivided into early (4.3-3.4 Ga), middle (3.4-3.15 Ga) and late (3.15-2.5 Ga) Archaean. At the early differentiation stages of the upper envelope of the Earth, simultaneously with the DM-1 domains, a complementary, EM-1 enriched mantle domain was formed. It includes the protocrust and the underlying mantle zones enriched in incompatible elements with the following parameters: εNd(T)= –52 (±0.9)+12.6 T, f= –0.5, 4300-4130 Ma. At the interval of 3.4-0 Ga, secondary EM mantle domains were formed; these were grouped in the lithosphere part of the mantle (EM-2) and in deep mantle zones (EM-3). The EM-2 domains are largely a result of mantle metasomatism, which was acting simultaneously with the generation of mantle magmas, as suggested by coinciding ages of mantle xenoliths and early stages of tectonic-magmatic cycles, based on the new detailed geochronologic scale of magmatic activity in the Baltic Shield. These stages were 40-60 Ma long and took place during early stages in the development of the cycles, the duration of which is 70-150 Ma. These stages resulted in the origination of mantle magmatic complexes, which were forming new crust in each tectonic-magmatic cycle, mostly from DM mantle domains. In rare cases, magmas appear to have been generated from the initial EM-1 domain that was preserved in the lithospheric zone of the mantle. Granulite metamorphism in the lower crust coincided with the intervals of mantle magmatism of the tectonic-magmatic cycles. As a result, in the vertical mantle-crust section, processes of magma-generation, mantle metasomatism and crustal metamorphism in each cycle were acting almost simultaneously. The secondary domains EM-3 resulted from Proterozoic subduction of crustal material into the mantle. K-enriched alkaline magmas are associated with these domains. The genesis of N- and E-types of MORB is related to DM-3 and DM-4 domains, respectively.

Keywords: mantle, crust, isotopic heterogeneity, differentiation stages, rock genesis

INTRDUCTION

Among various ways of studying the history of the Earth's envelopes, the combination of geochronologic and isotope-geochemical methods is notable for the opportunity to employ a quantitative approach to solving this fundamental problem. $^{143}Nd/^{144}Nd$ and $^{87}Sr/^{86}Sr$ isotope ratios, Pb isotopes and rare-earth elements (REE) compositions of mantle and crustal rocks are important geochemical signatures reflecting the conditions of genesis, evolution and transformation of these rocks. The most stable of these indicators are the $^{143}Nd/^{144}Nd$ ratios. Fundamentals of the Nd-isotope systematics were described in a number of papers [24, 25, 35-37, 60, 71 and others], which gave a definition to the main parameters, presented a model of a vertically layered mantle and suggested that most magmas originated from depleted mantle domains (DM). Extensive Sm-Nd isotope investigations of the 80s and 90s brought forward some data that were in disagreement with the previously accepted models of mantle structure and evolution, and the models had to be revised. However, the proposed versions did not change the essence of the models and did not account for the growing body of information about rocks with "anomalous" Nd-isotope characteristics.

This paper is a result of more than a 20 years long work of the author, who was engaged in the study of REE geochemistry and tried to systematize Rb-Sr and Sm-Nd isotope data in order to find objective identification criteria of major boundaries in the development of the Earth's mantle and crust [4-8, 10]. The paper presents a radically new model of development of heterogeneity in the upper mantle; the model reflects all types of Nd isotope ratios in mantle and crustal rocks and is in agreement with the geochronological data about cycles of crustal growth, with the geologic-tectonical concepts about major boundaries in the evolution of continental crust composition, and with the presumed changes in the energetics of the Earth that were manifested in alternating types of convection in the mantle. This model is far from being complete: it discusses only the results from analysis and systematization of Sm-Nd isotope data. Preliminary results obtained in the course of elaboration of the model were published in [14, 15]. A further study of other isotope and trace element systems is supposed.

BACKGROUND

The isotope-geochemical concepts about the composition and structure of the upper mantle have gone through several important stages in their development. First evidence of mantle heterogeneity was provided by REE distributions: oceanic basalts (MORB) were found to be deficient in LREE, whereas the continental crust is enriched in these elements, relative to the REE composition in chondrites [4-6, 31 and others]. The next stage was initiated by attempts to systematize $^{143}Nd/^{144}Nd$ values of mantle rocks [24-27, 60 and others], which served as a basis for recognizing and examining the geochemical and isotope heterogeneity of the mantle. All this made it possible to identify MORB sources as depleted mantle zones and propose a tentiative model of DM with $\varepsilon Nd(o)=+10$ and $\varepsilon Nd(4500\ Ma)=0$. In fact, these investigations posed a question as to "when" and "how" the mantle heterogeneity appeared and developed. Having perfomed a detailed Sm-Nd study of the Proterozoic magmatism in western States of the USA, DePaolo [27] amended the DM model by formulating the principle for estimating maximum values by an empirical equation $\varepsilon Nd(T)=0.25T^2-3T+8.5$, where the $\varepsilon Nd(o)$

value was taken to be equal to the average value for island arc volcanics. Presently, this model holds sway and is widely used for calculating T(DM) values and interpreting Sm-Nd data. Later, due to the growing body of information obtained, this model was updated [28] and expressed by a linear dependence: $\varepsilon Nd(T)=+8.6-1.9T$, where the initial εNd(o) is taken to be equal to the average for MORB. Advances in Sm-Nd studies in the 80-90s resulted in creation of other models, which either take into account the increasing evidence about high-εNd(o) MORB and assume an earlier differentiation of the mantle [40, 49], or account for the regional nature of the situation beneath of the Baltic Shield [51], or accept lower εNd(T) for deep-seated alkaline magmas [17], or discuss other possible ways [55, 57]. Elaboration of such models can be considered as an initial stage in the study of upper mantle heterogeneity. All these models bear inherent contradictions:

(1) The existence of different models impedes an assessment of the maximum age of protoliths, T(DM), for Precambrian metamorphosed rocks and makes it difficult to interpret unambiguously the increased εNd(T) values found in rocks of the mantle and crustal genesis, because depending on the model chosen, some of the data can be considered either as corresponding to the DM level or as exceeding it. Such εNd(T) values do not fit the models mentioned above, and this casts doubt on the correctness of the models.

(2) These models use a given **average value** of εNd(o) in MORB or other rocks, whereas the estimation of T(DM) is performed using **maximum values** of εNd(T) in the chosen model.

(3) All the DM models mentioned have one feature in common: they imply that the mantle parameter εNd(T) supposedly changed linearly or almost linearly (in 2 stages) over the entire geological time. A linear variation of the parameter εNd(T) is difficult to reconcile with the pulsational nature of Earth's energy release [68], or with a non-linear cooling [70] and the assumed alternation of the types of convective processes which must have altered the mantle composition [68], or with the existence of major tectonic-magmatic epochs in the development of the continental crust and tectonosphere, which reflect the energetic and compositional evolution of processes acting in the crust and the mantle [7, 13, 32 and others].

(4) These models sidestep such questions as the time of primary differentiation of the mantle and formation of the DM domains; whether this process was instantaneous or there were several stages and, correspondingly, several depleted mantle reservoirs; what kind of magmatic processes determined fractionation of the upper mantle and controlled the formation of DM domains. Depending on the opinion of the authors of the models, the beginning of mantle differentiation (the DM generation) is referred to as 4.6, 4.5 or 4.55 Ga, and on the whole for various rock types it varies from 1.75 to 4.6 Ga. These discrepancies indicate a lack of reliable criteria for estimating the time of the beginning of such processes.

(5) No consideration is given to the problem of identification of the mantle and crustal reservoirs that are enriched in incompatible and low-melting elements and are complementary to DM mantle domains. In this context we should note that in the proposed DM models the εNd(T) values of mantle-derived rocks, that plot considerably below DM lines and fall in the field of negative values, are considered mostly as a result of crustal contamination, and practically no account is taken of the mantle sources originally enriched in incompatible elements. However, the increasing amount of isotopic information has revealed that negative εNd(T) may be found in xenoliths of eclogite and ultrabasic rocks of the lithospheric zone of the mantle, in some Palaeoproterozoic intrusions, volcanic rocks, tonalites and metasediments of the Early Archaean. Consequently, the question about a role of enriched mantle domains (EM) in magma generation has acquired a fundamental importance for gaining an insight into the evolution of mantle and crust composition and ore deposition.

Thus, in the recent years, there have appeared a number of contradictions and unsolved questions in our understanding of evolution of mantle heterogeneity and genesis of mantle magmas, and this demands a revision of the previously accepted models.

Sm-Nd SYSTEMATICS OF MANTLE AND CRUSTAL ROCKS

Published data (over 4000 analyses) on εNd(T) values and U-Pb or Rb-Sr and Sm-Nd isochron ages have been plotted on diagrams of εNd(T) versus T for rocks of mantle and crustal genesis (Fig.1). MORB, island-arc volcanic and intrusive rocks, greenstone belts and intracratonic rifts, dolerite dykes and sills, xenoliths of eclogite, peridotite and pyroxenite, crustal xenoliths of basic composition, and alkaline assemblages (kimberlites, carbonatites, basalts, lamproites, leicitites, nephelinites, *etc.*) are incorporated in the mantle group; excluded was only a small set of data on metasomatically altered eclogites that have extremely high εNd(T), from +40 to +264 [41, 54]. The crustal group comprises various granitoids, gneisses, continental and oceanic sediments and their metamorphic analogues, shallow-depth amphibolite-facies crustal xenoliths of acid composition, and xenoliths of lower crustal granulites. When systematizing the Sm-Nd data, all the εNd(T) values were calculated to average mantle parameters $^{147}Sm/^{144}Nd=0.1967$ and $^{143}Nd/^{144}Nd=0.512638$ (0.511847) according to [35, 37, 71, 60].

Megacycles of mantle magmatic activity

As seen from Figure 1, there were synchronous periodical changes of εNd(T) in the fields of positive and negative values. It is possible to identify age intervals of strongly-(L) and poorly-(S) differentiated mantle magma-generating reservoirs and the related crustal rocks. Durations of the L and S epochs are different, being successively decreasing from Archaean to Phanerozoic: 0.85-0.7-0.55-0.4-0.35-0.2 Ga for L epochs, and 0.28-0.35-0.3-0.3-0.25-0.15 for S epochs. The synchronous changes of εNd(T) in different rock groups suggest a correlation between mantle and crustal processes throughout the entire geological time, and this lets us combine the isotope information (Fig. 1). The long duration of L and S intervals makes it possible to consider these intervals as megacycles which should incorporate several tectonic-magmatic cycles. The exception is the early epoch of S-1, when there was no differentiation.

When elaborating a detailed scale of Precambrian mantle and crustal events in the Baltic Shield, it has been established that the duration of tectonic-magmatic cycles does not exceed 150-70 Ma. Correspondingly, five tectonic-magmatic cycles are distinguished in the Late Archaean, and eight in the Palaeoproterozoic [13]. This scale, with supplementary information on the Riphean, is illustrated in Table 1. By comparing the data from this scale and the mentioned gradation of the megacycles, we can specify the boundaries between the megacycles. For instance, in the Palaeoproterozoic, there is a prominent difference between magmatism of Sumian-Upper Jatulian (2540-2227 Ma) megacycle S3, and Ludicovian-Vepsian (2110-1750 Ma) megacycle L3. This difference is manifested in characteristics of magma-generating mantle sources and it basically reflects the change of geotectonic conditions in time – a transition from rifting to marginal palaeo-island arc environments on the Baltic Shield.

It is also evident that the outlined boundaries between megacycles demarcate the major periods of maximum crustal growth that have been previously established by geological and geochronological methods [7, 55, 13, 15 and others] and suggested by the results from studies of the Earth's energetics [68]. For instance, if we take the total of maximum positive and negative εNd(T) values, we shall see that the Archaean differs from Proterozoic-Phanerozoic megacycles. The boundary between Archaean and Proterozoic is fairly distinct at about 2.5 Ga (megacycle S3). Within the Proterozoic, megacycles S4 characterizes the transition from Palaeoproterozoic to Riphean, and megacycle S5 – from Riphean to Palaeozoic.

It should be mentioned that the International Subcommittee on Precambrian Stratigraphy in Edinburgh (April, 1991) failed to establish criteria for subdividing the Archaean and recommended to subdivide it into conventional eras with the following boundaries: *ca.*3.6, 3.2 and 2.8 Ga, without constraining the initial and terminating limits of the Archaean [66]. The Sm-Nd systematics is the first to offer objective criteria for subdividing the Archaean – by the degree of differentiation of the upper mantle. From Fig.1 it is evident that the Archaean stage of the Earth's life should be subdivided into three megacycles; L-1 – Early Archaean (>4.0-3.4 Ga), S2 – Middle Archaean (3.4-3.15 Ga) and L2 – Late Archaean (3.15-2.5 Ga).

The "pulsational" variation of the εNd(T) parameter is thus an important tool for refining geochronologic boundaries in the geological time scale of the Earth's evolution. Certainly, these boundaries require some correction with the use of geochronological data. But even this would not change the noticed megacyclicity in the planet evolution.

Depleted and enriched mantle domains

If we look at the intensity of εNd(T) variations in different megacycles, we shall be able to identify lines of maximum εNd(T) values that correspond to the stages of maximum fractionation in the upper mantle and the formation of independent depleted (DM) or enriched (EM) domains. It should be noted that the evolution lines of DM-1,2 and 3 domains considerably exceed the maximum DM level adopted in previously published models. The data used for constructing these lines are summarized in Table 2 and illustrated in Fig.1. Calculation of Sm-Nd parameters for these lines was performed by the ISOPLOT program [46].

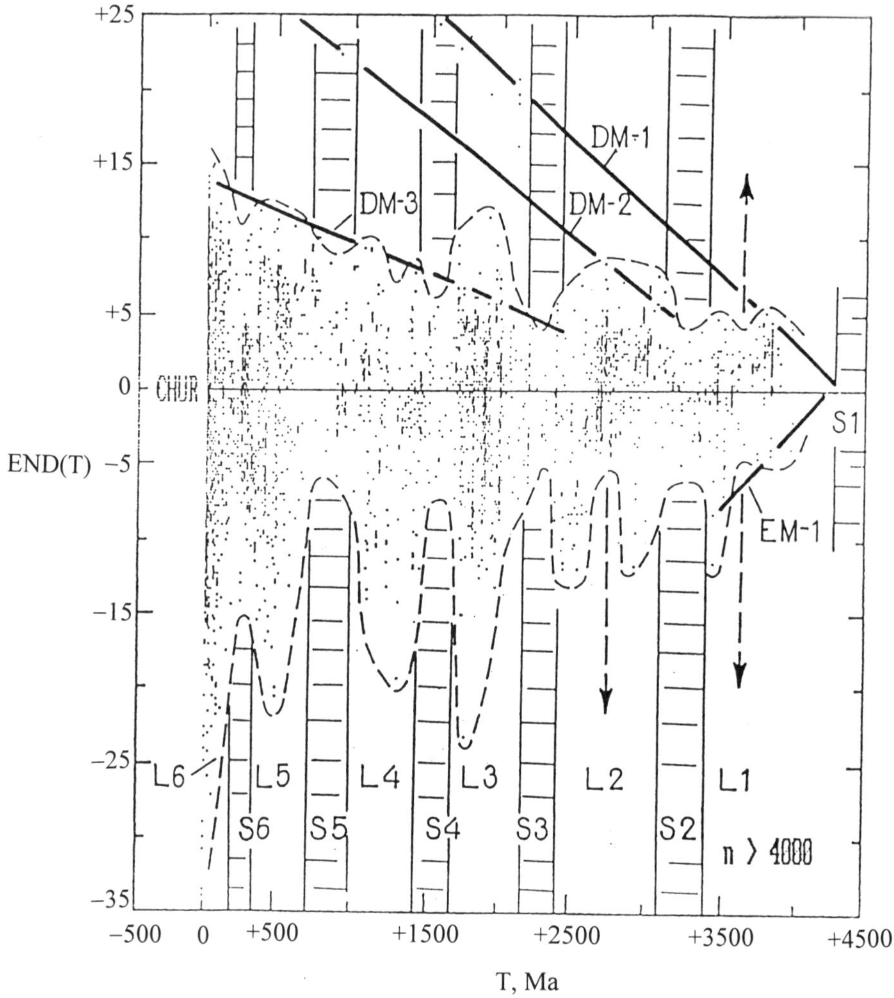

Figure 1. Periodic variation of εNd(T) of mantle and crustal rocks in geologic time. Vertical columns show the alternation and approximate intervals of L(1-6) and S(1-6) megacycles. A tentative, more detailed subdivision of the Late Archaean megacycle (L-2) into three intervals is indicated by a vertical arrow. Vertical arrows in the Early megacycle L-1 separate the intervals of maximum εNd(T) values in rocks of Gondwana and Laurasia. Bold lines mark the field of maximum εNd(T) values in the rocks that have been used for identification of DM-1,2,3 and EM-1 mantle magma-generating reservoirs.

Table 1. Detailed Precambrian geochronological scale of the Baltic Shield

Stratigraphic system	Mantle magmatism	Granitoids, detrital zircon, metamorphism	Beginning of magmatic cycle and stratigraphic interval A	H	Number of datings
Karatau	871 - 844	898 - 837			
	936 - 910	943 - 905			
	995 - 955	990 - 950	971± 8	981 ± 8	14
	-	1020 - 1002			
	1058 - 1040	1063 - 1030	1048 ± 6	1046 ± 8	11
Yurmatin	-	1144 - 1070			
	1184 - 1154	1180 - 1151	1168 ± 5	1170 ± 8	14
	-	1204 - 1192			
	1225 - 1213	1229 - 1210	1220± 5	1221 ± 4	12
	-	1241 - 1232			
	1299 - 1250	1293 - 1249	1279± 8	1281± 8	10
Burzyan	-	1362 - 1339			
	1433 - 1371	1422 - 1370	1413± 10	1417± 10	8
	-	1501 - 1444			
	1577 - 1508	1580 - 1512	1573± 3	1578± 5	8
	-	1617 - 1582			
	1667 - 1633	1672 -1630	1655± 5	1661± 6	13
PALAEOPROTEROZOIC					
Vepsian	-	1732 - 1694			
	1770 - 1756	1771- 1760	1770± 5	1766+ 10	10
Svecofennian	1815 ± 12	1815 - 1777			
	1836 ± 20	1843 - 1824			
	1898 - 1862	1898 - 1853			
	1910 - 1900	1915 - 1900	1905± 5	1907± 5	20
Kalevian	-	1959 - 1918			
	1980 - 1965	1984 - 1964	1968± 5	1972± 8	27
Ludicovian	-	2066 - 1990			
	2090 - 2078	2088 - 2074			
	2115 - 2105	2118 - 2110	2113± 5	2113± 5	10
Upper Jatulian	2220 - 2180	2233 - 2211	2227± 10	2214± 10	11
Lower Jatulian	-	2294 - 2243			
	2332 - 2324	2330 - 2316	2324 ± 5	2325± 5	7
Sariolian	2356 ± 4	2374 - 2346			
	2405 - 2398	2420 -2396	2405± 10	2408± 15	6
Sumian	2457 - 2435	2450 - 2433			
	2505 - 2502	2512 - 2468			
	2555 - 2526	2540 - 2530	2540 ± 20	2539± 20	7
LATE ARCHAEAN					
Tundrian	2580 ± 40	2595 - 2582			
	-	2640 - 2632			
	-	2660 - 2646			
	-	2693 - 2670			
	2703 - 2690	2707 - 2696	2701± 10	2699± 10	9
Upper Lopian	-	2741 -2707			
	2770 - 2744	2784 -2744			
	2807 - 2795	2813 -2785	2805± 15	2799± 15	14
Lower Lopian	-	2836 -2820			
	2860 - 2840	2881 -2840			
	2930 - 2903	2932 -2895	2917± 15	2921± 20	12
Saamian	2966 - 2960	2977 -2971			
	3020 - 2987	3029 - 2987	2998± 25	3005± 30	7
Semchenian	-	3115 -3050			
	3130 ± 30	3170 - 3136	3151± 20	3149± 25	6

Notes: A - the weighted average, H - estimated from histogram [46].

The formula expression of DM-1 is εNd(T)=38.6(±0.9) – 9.01 T (MSWD=2.58), which corresponds to the age 4280±60 Ma – the beginning of the early intensive differentiation of the upper part of the Earth. This age should be considered as the minimum estimate, since Sm-Nd data on the Early Archaean include the results on tonalititc gneisses, the nature of which has not been clarified yet. According to the current knowledge, the gneisses derived from melting of the mafic basement [7, 74]. Correspondingly, the age of the protolith could be older by 20-30 mln.y., and this should affect the estimate of the onset of mantle differentiation. Therefore, it is possible to take an age of 4.3 Ga as a more probable estimate of the onset of mantle differentiation. This age within precision limits coincides with the U-Pb age of the most ancient detrital zircon (4.28 Ga) found in the Yilgarn Block quartzite in West Australia [23] and with the time of formation of the most differentiated ferruginous anorthosite formation – Luna KREEP-basalt (*ca.*4.3 Ga) [7, 33]. If the time of meteorite agglomeration (4565 Ma) is taken to be the age of the Earth [1], then the interval when the upper Earth's envelope was heating up before melting and differentiation lasted 265-285 (±60 Ma), which is similar to the estimated duration of formation of the Earth and Luna mantle (238±56 Ma) obtained on the basis of the ^{146}Sm-^{142}Nd system [58, 59]. The calculated value of *f(Sm/Nd)* for the DM-1 stage of the Earth's mantle is 0.36, which is equal to the value obtained for the Lunar mantle (*f*=0.37)[18]. This is an average and not to expel rare extreme mantle fractionation [21].

In order to discuss the development of other strongly depleted mantle sources, let us turn back to the analysis of previously proposed DM models. If we exclude anomalously high εNd(T) values from consideration, the majority of moderately depleted εNd(T) of Phanerozoic, Proterozoic and Archaean would fit the range that is limited by the evolution line of initial εNd(o)=+8.6(±0.5) according to [28]. DePaolo *et al.* considered this value as an average for MORB. It is possible to suggest an alternative interpretation: the value +8.6 is the boundary between N-type and E-type MORB, which are different in time of formation and nature of mantle sources. The assumption about the difference in the nature of sources for N- and E-types of MORB can not be substantiated on the basis of Sm-Nd systematics of oceanic basalts because of the similarity of Sm and Nd distribution coefficients in rock-forming minerals of mantle peridotite and the fact that εNd(o) gradually changes in MORB. In this respect, the Rb-Sr isotope system seems to be more promising, although it also requires additional geochemical parameters to be involved. Specifically, both the MORB types show considerable differences in K, Rb and Cs concentrations and ratios. Taking into account the correlation of these values with the ^{87}Sr/^{86}Sr ratios in rocks of the mantle genesis [7], the difference between the mantle sources for the two types of MORB becomes quite obvious, and the values of ^{87}Sr/^{86}Sr <0.7028 almost without exception give us the benchmark to distinguish the N-type MORB. The corresponding values of εNd(o) are >+8.6. It should be emphasized that the values ^{87}Sr/^{86}Sr >0.7028 and εNd(o)<+8.6 are typical of most arc-rocks and alkaline basalts. Therefore, εNd(o)=+8.6(±0.5) should be considered as the real boundary dividing genetically different groups of mantle rocks of various petrochemical composition. This boundary and the evolution line, corrected for the time of the onset of differentiation of the uppermost envelope of the Earth, – 4280 Ma (line DM-4) – will be used below. The line has the following parameters: εNd(T)=8.6(±0.5) - 2.01 T, *f(Sm/Nd)*=0.08.

The DM-2 evolution line is described by the equation εNd(T)=29.7(±0.8) - 7.89 T (MSWD=0.15). Addition of data on amphibolites from Hebei Province [40] and basalts from Aldan [62] will not change the parameters of this line. Anomalous εNd(T) values are confined predominantly to the Late Archaean, the stage when the Earth's original mantle had already underwent initial differentiation and, possibly, partial homogenization [18]. Parameters of this moderately depleted mantle are taken to be equal to those of line DM-4. It should be noted that the maximum level of εNd(T) values for megacycle S2 coincides with this line. Therefore, the beginning of formation of DM-2 zones started in the interval of 3.4-3.5 Ga of Middle Archaean, and *f(Sm/Nd)*=0.32.

Table 2. Sm-Nd data used for constructing DM-1, 2, 3 and EM-1 models.

Rocks; Region	Age	ε Nd(T)	Authors
DM-1			
Gt-lherzolite, xenolith; Mir	1540±10	24.5±0.2	[50]
Gt-lherzolite, xenolith; Tanzania	1752±14	23.67±0.6	[39]
Gt-lherzolite, xenolith; Mir	1700±100	23.0±1.6	[75]
Eclogite,xenolith; S.Africa	2104±100	19.2±1.5	[67]
Tonalitic gneiss; E India	3775±89	5.1±1.4	[16]
Acid Boulder, E; Isua	3807±1	4.12±0.33	[53, 22]
Schist, B-2/3; Isua	3810±1	4.01±0.53	[38]
Gneisses; Amitsoq	3812±4	4.5±0.2	[18]
Gneisses; Amitsoq	3812±4	4.2±0.2	[18]
Gneisses; Amitsoq	3812±4	4.0±0.2	[18]
DM-2			
Gt-lherzolite, xenolith; Mir	910±10	22.5±0.5	[75]
Picrite; W Kolar, India	2715±110	8.46±0.4	[3]
Tonalitic gneiss; Garsjo, N Norway	2840±35	7.15±0.5	*[44]
Gneisses; Gruinard, Scotland	3064±91	5.7±0.8	[73]
DM-3			
MORB; Kane FZ (24 N), Atlantic	1±10	14.2±0.5	[47]
MORB; Kane FZ (24 N), Atlantic	1±10	13.7±0.5	[47]
Eclogite; W Alps	500±50	11.2±0.5	[61]
Sp-peridotite, xenolith (B-2-1); Arizona	615±50	10.9±1.4	[30]
Amphibolite; Pennine, Alps	1071±43	9.37±0.8	[69]
Sp-lherzolite, xenolith; Isl. Zhohov	1107±61	9.36±0.36	[19]
Ultramafic rocks, Precos, Colorado	1720±50	6.7±0.4	[55]
Komatiites; Kostomuksha, Karelia	2843±39	3.1±0.3	[63]
EM-1			
Tonalite gneiss; Acasta, Canada	3960±100 ?	2.0±0.5	[20]
Fe-quartzite, 167642, BIF; Isua	3813±5	−4.23±0.31	[52, 2]
Basaltic komatiite; Labrador	3810±46	−4.12±0.6	[21]
Gneiss; Amitsoq	3729±13	−4.6 ± 0.13	[18]
Gneiss; NWT, Canada	3600±100 ?	−7 ± 0.5	[20]
Light gray chert; Upper Hooggenoeg	3438± 12	−8.28 ± 0.39	[72, 42]
Eclogite, SM-2G; Snake River Plain	1±10	−52 ± 0.23	[43]

* [44] and V.Vetrin, pers.com., 1996.

For the DM-3 stage, the empirical equation is written as εNd(T)=13.7(±0.5) – 3.8 T (MSWD=1.77), *f(Sm/Nd)*=0.16, and the time of origination of an anomalously depleted source is close to 3.0-2.8 Ga, if calculated relative to the intersection with the DM-4 line. As the maximum εNd(o) values in this case correspond to the optimum depleted MORB, we can state that the sources for N-type MORB were formed in the mantle only in the Late Archaean, whereas the E-type MORB must have been generated from more ancient depleted domains of the mantle.

Negative εNd(T) values are recorded in some rocks of the Early Archaean sedimentary cover: in BIF formations and metamorphosed schists. Some tonalitic gneisses also have negative εNd(T), similar to those of BIFs. These gneisses can be considered as metamorphosed products that have been molten from an enriched lowercrustal source [74]; the source was represented in the Early Archaean by basitic and komatiitic volcanics. And finally, it is not inconceivable that products of protocrust or products of magmatism from an EM source may have been partially preserved in the lower crust, provided that it was covered by younger Archaean rocks. Therefore, one more identification criterion of initially enriched associations is the lowercrustal xenoliths [8].

The data on various Early Archaean rocks with maximum negative εNd(T) values are summarized in Table 2. Note that only whole-rock data have been selected for this Table, because calculations performed on rock fragments or fractions give a great scatter of εNd(T) values [52,72]. The evolution line can be expressed as εNd(T)= – 50.7(±9.1)+12.3 T (MSWD=6.83). Its intersection with CHUR corresponds to an age of 4136±92 Ma. In the case of the lowercrustal xenolith (SM-G2) from the Snake River region [43], which has εNd(o)= –52, the intersection with CHUR gives an age of 4144 Ma. If we add this sample to the Archaean set of data, the parameters of the evolution line for the early differentiation of the mantle will not change, being described by the equation εNd(T)= – 52(±0.9)+12.6 T (MSWD=6.77). The intersection with CHUR corresponds to an age of 4125±50 Ma; *f(Sm/Nd)*= –0.50. Even though the age of intersection with CHUR is lower than the age estimated from line DM-1, still this result can be interpreted to-date as the closest approximation to the assessment of characteristics of the maximum enriched (EM-1) initial mantle domains of the Earth.

Among mantle and lowercrustal xenoliths there are examples with EM-characteristics which are considerably different from EM-1. However, it is possible to calculate their model T(EM-1) ages and εNd(T) values for these ages – εNd(T) (EM-1) (Table 3). Apparently, the characteristics of the initial enriched mantle source were conserved in the lower crust or in the lithospheric part of the mantle in a number of regions. Note that among upper crustal rocks, only the Stillwater intrusion [26] features analogous Nd parameters (Table 3).

Calculation of model ages for other xenoliths with negative εNd(T) makes these xenoliths fall into the depleted mantle field. Xenoliths of this type predominate, and most of them plot near the CHUR line (field EM-2 with the following parameters: εNd(T)= –19+5.57 T).

When analyzing the information on these xenoliths, of fundamental importance is the estimate of the time when the lithosphere started to be intensely reworked by mantle metasomatic processes. We should note here that the age of all studied mantle and crustal xenoliths does not exceed 2900 Ma. An older U-Pb dating has been obtained for peridotite xenoliths in diamonds (3320 Ma) [45]. Therefore, the maximum age of the xenoliths is constrained by the Middle Archaean. It can be concluded that recrystallization, metasomatism and other processes responsible for the generation of secondary EM mantle domains started to act in the lower crust and the lithospheric part of the mantle at that time. This conclusion is in agreement with the estimated time when a strongly depleted mantle domains (DM-2) started to form in the Middle Archaean.

Table 3. Mantle and crustal xenoliths and the rocks of the Stillwater Complex with negative εNd(T) and the calculated εNd(T)(EM-1)

Rock; Region	Age	εNd(T)	T(EM-1)	εNd(T) (EM-1)	f	Sm/M*	Authors
Amphibolite; Sierra Nevada	1**	−0.55	3440	−11.4	0.126	33.8	[29]
Pyroxenite; Sierra Nevada	1	−6.13	3750	−7.2	0.013	7.93	[29]
Eclogite; Angola	1	−0.55	1615	−35.4	0.92	5.42	[67]
Eclogite; Camp Creek, Arizona	30	−4.6	3700	−8	0.032	-	[30]
Eclogite; Camp Creek, Arizona	30	−4.74	3970	−4.3	0.0035	-	[30]
Eclogite; Camp Creek, Arizona	30	−3.88	4040	−3.5	0.0035	-	[30]
Eclogite; Camp Creek, Arizona	30	−3.47	4300	0.01	-0.034	-	[30]
Eclogite; Camp Creek, Arizona	30	−3.18	4190	−1.65	-0.014	-	[30]
Eclogite; Camp Creek, Arizona	30	−5.46	3800	−6.8	0.032	-	[30]
Garnetite; Camp Creek,Arizona	30	−1.76	1475	−37.5	0.988	-	[30]
Mafic, K88, Calcutteroo, S. Australia	1	−0.53	2500	−23.9	0.37	5.28	[48]
Granulite, 83-162; McBridge, Queensland, Australia	376	−6.5	4130	−2.3	-0.041	16.54	[65]
Mafic granulite; Chudliegh, Queensland, Australia	376	0.08	3220	−14.5	0.218	6.94	[64]
Gabbro, 150; Stillwater, Montana	2701	−2.07	3850	−2.7	0.0469	0.957	[26]
Norite, 6; Stillwater, Montana	2701	−1.97	4080	−0.6	-0.0392	0.844	[26]
Pyroxenite, 25; Stillwater, Montana	2701	−2.05	3655	−5.9	0.1592	0.51	[26]

* The Sm/M ratio indicates how much Sm was accumulated relative to the average Sm content in the mantle (Sm=0.342 ppm) [7]. ** The age has not been determined, εNd(T)= εNd(o)

However, mantle and crustal xenoliths have greatly varying ages. Could this mean that processes of recrystallization, metamorphism and metasomatism wer autonomous?In the geochronological scale of tectonic-magmatic cycles of the Baltic Shield, covering the Late Archaean to Riphean (Table 1), the intervals of mantle volcanic and plutonic activity in each tectonic-magmatic cycles are confined to first 40-60 mln. years. It turns out (Fig.2) that age datings of crustal and mantle xenoliths strictly correspond to these intervals of magamtic activity. Therefore, in the vertical mantle-crust section, processes of magma-generation, mantle metasomatism and crustal metamorphism in each cycle were acting almost simultaneously.

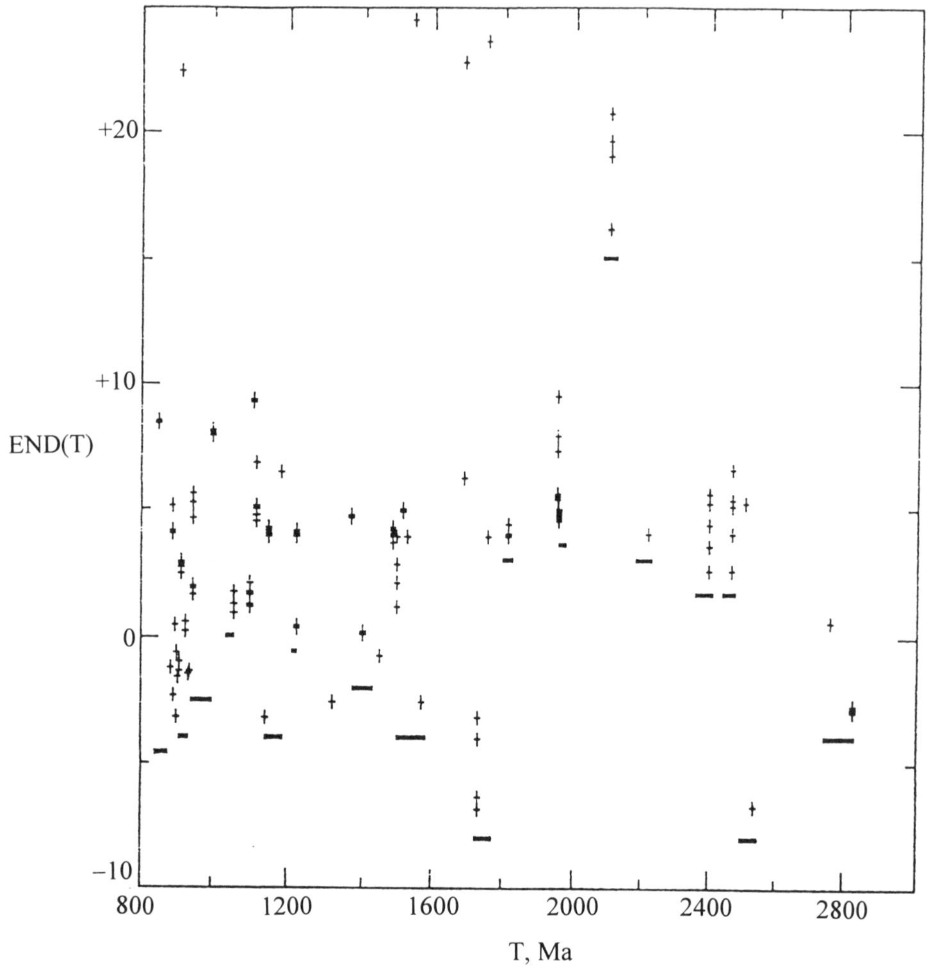

Figure 2. Coincidence of the ages of mantle and crustal xenoliths with the epochs of mantle magmatism within some tectonic-magmatic cycles of the Late Archaean-Riphean. Crosses indicate the age and εNd(T) values of the xenoliths. Horizontal bold lines – the duration of mantle magmatism according to the detailed geochronologic scale of the Baltic Shield (Table 1). For the interval of 1090-1120 Ma, the data on the mantle magmatism of the Baltic Shield have been not determined, but tholeiites, mangerites and syenodiorites of this age are known in the Grenville Province of the Canadian Shield.

In the recent years, many researchers started to interpret negative εNd(T) values in basic and ultrabasic massifs and xenoliths as indicators of generation from an enriched mantle [9 and others]. For instance, the layered basic-ultrabasic intrusions and komatiites of the Proterozoic ages (2505-2390 Ma) of the Baltic Shield, being confined to rift structures in the Archaean basement and located tens and hundreds of kilometers from each other, reveal a striking similarity of negative εNd(T) values: from −1.6±0.6 to −2.3±0.4 [9, 12, 34]. In addition, the initial ratio $^{87}Sr/^{86}Sr$ in these rock formations is 0.7022-0.7026, which is considerably higher than the average $^{87}Sr/^{86}Sr$ in the UR mantle of the same age (0.7015-0.7017), and their REE composition is characterized by a predominance of LREE. This makes it possible to suggest that such rock formations were derived from an enriched mantle domains [9] which formed in the course of mantle metasomatism. The T(DM-4) values for the Palaeoproterozoic layered intrusions and komatiites of the Baltic Shield fall into the interval 2760-3020 Ma, which coincides with the time of formation of the Late Archaean basement in the northeastern part of the Shield [11]. It can be therefore believed that processes of crust generation and mantle metasomatism in the lithosphere were acting simultaneously.

And finally I would like to note that the genesis of high-K alkaline rocks is related to melting of subducted parts of the crust beneath deep zones of the upper mantle, which represented an independent, secondary-enriched mantle domains (EM-3) [56].

DISCUSSION

The age of the Earth and the beginning of its differentiation is a hotly debated problem. It seems that we shall have to revise the current concepts that the planet surface was heated up already at the stage of accretion, and a mantle ocean was formed at that time. Various modern approaches to estimate the time when the Earth's material started to differentiate have given fairly close results: *ca.* 4.30 Ga. Possibly, heating of the upper part of the Earth at that stage can be adequately explained by the model [68].

Another important problem is to determine possible composition of initial magmas of the early differentiation stage (megacycle L-1) and to find the domains where excess of incompatible elements were buried. A considerable amount of incompatible elements must have been removed from the upper mantle at that stage, as evidenced by the identified anomalously depleted domains DM-1 and by the prevalence of depleted mantle products in younger megacycles. The questions regarding the place of concentration of these elements and the volume of the enriched reservoirs still remain to be solved. The existence of an enriched domain EM-1 has been now proved. It may be suggested that this reservoir included not only the protocrust, but also some portion of the proto-lithospheric part of the mantle. However, according to the opinion of different authors, the Early Archaean crust was not large in volume. On the other hand, the inferred higher temperature [70] and the enrichment in volatiles (water, *etc.*) of the initial material of the Earth are interpreted as an indicator of intense magmatism in the Early Archaean. We are forced to assume that our understanding of the nature and amount of mantle products generated at the earliest differentiation stage is contradictory. It can be conjectured only that either the magmas were petrochemically more diverse, up to the presence of alkaline magmas [50], or some komatiitic magmas were more strongly enriched in incompatible and volatile elements. The difficulty associated with identification of initial EM domains may be caused by the influence of later recrystallization processes, which "rejuvenated" the age characteristics and changed the composition of the ancient enriched mantle domains. The occurrence of eclogites with anomalous depleted parameters indicates that such secondary processes existed.

The other aspect of this problem is the relationship of differentiation and secondary homogenization. The degree of mantle differentiation was changing at the transition from the Early to Late Archaean; the change can be accounted for by mixing of previously formed mantle zones with contrasting isotopic and geochemical characteristics. In this respect, we must note that the result from mixing corresponded to the DM-4 line already in the S-2 megacycle of the Middle Archaean. Thus, the DM-4 domain could well be formed at that time, but not in the beginning of the L-1 megacycle, as it was assumed for the analysis of DM-2 and DM-3 domains. The mantle temperature was likely to be high enough for the secondary homogenization to take place in the Early Archaean. However, it appears that cooling of the mantle should cause a decrease in the thickness of the molten mantle ocean, if it ever existed. Possibly, this "ocean" had properties of the asthenosphere. If we accept the estimated rate of cooling of the Earth in the Early and Middle Archaean as it is given in [70], the relatively "thin" asthenosphere and "thick" lithosphere were formed in the Middle Archaean. This apparently was another reason why remaining EM-1 and DM-1 domains were removed from the magma-generating sphere and why there occurred a change-over from mantle melting to mantle metasomatism. This is confirmed by the fact that the beginning of active mantle metasomatism and processes of lithosphere recrystallization (generation of DM-2 domains) started in the Middle Archaean. However, from the cooling model according to [70], it is impossible to understand how thick crustal growth and generation of DM-3 domains could proceed in the Late Archaean. In this respect, the model proposed by [68] seems to be more promising, as it considers additional heating related to the formation of the core some time in the Late Archaean.

Acknowledgements

I thank F. P.Mitrofanov, Director of the Geological Institute KSC RAS, corr.-memeber RAS, for the opportunity to use the Institute's facilities to carry out the research on the problems of mantle heterogeneity, S. Yu. Delenitsina for translating the first version of the manuscript into English, and Prof. Wang Hongzhen, Member of Chinese Academy of Sciences, for the kind offer to include this paper in the 30th IGC Proceedings Volume.

REFERENCES

[1] C. J. Allegre, G. Manhes and C. Gopel. The age of the Earth, *Terra Abstr.Suppl. 1 to Terra nova* **7**, 97 (1995).

[2] H. Baadsgaard, A. P. Nutman, D. Bridgwater, M. Rosing, V. R. McGregor and J. H. Allaart. The zircon of the Akilia association and Isua supracrustal belt, West Greenland, *Earth and Planet. Sci. Lett.* **68**, 221-228 (1984).

[3] S. Balakrishnan, G. N. Hanson and V. Raiamani. Pb and Nd isotope constraints on the origin of high Mg and tholeiitic amphibolites, Kolar Schisy Belt,South India, *Contrib.Miner.Petrol.* **107**, 279-292 (1991).

[4] Yu. A. Balashov. Regularities of REE distribution in the crust of the Earth. *Geokhimiya* **2**, 99-114 (1963).

[5] Yu. A. Balashov and G. V. Nesterenko. REE distribution in trapps of the Siberian Plaform, *Geochim. Intern.* **3**, 672- 678 (1966).

[6] Yu. A. Balashov. *Geochemistry of rare earth elements.* Nauka, Moscow, USSR (1976).

[7] Yu. A. Balashov. *Isotope- geochemical evolution of the Earth mantle and crust.* Nauka, Moscow, USSR (1985).

[8] Yu. A. Balashov. Evidence of most ancient Earth crust material in metamorphic rocks of the Archaean, *Izv. AN SSSR.* **12**, 126-128 (1990).

[9] Yu. A. Balashov, T. B. Bayanova and F. P. Mitrofanov. Sources of the layered intrusions. In: *Geochronology and genesis of layered basic intrusions, volcanites and granite-gneisses of the Kola Peninsula.* pp. 24-30. Preprint, Apatity (1990).

[10] Yu. A. Balashov, A. N. Vinogradov and F. P. Mitrofanov. Isotope-geochemical and petrologic data on the formation and transformation of the protocrust. In: *Early crust: its composition and age.* pp. 102-112. Moscow, Nauka (1991).

[11] Yu. A. Balashov, F. P.Mitrofanov and V. V. Balagansky. New geochronological data on Archaean rocks of the Kola Peninsula. In: *Correlation of Precambrian formations of the Kola-Karelian Region and Finland.* pp. 13- 34. Apatity (1992).

[12] Yu. A. Balashov, T. B. Bayanova and F. P.Mitrofanov. Isotope data on the age and genesis of layered basic- ultra-basic intrusions in the Kola Peninsula and northern Karelia, northeastern Baltic Shield, *Precambrian Res.* **64**, 197-205 (1993).

[13] Yu. A. Balashov. Detailed AR2-PR1- geochronological scale of the Baltic Shield, *Doklady RAN* **343**, 513-516 (1995).

[14] Yu. A. Balashov. Processes of differentiation and mixing in the upper mantle: a new model based on Sm-Nd isotope data, *Doklady RAN* **347**, 81-85 (1996).

[15] Yu. A. Balashov. Palaeoproterozoic geochronology of the Pechenga-Varzuga Structure, Kola Peninsula, *Petrology* **4**, 1-22 (1996).

[16] A. R. Basu, S. L. Ray, A. K. Saha and S. N. Sarkar. Eastern Indian 3800-Million-Year-old crust and early mantle differerntiation, *Science* **212**, 1502-1506 (1981).

[17] K. Bell and J. Blenkinsop. Carbonatites and the sub-continental upper mantle, *Geol. Assoc. Can. Mineral.Assoc. Can./ Can. Geophys. Union* Progr. with Abstrs. **11,** 44 (1986).

[18] V. C. Bennett, A. P. Nutman and M. T. McCulloch. Nd isotopic evidence for transient, highly depleted mantle reservoirs in the early history of the Earth, *Earth and Planet.Sci.Lett.* **119,** 299-317 (1993).

[19] O. G. Bogdanovsky, S. A. Silant'ev, S. F. Karpenko, S. D. Mineev and L. A. Savostin. Ancient mantle xenoliths in young effusions of Zhohov Island (DeLong Archipelago), *Dorkfdy RAN* **330**, 750-753 (1993).

[20] S. A. Bowring, D. S. Coleman and T. B. Housh. The 4.0 Ga Acasta gneisses: constraints on the growth and recycling of continental crust. *Program and Abstr. " Precamb.95", Montreal, Can.*, 275 (1995).

[21] K. D. Collerson, L. M. Campbell, F. L. Weaver and Z. A. Palacz. Evidence for extreme mantle fractionation in early Archaean ultramafic rocks from northern Labrador, *Nature* **349**, 209-214 (1991).

[22] W. Compston, P. D. Kinny, I. S. Williams and J. J. Foster. The age and Pb loss behaviour of Isua zircons as determined by ion microprobe, *Earth and Planet.Sci.Lett.* **60**, 71-81 (1986).

[23] W. Compston and R. T. Pidgeon. Jack Hills, evidence of more very old detrital zircons from Western Australia, *Nature* **321**, 766-769 (1986).

[24] D. J. DePaolo, G. J. Wasserburg. Nd isotopic variations and petrogenic models, *Geophys. Res. Letters* **3**, 249-252 (1976).

[25] D. J. DePaolo, G. J. Wasserburg. Inferences about magma sources and mantle structure from variations of $^{143}Nd/^{144}Nd$, *Geophys. Res. Letters* **3,** 743-746 (1976).

[26] D. J. DePaolo and G. J. Wasserburg. Sm-Nd age of the Stillwater complex and the mantle evolution curve for neodymium, *Geochim. Cosmochim. Acta* **43**, 999-1008 (1979).

[27] D. J. DePaolo. Neodymium isotopes in the Colorado Front Range and crust-mantle evolution in the Proterozoic, *Nature* **291,** 193-196 (1981).

[28] D. J. DePaolo, A. M. Linn and G. Schubert. The continental crustal age distribution: Methods of determining mantle separation ages from Sm-Nd isotopic and application to the Southwestern United States. *J. Geophys. Res.* **96:2**, 2071-2088 (1991).

[29] M. A. Domenick, R. W. Kistler, F. C. W. Dodge and M. Tatsumoto. Nd and Sr isotopic study of crustal and mantle inclusions from the Sierra Nevada and implications for batholith petrogenesis, *Geol. Soc. Amer. Bull.* **94**, 713-719 (1983).

[30] S. Esperanca, R. W. Carlson and S. B. Shirey. Lower crustal evolution under central Arizona: Sr, Nd, and Pb isotopic and geochemical evidence from the mafic xenoliths of Camp Creek, *Earth and Planet. Sci. Lett.* **90**, 26-40 (1988).

[31] F. A. Frey and L. Haskin. Rare earths in oceanic basalts, *J. Gephis. Res.* **69**, 775-778 (1964).

[32] B. E. Hain, K. B. Seslavinsky. Global rhythms in Phanerozoic endogenic activity of the Earth, *Stratigraphy. Geol. Correlation* **2,** 40-63 (1994).

[33] B. Holmberg and M. J. Rutherford. An experimental study of KREEP basalt evolution, *Lunar and Planet. Sci. 25. Abstr. Pap. 25th Lunar and Planet.Sci.Conf.*, March 14-18, 1994. Pt.**2**, Houston (Tex.), pp. 557-558 (1994).

[34] H. Huhma, T. Mutanen, E. Hanski, J. Rasanen, T. Mannien, M. Lehtonen, P. Rastas and H. Juopperi. Sm-Nd isotopic evidence for contrasting sources of the prolonged Palaeoproterozoic mafic-ultramafic magmatism in northern Finland, *Program and Abstrs. IGCP Progect 336 Symp. In Rovaniemi*, Finland, August 21-23, 17 (1996).

[35] S. B. Jacobsen and G. J. Wasserburg. Sm-Nd isotopic evolution of chondrites, *Earth and Planet. Sci. Lett.* **50,** 139-155 (1980).

[36] S. B. Jacobsen and G. J. Wasserburg. A two-reservoir recycling model for mantle-crust evolution, *Proc. National Academy of Sci. USA* **77,** 6298-6302 (1980).

[37] S. B. Jacobsen and G. J. Wasserburg. Sm-Nd evolution of chondrites and achondrites, II. *Earth and Planet. Sci. Lett.* **67**, 137-150 (1984).

[38] S. B. Jacobsen and R. F. Dymek. Nd and Sr isotope systematics of clastic sediments from Isua, West Greenland: Identification of pre-3,8 Ga differentiated crustal components, *J. Geophys. Res.* **93**, B1. 338-354 (1988).

[39] E. Jagoutz. Nd and Sr systematics in an eclogite xenolith from Tanzania: evidence from frozen mineral equilibria in the continental lithosphere. *Geochim. Cosmochim. Acta* **52**, 1285-1293 (1988).

[40] B. M. Jahn, B. Auvray, J. Cornichet, Y. L. Bai, Q. H. Shen and D. Y. Liu. 3.5 Ga old amphibolites from Eastern Hebei Province, China: Field occurrence, petrography, Sm-Nd isochron age and REE geochemistry, *Precambrian Res.* **34**, 311-346 (1987).

[41] B. M. Jahn, J. Cornichet, B. Cong and T. F. Yui. Ultrahigh-E Nd eclogites from an ultrahigh-pressure metamorphic terrane of China, *Chem.Geol.* **127**, 61-79 (1996).

[42] A. Kroener and A. Tegtmeyer. Gneiss- greenstone relationships in the Ancient Gneiss Complex of southwestern Swaziland, southern Africa, and implications for early crustal evolution, *Precambrian Res.* **67**, 109-139 (1994).

[43] W. P. Leeman, M. A. Menzies, D. J. Matty and G. F. Embree. Strontium, neodymium and lead isotopic composition of deep crustal xenoliths from the Snake River Plain: evidence for Archaean basement, *Earth and Planet.Sci.Lett.***75**, 354 - 368 (1985).

[44] O. A. Levchenkov, L. K. Levsky, O. Nordgulen, L. F. Dobrzhinetskaya, N. Lars-Petter, V. R. Vetrin and B. A. Sturt. New geochronological data from the Sor-Varanger district of Finnmark, Norway, and Kola Peninsula, Russia. In: *1st Intern.Barents Symp.Norwegian-Russian Collaboration Programme «North Area»,* Abstrs, 21-24 october 1993, Kirkenes, Norway (1993).

[45] D. Lowry, D. P. Mattey, C. G. Macpherson and J. W. Harris. Oxygen isotope variations among peridotitic and eclogitic syngenetic inclusions in diamond. In 7th Meet. Eur. Union. Geosci. Strabourgh, Ap.4th-8th,1993:EUG VII. *Terra nova* **5:1**, 24 (1993).

[46] K. R. Ludwig. ISOPLOT for MS-DOS: A plotting and regression program for radiogenic-isotope data, for IBM-PC compatible computers, Version 2.00. *USGS Open-File Report 88-557.* 1-38 (1990).

[47] N. Machado, J. N. Ladden, C. Brooks and G. Thompson. Fine-scale isotopic heterogeneity in the sub-Atlantic mantle, *Nature* **295**, 226- 228 (1982).

[48] M. T. McCulloch, K. J. Arculus, B. W. Chppell and J. Ferguson. Isotopic and geochemical studies of nodules in kimberlite have implications for the lower continental crust, *Nature* **300**, 166-169 (1982).

[49] M. T. McCulloch and L. P. Black. Sm-Nd systematics of Enderby Land granulites and evidence for the redistribution of Sm and Nd during metamorphism, *Earth Planet Sci.Lett.* **71**, 46-58 (1984).

[50] M. T. McCulloch. Sm-Nd systematics in eclogite and garnet peridotite nodules from kimberlites: implications for the early differentiation of the Earth. In: *Fourth Intern. Kimberlite Conf. Extended Abstrs.* pp. 285-287. Perth, W. Australia, August 11-15 (1986).

[51] E. W. Mearns, T. Andersen, M. B. E. Mork and R. Morvik. $^{143}Nd/^{144}Nd$ evolution in depleted Balticscandian mantle, *Terra Cognia* **6,** 247 (1986).

[52] R. G. Miller and R. K. O'Nions. Source of Precambrian chemical and clastic sediments, *Nature* **314**, 325-330 (1985).

[53] S. Moorbath, P. N. Taylor and N. W. Jones. Dating the oldest terrestrial rocks - fact and fiction, *Chem. Geol.* **57**, 63-86 (1986).

[54] C. R. Neal, L. A. Taylor, J. P. Davidson, P. Holden, A. N. Halliday, P. H. Nixon, J. B. Paces, R. N. Clayton and T. K. Mayeda. Eclogites with oceanic crust and mantle signatures from the Bellsbank

kimberlite, South Africa, part 2: Sr, Nd and O isotope chemistry, *Earth and Planet.Sci.Lett.* **99**, 362 - 379 (1990).

[55] B. K. Nelson, D. J. DePaolo. Rapid production of continental crust 1.7-1.9 Gy ago: Nd and Sr isotopic evidence from the basement of the North American midcontinent, *Geol. Soc. Am. Bull.* **96,** 746-754 (1985).

[56] D. K. Nelson. Isotopic characteristics of potassic rocks: evidence for the involvement of subducted sediments in magma genesis, *Lithos* **28**, 403-420 (1992).

[57] D. I. Norman, K. C. Condie and R. W. Smith. Geochemical and Sr and Nd isotopic constraints on the origin of late Proterozoic volcanics and associated tin-bearing granites from the Franklin Mountains, west Texas, *Can. J.Earth Sci.* **24**, 830-839 (1987).

[58] L. E. Nyquist, H. Wiesmann, B. M. Bansal and C. Y. Shih. New data supporting a ^{144,147}Sm-^{142,143}Nd formation interval for the lunar mantle, *Lunar and Planet. Sci. 25. Abstr.* pp. 1017-1018. Pap 25th Lunar and Planet. Sci. Conf., March 14-18, 1994, Houston , Texas (1994).

[59] L. E. Nyquist, H. Wiesmann, B. M. Bansal, C. Y. Shih, J. E. Keith and C. L. Harper. ^{146}Sm-^{142}Nd formation interval for the lunar mantle, *Geochim. Cosmochim. Acta* **59**, 2817-2837 (1995).

[60] R. K. O'Nions, P. J. Hamilton and N. M. Evensen. Variations in ^{143}Nd/^{144}Nd and ^{87}Sr/^{86}Sr ratios in oceanic basalts, *Earth Planet. Sci. Lett.* **34**, 13-22 (1977).

[61] J. L. Paquette, R. P. Menot and J. J. Peucot. REE, Sm-Nd and U-Pb zircon study of eclogites from the Alpine External Massif (Western Alps): evidence for crustal contamination, *Earth and Planet. Sci. Lett.* **96,** 181-198 (1989).

[62] I. S. Puchtel, O. A. Bogatikov and A. K. Simon. Early Precambrian evolution of crust-mantle system of Olekminsky gneiss-greenstone region (the Aldan Shield), *Petrology* **1**, 499-523 (1993).

[63] I. S. Puchtel, A. V. Samsonov, A. A. Shchipansky and V. N. Furman. Accretional tectonics within Karelian granite-greenstone terrain: evidence from the field and isotope-geochemistry data for two distinct terrains within the Kostomuksha greenstone belt. In: *The 1st Inern. Conf. "Fennoscandian Geological Correlation Abstrs."* pp. 163-165. 8-11 September, St.Peterburg (1996).

[64] R. A. Rudnick, W. F. McDonough, M. T. McCulloch and S. R. Taylor. Lower crustal xenolith from Qeensland, Australia: evidence for deep crustal assimilation and fractionation of continental basalts, *Geochim. Cosmochim. Acta* **50**, 1099-1115 (1986).

[65] R. A. Rudnick. Nd and Sr isotopic compositions of lower-crustal xenoliths from north Queensland, Australia: implications for Nd model ages and crustal growth processes. *Chem.Geol.* **83**, 195-208 (1990).

[66] M. A. Semikhatov. Chronometric scale of Archaean (Proposal of the International Subcommittee on Stratigraphy of the Precambrian), *Izv. RAN, cer. geologiya* **9,** 153-160 (1992).

[67] J. W. Shervais, L. A. Taylor, G. W. Lugmair, R. W. Clayton, T. K. Mayeda and R. L. Korotev. Early Proterozoic oceanic crust and the evolution of subcontinental mantle: eclogites and related rocks from southern Africa, *Geol. Soc. Amer. Bull.* **100**, 411-423 (1988).

[68] O. G. Sorokhtin., S. A. Ushakov. *Global evolution of the Earth.* Moscow Univ. Press (1991).

[69] P. Stille and M. Tatsumoto. Precambrian tholeiitic-dacitic rock-suites and Cambrian ultramafic rocks in the Pennine nappe system of the Alps: evidence from Sm-Nd isotopes and rare-earth elements, *Contrib. Mineral. Petrol.* **89**, 184-192 (1985).

[70] N. J. Vlaar, A. P. van Keken and A. P. van den Berg. Cooling of the Earth in the Archaean: Consequences of pressure-release melting in a hotter mantle, *Earth and Planet. Sci. Lett.* **121,** 1-18 (1994).

[71] G. J. Wasserburg, S. B. Jacobsen, D. J. DePaolo, M. T. McCulloch and T. Wen. Precise determination of Sm/Nd ratios, Sm and Nd isotopic abundances in standard solutions, *Geochim. Cosmochim. Acta* **45**, 2311-2323 (1981).

[72] D. Weis and G. J. Wasserburg. Rb-Sr and Sm-Nd isotope geochemistry and chronology of cherts the Onverwacht Group (3.5 AE), South Africa. *Geochim. Cosmochim.Acta* **51**, 973-984 (1987).

[73] M. J. Whitehouse. Sm-Nd evidence for diachronous crustal accretion in the Lewisian complex of northwest Scotland, *Tectonophysics* **169,** 245-256 (1989).

[74] K. T. Winther. An experimentally based model for the origin tonalitic and trondhjemitic melts, *Chem. Geol.* **127**, 43-59 (1996).

[75] A. Z. Zhuravlev, E. E.Laz'ko and A. I. Ponomarenko. Radiogenic isotopes and rare-earth elements in minerals from xenoliths of garnet peridotite from the kimberlite pipe Mir. *Geokhimiya* **7**, 982-994 (1991).

Proc. 30th Intern. Geol. Congr. Vol. 1, pp. 97-110
Wang *et al.* (Eds)

Geochronology, Accretion and Tectonic Framework of Precambrian Continental Crust of Eastern China

SUN DAZHONG, LI XIANHUA and QIU HUANING
Guangzhou Institute of Geochemistry, Chinese Academy of Sciences, P.O. Box 1131, Guangzhou 510640, P. R. China

Abstract

Eastern China consists of three major Precambrian continental blocks: the North China Block (NCB), the Yangtze Block (YB) and the Cathaysia Block (CB). The NCB is composed mainly of Archaean basement with the oldest sialic rocks as old as early Archaean (*ca.* 3.8 Ga). The most important crustal growth in the NCB took place at 3.0-2.8 Ga. Two periods of crustal reworking accompanied by some juvenile crustal growth are the latest Archaean (*ca.* 2.5 Ga) and early Palaeoproterozoic (2.4-2.0 Ga), respectively. It was finally cratonized following the late Palaeoproterozoic Lüliang Orogeny of *ca.* 1.8 Ga. The oldest rocks in the YB is of late Archaean age (2.8-2.5 Ga), probably representing the crustal nucleus of the YB. Rapid crustal growth and extensive reworking during the late Palaeoproterozoic to Mesoproterozoic resulted in the formation of the continental basement of the YB. The basement rocks in the CB are of Palaeoproterozoic age of *ca.* 1.8 Ga, and the major crustal growth in the CB took place during the Proterozoic. The unexposed late Archaean crust, however, is postulated to exist based on the available radiometric age data. The collision of the CB to the YB took place during the early Neoproterozoic Jinning Orogeny when both oceanic and continental arcs accreted to the southern margin of the YB, accompanied by extensive crustal remelting and some newly mantle input. Whereas, the timing of the NCB-YB collision is still controversial, the Early Palaeozoic (Caledonian-age) and the Triassic (Indosinian) collision models being the most probable.

Keywords: eastern China, North China, Yangtze, Cathaysia, Precambrian crust, geochronology, crustal accretion, tectonic framework

INTRODUCTION

Eastern China consists of three major Precambrian continental blocks: The North China Block (NCB), the Yangtze Block (YB), and the Cathaysia Block (CB). Complicated geological and tectonic evolution including multiple stages of crustal accretion and reworking has been recorded in all blocks. Several important problems and controversies about the Precambrian crustal growth and tectonic evolution still pose a challenge to geoscientists, including: (1) When and how did these Precambrian crusts grow and get reworked? (2) When and how did they accrete to each other? (3) Do they have similar or distinct mean crustal residence ages? In the last decade numerous radiometric and isotopic data have accumulated for various rock types from eastern China. Based on available U-Pb zircon ages and Sm-Nd isotopic data in the literature and our new data, this paper gives a synthesis which attempts to unravel the history of the crustal growth and reworking, and to give a tectonic outline of the region in the Precambrian time.

NORTH CHINA BLOCK (NCB)

Archaean: Crustal Nuclei and Crustal Basement

The Archaean basement rocks crop out extensively in six broad regions in the NCB, including (1) the eastern Hebei Province; (2) the eastern Liaoning and southern Jilin Provinces, (3) the western Shandong Province; (4) Taihang-Wutai mountains in the northeastern Shanxi and the western Hebei Provinces; (5) the Songshan and Zhongtiao mountains in the western Hebei and southern Shanxi Provinces, and (6) the mid-southern Inner Mongolia (Fig. 1). These high-grade belts are dominated by granulites and gneisses, including supracrustal sequences, orthogneisses, paragneisses, grantitic intrusions and migmatites. Metamorphosed mafic-ultramafic layered bodies occur occasionally in the region. Orthogneisses occupy about 60% of the total outcrop.

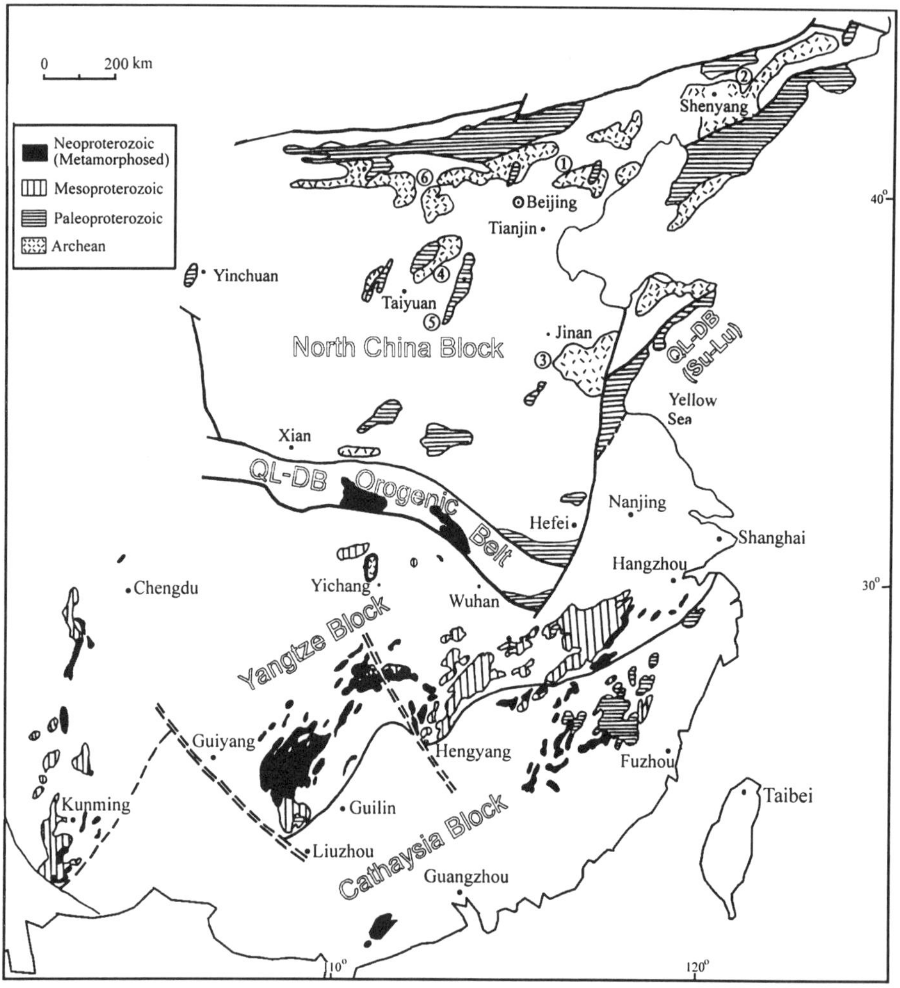

Figure 1. Simplified map showing the basement distribution in eastern China.

Numerous radiometric age data for the Archaean rocks in the NCB have accumulated during the last decade, with three age groups of 3.8 to 3.4 Ga, 3.2 to 2.8 Ga and 2.7 to

2.5 Ga. The oldest early Archaean rocks crop out in the Caozhuang area, eastern Hebei Province, including granulite and upper amphibolite facies gneisses with variable proportions of supracrustals and granites. Amphibolites, fuchsite quartzites, paragneisses, BIF and marbles occur as enclaves or lenses embedded in grey gneisses. Sm-Nd ages of about 3.5 Ga were obtained for the amphibolite enclaves in the region [21, 22]. If the amphibolite enclaves are remains of ancient mafic crust, the age of 3.5 Ga may record the earliest crustal formation [22]. Recently, U-Pb and Pb-Pb ages of 3.85 to 3.55 Ga have been obtained for detrital zircons from the fuchsite quartzite by means of SHRIMP and single-grain zircon stepwise evaporation technique [30, 31]. The age of 3.55 Ga was interpreted as the sedimentation age of the quartzite, and the age of 3.85 Ga as representing the age of its protolith – the oldest sialic crust. Early Archaean ages of *ca.* 3.8 Ga and *ca.* 3.4 Ga were also obtained recently by means of SHRIMP zircon U-Pb dating for the mylonitized granitic gneiss and the granites from the Anshan Group, eastern Liaoning and southeastern Jilin Provinces respectively [30, 36]. These age data suggest that the crustal nuclei in the NCB was most likely formed since the early Archaean.

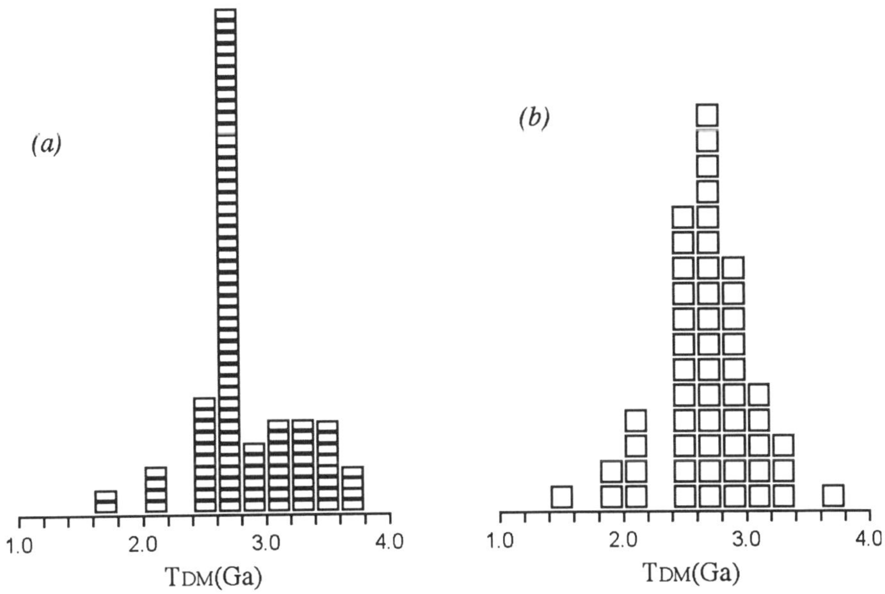

Figure 2. Histograms of T_{DM} in the NCB. (a) charnockite and TTG; (b) meta-sedimentary rocks

Numerous granitic gneisses and amphibolites of 3.0-2.8 Ga from eastern Hebei, eastern Liaoning, southeastern Jilin, and Shandong Provinces have been determined by zircon U-Pb, Sm-Nd, Rb-Sr and ^{39}Ar-^{40}Ar methods [20, 30, 33, 44, 46, 54]. These data suggest that a rapid crustal growth occurred in the NCB at 2.8-2.6 Ga, during which time some crustal basements may have been formed. Histogram of Nd model ages (T_{DM}) for various rock types in the NCB (Fig. 2) shows a peak in 2.8-2.6 Ga. Mafic granulites, charnockites and TTG suites have a main population of T_{DM} ages clustering ground 2.8-2.6 Ga, implying an crustal growth event at that time. Another minor group of T_{DM}

ages is around 2.4 to 2.6 Ga. Although some T_{DM} ages of >3.0 Ga are also identified, the rocks formed at that time are still quite limited.

Another important event of crustal reworking accompanied with some crustal growth took place during the latest Archaean age of 2.6-2.45 Ga [39, 39]. Various rock types including acidic, intermediate and mafic intrusives and extrusives were formed throughout the NCB by either extensive crustal remelting or by mantle magmatism. It is noteworthy that the generation of extensive peraluminous granites of *ca.* 2.5 Ga was closely related to the intracrustal remelting or reworking, rather than to the juvenile crustal growth. Contemporaneous granulite-facies metamorphism and tectonic deformation also occurred at this time, resulting in the consolidation of the Archaean basement. The unified North China Block was considered to have been formed at the end of Archaean.

Palaeoproterozoic: Rapid Crustal Growth and Extensive crustal Reworking

Two kinds of Palaeoproterozoic mobile belts were developed in the NCB. One occurred around the continental margin, including the Yinshan-Yanshan Belt along the northern margin and the Qingling-Dabie Belt along the southern margin of the NCB. The tectonic evolution of these two mobile belts has been controversial due to extensive and complicated Phanerzoic reworking. The other is the intracratonic mobile belts, including the Jiao-Liao belt, the Qinglong Belt, the Wutaishan-Lulliangshan Belt, the Taihangshan Belt, the Zhongtiaoshan Belt and the Songshan Belt, *etc.* These intracratonic mobile belts are composed of two sequences of rock association, with the lower sequence of volcanic rocks and the sequence of clastic, pelitic and carbonate sedimentations. Among these mobile belts, systematic dating work has been carried out for the Zhongtiaoshan Belt, and the chrono-tectonic framework and the chrono-crustal structure have been established [40, 42]. Both SHRIMP and single-grain zircon U-Pb analyses gave consistent ages of *ca.* 2.35 Ga for the intermediate and basic volcanics and tonalitic intrusions and *ca.* 2.15 Ga for the acidic volcanics and granitic intrusions within the lower sequence. Zircon U-Pb ages of 2.05 Ga have been obtained for intermediate and acidic tuffites within the sediments of the upper sequence. It is therefore concluded that the extensive extrusive and intrusive magmatism were widespread within the NCB during the early Palaeoproterozoic of 2.4-2.0 Ga, significantly differing from most other continents with a quiescence of magmatism at this time.

The early Palaeoproterozoic intracratonic mobile belts with development of rifts ended in the late Palaeoproterozoic Luliang Orogeny (*ca.* 1.8 Ga), and were accompanied by extensive greenschist to lower amphibolite facies metamorphism, deformation and granitic intrusions [37, 41]. Therefore, the NCB was further reworked and finally cratonized after the Luliang Orogeny. The late Palaeoproterozoic (*ca.* 1.8 Ga) to the uppermost Neoproterozoic sequences form the cover strata on the North China continental basement.

YANGTZE BLOCK (YB)

Archaean: the Crustal Nuclei

The oldest rocks in the YB are known as the late Archaean, Kongling Group located near the Yangtze Gorge of Hubei Province. The Kongling Group is composed mainly of granitic gneisses, amphibolites and metasediments. Zircon U-Pb ages of about 2.8 Ga for the granitic gneiss have been reported [51, 53]. It is still uncertain whether the entire Kongling Group or a part of it is of Archaean age. The Neoproterozoic Huangling Granite was intruded into the Kongling Group, which has T_{DM} ages of 2.2 to 3.0 Ga (Fig. 3), probably indicating an old Archaean protolith. Late Archaean T_{DM} ages of ~2.5 Ga have also been obtained from the meta-volcanic and sedimentary rocks within the Puling Formation in southern Anhui. A few of detrital zircon U-Pb ages of 2.5 to 2.8 Ga have been reported from the Mesoproterozoic amphibolites of the Xingzi Group nearby Lushan of Jiangxi Province and some Neoproterozoic granites along the southern belt of the YB. We suggest that an Archaean crustal nuclei may exist in the YB on account of the available geological and radiometric age data.

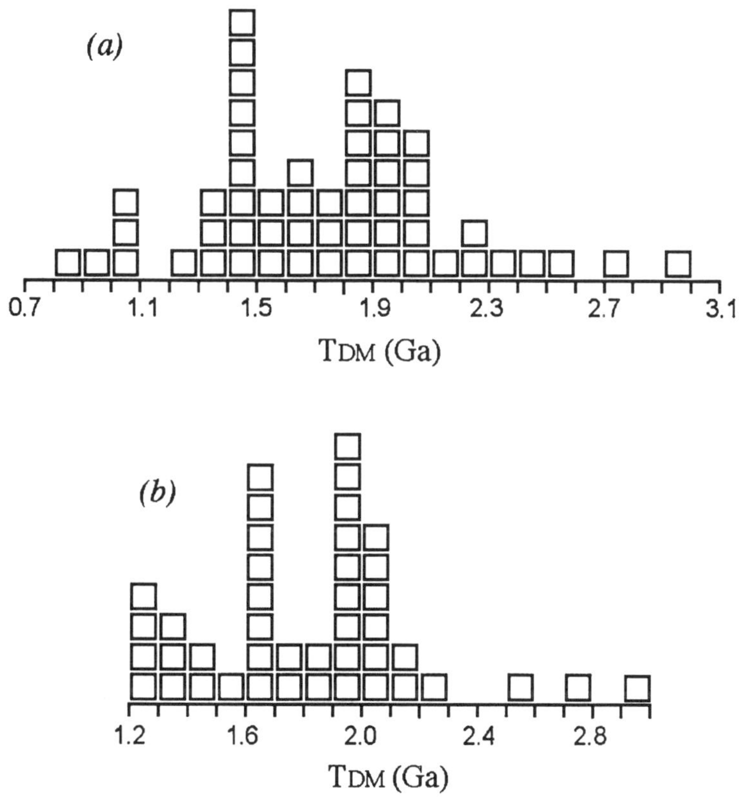

Figure 3. Histograms of T_{DM} in the YB. (a) granitiod rocks; (b) sedimentary rocks.

Palaeoproterozoic and Mesoproterozoic: Rapid Crustal Growth and Formation of Crustal Basement

Late Palaeoproterozoic to Mesoproterozoic low-grade metamorphic sedimentary rocks with intercalated mafic and ultramafic volcanics occur widely around the YB, including the Kunyang Group in western Yunnan, the Sibao Group in northern Guangxi and southern Guizhou, the Lengjiaxi Group in western Hunan, the Jiuling Group in northern Jiangxi and the Shuangqiaoshan Group in northeastern Jiangxi and northwestern Zhejiang. Ages of 1.6-1.7 Ga have been reported recently, including Pb-Pb whole-rock isochron ages for carbonates and slates [3, 32], and zircon U-Pb ages for volcanic rocks [14]. Fine-grained sediments from the late Palaeoproterozoic to early Mesoproterozoic sediments have quite uniform T_{DM} ages of 1.8-1.9 Ga [24]. The very small difference between the stratigraphic and T_{DM} ages of the Sibao sediments (less than 0.3 Ga) indicates that the sediments were derived dominantly from immature continental materials, and were probably the product of the first stage of sediment recycling from these materials. Nd model ages of 1.8-1.9 Ga were also obtained from the Neoproterozoic S-type granites around the YB. As the product of crustal anatexis, T_{DM} ages of these S-type granites may imply that the southern YB was mostly formed since Palaeoproterozoic.

Relative high-grade basement rocks called the Kangding Group in southern Sichuan Province occur mainly in the northwestern of the YB. It was formed during the late Palaeoproterozoic to Mesoproterozoic, mainly through amphibolite facies metamorphism. Basement rocks with amphibolite facies metamorphism, the Xingzi Group, are recently identified in the northern margin of the YB. The Yangtze Block was finally cratonized following the early Neoproterozoic Jinning Orogeny. The late Neoproterozoic Sinian sequence forms the cover stata on the folded basement.

CATHAYSIA BLOCK (CB)

Archaean: Probable Unexposed Crustal Nuclei or Basement

Although a few of Sm-Nd whole-rock isochron age of 3.1-2.7 Ga were reported for amphibolites from northwestern Fujian and southeastern Zhejiang [8, 55], there has no compelling evidence for any outcrops of the Archaean basement rocks in the CB. In fact, the reliability and interpretation of these Archaean Sm-Nd isochron ages are problematic. On the other hand, Archaean ages have been frequently obtained for inherited zircons from granitic and meta-volcanic rocks in the CB. For instance, inherited zircons with ages of 2.7-2.5 Ga have been recognized within the Early Palaeozoic granites from Tanghu in the southeastern Hunan Province [25], Mesozoic granites from Yangkou in Fujian, and Palaeoproterozoic meta-volcanic rocks from Fujian [11]. Most Palaeoproterozoic granitic gneiss from southwestern Zhejiang and northeastern Fujian gave Archaean Nd model ages of 3.0-2.5 Ga (Fig. 4) [19, 47], implying a derivation from ancient protoliths with Archaean mean crustal residence ages. We tentatively suggest, therefore, that the unexposed Archaean crustal nuclei or basement may exist in the CB.

Palaeoproterozoic and Mesoproterozoic: Rapid Crustal Growth and Extensive Crustal Reworking

The Proterozoic basement rocks with amphibolite-facies metamorphism crop out widely in Zhejiang and Fujian of the CB. The oldest rocks in the CB is Palaeoproterozoic, as demonstrated by zircon U-Pb ages of about 1.8 Ga for both granitic gneisses and amphibolites of the Badu Group in southwestern Zhejiang and the Mayuan Group in northwestern Fujian [11, 19]. With the chemical characteristics of tholeiitic basalts, the amphibolites show general geochemical similarities to MORB and within-plate basalts. They are most likely to have been formed in an environment of back-arc basin floored by continental crust. The granitic gneisses, however, have T_{DM} ages of 2.8-2.5 Ga, indicating a derivation from the Archaean protolith. Thus, we postulate that there was an important period of crustal growth and reworking in the CB around 1.8 Ga as based on the spatial and temporal association of amphibolites and granitic gneisses, with new input of mantle-derived magma and remelting of ancient crustal materials.

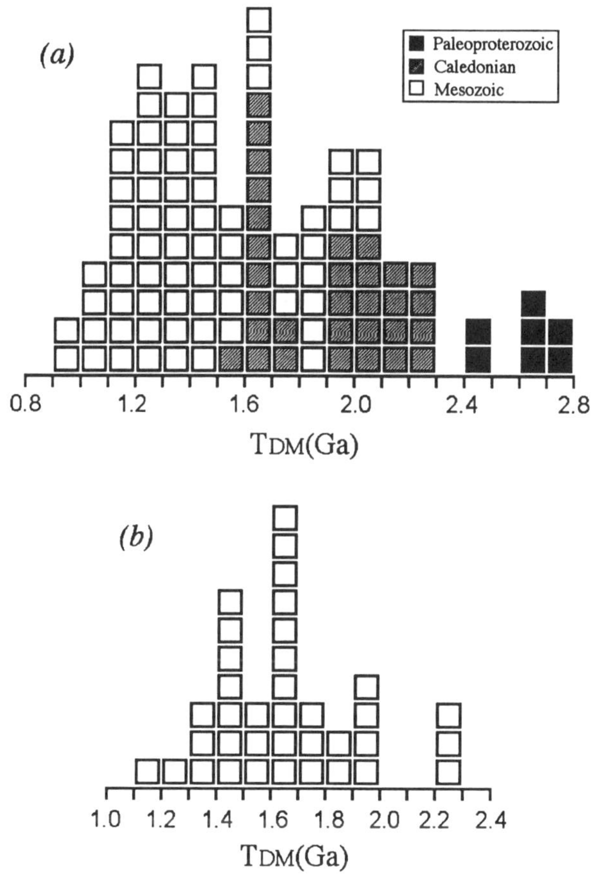

Figure 4. Histograms of T_{DM} in the CB. (a) granitiod rocks; (b) sedimentary rocks.

The Mesoproterozoic metamorphic basement, consisting of the Chencai Group in Zhejiang and the Mamianshan Group in Fujian, has also been identified. Both Sm-Nd isochron and zircon U-Pb ages of 1.4-1.1 Ga have been obtained for the intercalated meta-volcanic rocks [9, 19]. Contemporaneous granitic intrusives are absent in this region. These evidences may suggest an episode of Mesoproterozoic crustal growth, but no crustal reworking, in the CB. The mechanism of crustal growth during this time is however, not known.

Most Palaeozoic to Mesozoic granitoids and Phanerzoic sediments from the CB give T_{DM} ages ranging from 1.0 to 2.3 Ga (Fig 4), in contrast to the Palaeoproterozoic granitic gneisses with consistent Archaean T_{DM} ages. Interpretation of these ubiquitous Proterozoic Nd model ages is not unique. It is noted that the Mesozoic granitoids display a much wider spectrum of T_{DM} ages ranging from 0.5 to 2.3 Ga, in contrast to the early Palaeozoic (Caledonian) granitoids with T_{DM} ages of 1.6 to 2.3 Ga. If any granitoids were produced by mixing of crustal materials and the new mantle input, the Nd model ages are only the "weighted" mean ages of different components. This is likely the process for generation of some Mesozoic (Indosinian and Yanshanian) granitoids from the coastal provinces and along the major faults in the interior. They usually have T_{DM} ages younger than 1.6 Ga. On the other hand, if any granitoids were produced by melting of the middle and upper crust, such as most early Palaeozoic granites, their Nd model ages can not reflect the crust residence ages of the lower crust which might be of Archaean age and were not involved in the granitoid generation. Alternatively, the presence of Archaean crust might have been completely obliterated during previous magmatism through mixing with younger mantle input. Nevertheless, except for a few Archaean model ages registered in *ca.* 1.8 Ga-old granitic gneisses, the absence of Archaean model ages in the Phanerozoic granitoids and sediments suggests that the Archaean crust may have not played an important role in the crustal evolution of the CB. The major crustal growth in the CB took place in the Proterozoic.

DISCUSSION

Neoproterozoic YB-CB Accretion

The tectonic evolution of the YB and CB has been controversial, with a debate on the age and tectonic affinity of the Banxi Group as well as the time of collision between the two blocks. Hsü and his coworkers proposed a Mesozoic collision model for the YB-CB tectonic evolution, with the Banxi Group being a tectonic melange [16, 17]. However, based on field observations, sedimentological analyses, Sm-Nd isotopes and radiometric dating, many recent studies demonstrated that the Banxi Group is a normal Neoproterozoic lithostratigraphic unit occurring along the southern margin of the YB [24, 10, 43]. Moreover, Geochronological and geochemical investigations of the ophiolites and granitoids suggest that the YB and CB represent a Neoproterozoic continent-arc-continent collision belt.

Three ophiolite suites crop out in Guangxi, Anhui and Jiangxi. Zircon U-Pb ages of 968 ± 23 Ma by SHRIMP for Jiangxi Ophiolite [27] and 977 ± 10 Ma by conventional single-grain zircon chemical-TIMS method for Guangxi Ophiolite [10] have been reported, which are indistinguishable within analytical errors from previously published Sm-Nd mineral internal isochron ages of *ca.*1.0 Ga for Jiangxi and Anhui Ophiolites [5, 58]. The three ophiolite suites are similar to each other in terms of geochemical features, with volcanic rocks displaying affinity to basalts forming in subduction-related environments, typical of the "supra-subduction zone (SSZ)" ophiolites. However, they differ significantly in their initial Nd isotopic compositions. In the southeastern margin of the YB, the Jiangxi ophiolite displays very uniform initial eNd values of +5.5 ± 1.2, implying derivation from a depleted upper mantle source without significant contamination by evolved crustal materials. In contrast, the Anhui suite shows a large range of initial eNd values (+4.5 to −1.0), indicating a binary mixing between a depleted mantle component and an evolved crustal component. In the southern margin of the YB, the Guangxi suite has very uniform but relatively less radiogenic initial Nd isotopic compositions (eNd(T) = +1.0). It is postulated, therefore, the CB and YB was separated during the early Neoproterozoic (*ca.* 1.0 Ga?) by a multi-arc (archipelago) oceanic basin in the southeastern margin of the YB, with both oceanic and continental margin island arcs being developed to Jiangxi and Anhui, and by a continental marginal basin in the Guangxi of the southern margin of the YB.

S- and I-types of Neoproterozoic granitoids were recognized to occur around the southern margin of the YB. The S-type granites are voluminously dominant and occur to the north of the suture zone, including a number of cordierite granites (such as Xucun, Xiuning, Shexian and Jiuling) and the Baiji two-mica granite along the southeastern margin of the YB, and a number of biotite and two-mica granites such as Sanfang, Bendong and Yuanbaoshan along the southern margin of the YB. Although the ages of these S-type granites are still controversial, with previously published ages ranging from 0.7 to 1.1 Ga [13, 18, 48, 49, 52, 56], high-precision zircon U-Pb ages of 0.82 Ga have been obtained recently for the granites from Guangxi. These S-type granites, with eNd(T) = −2 to −7 and $d^{18}O$ = +9.2 to +12.6, are considered as the products of crustal anatexis during the Neoproterozoic collision between the CB and YB.

The I-type granitoids mostly occur within the Jiangshan-Shaoxing Fault Zone. They are mainly intermediate dioritic rocks, with rock types including pyroxene diorite, diorite and quartz diorite, covering relatively small areas and intruded during the period of 0.80 to 0.89 Ga [35, 57]. Their initial Sr isotopic ratios range from 0.7023 to 0.7031, eNd(T) = +4 to +5 and $d^{18}O$ = +5.7 to +6.7, implying a juvenile source [6, 34, 45, 57]. These I-type granitoids are, therefore, the products of crustal accretion during the Neoproterozoic Jinning Orogeny.

The Neoproterozoic mantle input is also recorded in the sediments from the southern margin of the YB. Fine-grained sediments from the Mesoproterozoic and early Neoproterozoic strata including the Sibao Group and the lower Banxi Group and their equivalents have quite uniform T_{DM} ages of 1.8 to 1.9 Ga. The middle Neoproterozoic including early Sinian (0.90 to 0.77 Ga) sediments have, however, a dramatic decrease in T_{DM} ages from 1.8 Ga to 1.3 Ga [24]. It is demonstrated that this pronounced decrease in T_{DM} ages requires a large increase in the proportion of juvenile mantle-derived materials which had been incorporated in the sedimentary provenance after the collision between the CB and YB.

In summary, a pronounced crustal accretion and reworking took place in the YB during the Neoproterozoic Jinning Orogeny when the YB and CB collided with each other.

Debate on the Timing of the NCB-YB Accretion.
The Qinling-Dabie orogenic belt is generally considered as the suture zone formed by the collision between the NCB and the YB. Ultra high-pressure metamorphic (UHPM) rocks including coesite-bearing eclogites and mantle xenoliths are widely distributed in the Dabie and Su-Lu UHPM terranes. These UHPM rocks are generally considered as generated by the NCB-YB collision. Plenty of radiometric ages have been reported, but the ages of these UHPM rocks are, however, still an issue of hot debate. Most Sm-Nd mineral isochron and high-precision zircon U-Pb analyses yield a consistent Triassic age of about 210-220 Ma, implying that the NCB-YB collision most likely took place during the Triassic[1, 4, 23, 26], consistent with the palaeomagnetic results [7, 28]. On the other hand, U-Pb zircon and Sm-Nd isochron ages ranging from 400 Ma to >2.0 Ga have also been reported [2, 29, 50]. Amongst these ages, attention has been directed to the early Palaeozoic age of *ca.* 400 Ma. Sedimentological studies in the Qinling orogen to the west of Dabie Mountains indicate that the beginning of NCB-YB collision may have taken place earlier in Odovician about 400 Ma ago [12].

Further more, the crustal growth and tectonic evolution within the Qinling-Dabie orogenic belt are also controversial. Different rock types including igneous and metamorphic rocks have Nd T_{DM} ages ranging from 1.6 to 2.6 Ga. It is still unknown whether there exists any Archaean basement rocks in the belt. Further works are needed to unravel the complicated history of the Qinling-Dabie orogenic belt .

SUMMARY

The Precambrian continental crust of East China has been experienced multiple stages of crustal growth and reworking, which is registered by U-Pb zircon and Sm-Nd isotopic systematics in various rock types. The important crustal growth and reworking intervals are summarised in Table 1. The oldest sialic rocks are known to be as old as early Archaean (*ca.* 3.8 Ga) in the NCB. The most important crustal growth in the NCB took place at 3.0-2.8 Ga. Extensive crustal reworking, accompanied by some new mantle input during the latest Archaean resulted in consolidation of the Archaean basement. Rapid crustal growth and extensive crustal reworking took place during the early Palaeoproterozoic, with intracratonic mobile belts being developed. The NCB was finally cratonized following the late Palaeoproterozoic Lüliang Orogeny of *ca.* 1.8 Ga.

The oldest rocks in the YB is known to be late Archaean of 2.8 to 2.5 Ga, which occurs sporadically in the northwestern part of the block and may represent the crustal nucleus of the YB. Rapid crustal growth and extensive reworking during the late Palaeoproterozoic to Mesoproterozoic resulted in the formation of the continental basement of the YB.

Although the oldest outcropped crustal basement rocks in the CB are *ca.* 1.8 Ga in age, the unexposed late Archaean basement is postulated to exist on account of the relevant radiometric age data. Multiple stages of crustal reworking including extensive Phanerozoic granitic magmatism may have strongly obliterated the Precambrian basement. However, the massive Nd T_{DM} ages from the granitic and (meta)sedimentary rocks strongly indicate that most parts of the CB were formed since the Palaeoproterozoic.

Table 1. The important crustal accretion and reworking intervals of Precambrian continental crust in eastern China

North China		Yangtze		Cathaysia	
possible accretion	reworking	possible accretion	reworking	possible accretion	reworking
(Ga)		(Ga)		(Ga)	
		0.9	0.9	(YB-CB collision)	0.9
		1.4−1.1		1.4−1.1	
		1.8−1.5		1.8−1.6	
	1.9−1.8				
2.3−2.1					
2.5−2.4	2.5−2.4				
2.8−2.6		2.8−2.5		2.8−2.5	
3.2−3.0					
3.8−3.6					

The collision of the CB to the YB took place during the early Neoproterozoic Jinning Orogeny when both oceanic and continental arcs were accreted to the southern margin of the YB, accompanied by extensive crustal remelting and some newly input mantle. The timing of the NCB-YB collision is controversial, and both the early Palaeozoic (Caledonian) and Triassic (Indosinian) collision models have been proposed. This important issue still poses a challenge to geoscientists.

Acknowledgements

This study was supported by the special grant of the Ministry of Geology and Mineral Resources and the President Grant of the Chinese Academy of Sciences. We are grateful to Professor Wang Hongzhen and Professor Chen Jiangfeng for fruitful discussions.

REFERENCES

[1] L. Ames, G. R. Tilton and G. Zhou. Timing of collision of the Sino-Korean and Yangtze Cratons: U-Pb zircon dating of coesite-bearing eclogites, *Geology* **21**, 339-342 (1993).

[2] R. L. Cao and S. H. Zhu. A U-Pb and $^{40}Ar/^{39}Ar$ geochronological study Bixiling coesite-bearing eclogites from Anhui province, China (in Chinese with English Abstract), *Geochimica* **24**, 152-161 (1995).

[3] X. Y. Chang, B. Q. Zhu, D. Z. Sun, H. N. Qiu and R. Zou. Isotope geochemistry study of the Dongchuan copper deposits in the middle Yunnan province: 1, stratigraphic chronology and application of geochemical exploration by lead isotopes, *Geochimica* **26** (1997).

[4] V. Chavagnac and B. M. Jahn. Coesite-bearing eclogites from the Bixiling Complex, Dabie Mountain, China: Sm-Nd ages, geochemical characteristics and tectonic implications, *Chem. Geol.* **133,** 29-51 (1996).

[5] J. Chen, K. A. Foland, F. Xing, X. Xu and T. Zhou. Magmatism along the southeastern margin of the Yangtze block: Precambrian collision of the Yangtze and Cathaysia blocks of China, *Geology* **19**, 815-818 (1991).

[6] H. Cheng. A preliminary study of an early Late Proterozoic collisional orogenic belt in northwestern Zhejiang Province (in Chinese with English abstract), *Geol. Rev.* **37**, 203-213 (1991).

[7] R. Enkin, Z. Yang, Y. Chen and V. Courtillot. Palaeomagnetic constraints on the geodynamic history of the major blocks of China from the Permian to the present, *J. Geophys. Res.* **97**, 13953-13989 (1992).

[8] S. C. Fu, J. M. Chcn and W. S. Lin. Geological characteristics of upper-Archaeozoic Tianjingping Formation (Ar_2t) in western Jianning, Fujian Province (in Chinese with English abstract), *Geology of Fujian* **10**, 103-113(1991).

[9] X. C. Gan, H. M. Li, D. Z. Sun and J. M. Zhuan. Geochronology study of Precambrian basement in northern Fujian (in Chinese with English abstract), *Geology of Fujian* **12**, 17-32 (1993).

[10] X. C. Gan, X. H. Li, F. Q. Zhao and H. B. Huang. U-Pb zircon and Sm-Nd isotopic ages of spilites from Danzhou Group, Longsheng, Guangxi (in Chinese with English abstract), *Geochimica* **25**, 270-276 (1996).

[11] X. C. Gan, F. Q. Zhao, W. S Jin and D. Z. Sun. The U-Pb ages of early Proterozoic - Archaean zircons captured by igneous rocks in southern China (in Chinese with English abstract), *Geochimica* **25**, 112-120 (1996).

[12] R. L. Gao, B. R. Zhang, X. M. Gu, Q. L. Xie, C. L. Gao and X. M. Guo. Silurian-Devonian provenance changes of South Qinling basins: Implications for accretion of the Yangtze (South China) to the North China cratons, *Tectonophysics* **250**, 183-197 (1995).

[13] The Granitoid Research Group of the Nanling Project. *Geology of granitoids of Nanling Region and their petrogenesis and mineralisation* (in Chinese with English abstract). Geological Publishing House, Beijing (1989).

[14] F. Han, J. Z. Shen and F. J. Nie. Geochronological study on the Sibao Group in the southern margin of the Jiangnan Old land (in Chinese with English abstract), *Acta Geoscietia Sinica* **(1-2)**, 43-50 (1994).

[15] J. K. Hsü, J. L. Li, H. Chen, Q. Wang, S. Sun and A. M. C. Sengor. Tectonics of South China: key to understanding West Pacific geology, *Tectonophysics* **183**, 9-39 (1990).

[16] J. K. Hsü, S. Sun and J. L. Li. Huanan. Alps, not South China Platform, *Sci. China*, Ser. *B* **10**, 1107-1115 (1987).

[17] J. K. Hsü, S. Sun, J. L. Li, H. Chen, H. Pen and A. M. C. Sengor. Mesozoic overthrust tectonics in south China, *Geology* **16**, 418-421 (1988).

[18] S. L. Hu, S. S.Wang, H. Q Sang and J. Qu. The application of the $^{40}Ar/^{39}Ar$ fast neutron activation dating technique to the study of the emplacement ages of the Jiuling Granite, Jiangxi Province (in Chinese with English abstract), *Acta Petrol. Sinica*, **1**, 29-34 (1985).

[19] X. J. Hu, J. K. Xu, C. X. Tong and C. H. Chen. *The Precambrian Geology of Southwestern Zhejiang Province* (in Chinese with English abstract), Geological Publishing House, Beijing (1993).

[20] A. Q. Hu, B. Q. Zhu, S. K. Fan, Z. L. Cheng, Z. P. Pu and Q. F. Zhang. Study on geochronology of Huadian are in northeastern China and its implication for mineralization. *Abstract Volume of International Symposium on Metallogeny of the Early Precambrian*, Changchun, 139-140 (1985).

[21] X. Huang, Z. Bi, and D. J. DePaolo. Sm-Nd isotopic study of early Archaean rocks, Qianan, Hebei Province, China, *Geochom. Cosmochim. Acta* **50**, 625-631 (1986).

[22] B. M. Jahn, B. Auvray, J. Cornichet, Y. L. Bai and D. Y. Liu. 3.5 Ga old amphibolites from eastern Hebei Province, China: Field occurrence, petrology, Sm-Nd isochron age and REE geochemistry, *Precambrian. Res.* **38**, 311-346 (1987).

[23] S. Li, S. R. Hart, S. Zheng, D. Liou, G. Zhang and A. Guo. Timing of collision between the North and South China Blocks – Sm-Nd isotopic age evidence, *Sci. China, Ser. B* **32**, 1391-1400 (1989).

[24] X. H. Li and M. T. McCulloch. Secular variation in the Nd isotopic composition of Neoproterozoic sediments from the southern margin of the Yangtze Block: evidence for a Proterozoic continental collision in southeast China, *Precambrian Res.* **76**, 67-76 (1996).

[25] X. Li, M. Tatsumoto, W. R. Premo and X. Gui. Age and origin of the Tanghu Granite, southeast China: Results from U-Pb zircon and Nd isotopes, *Geology* **17**, 395-399 (1989) .

[26] S. Li, Y. Xiao D. Liou, Y. Chen, N. Ge, Z. Zhang, S. S. Sun, B. Cong, R. Zhang, S. R. Hart and S. Wang. Collision of the North China and Yangtze Blocks and formation of coesite-bearing eclogites: Timing and process, *Chem. Geol.* **109**, 89-111 (1993).

[27] X. H. Li, G. Q. Zhou, J. X. Zhao, C. M. Fanning and W. Compston. SHRIMP ion probe zircon age of the NE Jiangxi Ophiolite and its tectonic implications (in Chinese with English abstract), *Geochimica* **23**, 117-123 (1994).

[28] J. L. Lin, M. Fuller and W. Zhang. Preliminary polar wander paths for the North and South China blocks, *Nature* **313**, 444-449 (1985).

[29] R. Liu, Q. Fan, H. Li, Q. Zhang, D., Zhao and B. Ma. The nature of protolith of Bixiling garnet peridotite-eclogite massif in Dabie Mountains and the implication of its isotopic geochronology, *Acta Petrol. Sinica* **11**, 243-256 (1995).

[30] D. Y. Liu, A. P. Nutman, J. S. Williams, W. Compston, J. S. Wu, Q. H. Shen. The remnants of ≥3800 Ma crust in Sino-Korean Craton — The evidence from ion microprobe U-Pb dating of zircon (in Chinese with English abstract), *Acta Geoscientia Sinica* **(1-2)**, 3-13 (1994).

[31] D. Y. Liu, Q. H. Shen, Z. Q. Zhang, B. M. Jahn and B. Auvray. Archaean crustal evolution in China: U-Pb geochronology of the Qianxi complex, *Precambrian Res.* **48**, 223-244 (1990).

[32] H. C. Liu and B. Q. Zhu. Study on the ages of Banxi and Lengjiaxi Groups from western Hunan Province (in Chinese with English abstract), *Chin. Sci. Bull.* **39**, 148-150 (1994).

[33] S. N. Lu, C. L. Yang, M. M Yang, H. K. Li, H. M. Li. *Tracing of Precambrian Continental Crustal Evolution* (in Chinese). pp. 1-18. Geological Publishing House, Beijing (1996).

[34] Q. Qi, X. M. Zhou and D. Z. Wang. The origin of the Xiqiu Spilite-Keratophyre Series and the characteristics of associated mantle-derived granites in Zhejiang Province(in Chinese with English abstract), *Acta Petrol. Minerl.* **5**, 299-308 (1986).

[35] T. Shui, B. T. Xu, R. H. Liang and Y. S. Qiu. *Geology of the metamorphic basement in Fujian and Zhejiang Provinces* (in Chinese with English abstract). *China*. Beijing, Science Press (1988).

[36] B. Song, A. P. Nutman, D. Y. Liu and J. S. Wu. 3800 to 2500 Ma crustal evolution in the Anshan area of Liaoning Province, northeastern China, *Precambrian Res.* **78**, 79-94 (1996).

[37] D. Z. Sun. Tectonic and geochemical development of Archaean and Proterozoic mobile belts in eastern China. In: *Progress in Geosciences of China* (1985-1988). pp. 131-134. Geological Publishing House, Beijing (1988).

[38] D. Z. Sun. A subdivision of the Archaean of China, *Report to 9th Meeting of ISPS*, 1-21 (1991).

[39] D. Z. Sun. The Archaean in China, *Report to 10th Meeting of ISPS*, 1-17 (1995).

[40] D. Z. Sun and W. X. Hu (ed.). *Precambrian chronotectonic framework and chronocrustal structure of the Zhongtiao Mountains* (in Chinese). pp. 1-18. Geological Publishing House, Beijing (1993).

[41] D. Z. Sun and S. N. Lu. A subdivision of Precambrian of China. *Precambrian Res.* **28**, 137-162 (1985).

[42] D. Z. Sun, H. M. Li, Y. X. Lin, H. F. Zhou, F. Q. Zhao and M. Tang. Precambrian geochronology, chronotectonic framework and model of chronocrustal structure of the Zhongtiao Mountains, *Acta Geologica Sinica* **5**, 23-37.

[43] J. F. Tang. Field excursion note, Banxi Group, Hunan Province (in Chinese with English abstract), *J. SE China College of Geology* **16**, 420-424 (1993).

[44] Y. S. Wan and D. Y. Liu. Ages of zircons from mid-Archaean gneissic granite and fuchsite quartzite in the Gongchangling area, Liaoning, *Geol, Rev.* **39**, 124-129 (1993).

[45] Z. J. Wang. A study of various rock types within the Jiangshan-Shaoxing Fault Zone (in Chinese with English abstract), *Geology of Zhejiang*, **1**, 12-20 (1988).

[46] S. S. Wang, S. L. Hu, M. G. Zhai, H. Q. Sang and J. Qiu. $^{40}Ar/^{39}Ar$ age spectrum for biotite separated from Qingyuan tonalite, NE China (in Chinese with English abstract) *Scientia Geol. Sinica.* **21:1**, 97-99 (1987) .

[47] Y. X. Wang, J. D. Yang, L. Z. Guo, Y. S. Shi, X. J. Hu and X. Wang, 1992. The discovery of the lower Proterozoic granite in Longquan, Zhejiang Province, and the age of the basement, *Geol. Rev.* **38**, 525-231.

[48] S. Wu. Isotopic chronological study on the Precambrian Bendong granite from Guangxi (in Chinese with English abstract), *Geochimica* **8**, 187-194 (1979).

[49] F. M. Xing, X. Xu, J. F. Chen, T. X. Zhou and K. A. Foland. Late Proterozoic continental growth history in the southeastern margin of the Jiangnan Old Land (in Chinese with English abstract), *Acta Geologica Sinica* **66**, 59-72 (1992).

[50] Z. You, Y. Han, W. Yang, Z. Zhang, B. Wei and R. Liu. *The high-pressure and ultra-high-pressure metamorphic belt in the East Qinling and Dabie Mountains*. China University of Geosciences Press, Wuhan (1996).

[51] Z. C. Zhang and J. P. Zhu. $^{207}Pb/^{206}Pb$ age determination on single zircon grains by direct evaporation on double filament thermal emission mass spectrometer (in Chinese with English abstract), *Bull.*

Yichang Institute of Geology and Mineral Resources, Chinese Academy of Geological Sciences **17**, 153-161 (1991).

[52] Z. J. Zhao, D. Q. Ma, H. K. Lin, C. L. Yuan and X. H. Zhang. A study on the Precambrian granitoids from Bendong and Sanfang Massifs, Northern Guangxi (in Chinese with English abstract). In: *Selected Papers on Geology and Mineral Resource in Nanling Area (Vol. 1)*. Yichang Institute of Geology and Mineral Resources (ed.). pp. 1-27. Publishing House of Wuhan College of Geology, Wuhan (1987).

[53] W. Z. Zheng, G. L. Liu and X. W. Wang. New information on the Archaean Eon for the Kongling Group in northern Huangling anticline, Hubei (in Chinese with English abstract). *Bull. Yichang Institute of Geology and Mineral Resources, Chinese Academy of Geological Sciences* **16**, 97-106 (1991).

[54] F. D. Zhong. Geochronology study of Archaean granite-gneiss in Anshan area, Northeast China (in Chinese with English abstract), *Geochimica* **13:3**, 195-205 (1984).

[55] X. H. Zhou. *Geochemical constraints on tectonic evolution of South China. In: Isotopic Geochemical Research in China*. J. S. Yu (ed). pp. 98-117. Beijing, Science Press (1996).

[56] X. M. Zhou and D. Z. Wang. Peraluminous granites with low $^{87}Sr/^{86}Sr$ ratios in southern Anhui Province and their genesis (in Chinese with English abstract), *Acta Petrol. Sinica* **4**, 37-45 (1988).

[57] X. M. Zhou and Y. H. Zhu. Petrological evidence for a Late Proterozoic collision orogenic belt and suture zone in southeast China (in Chinese with English abstract). In: *The structure and geological evolution of the continental lithosphere in southeast China*. J. L. Li (ed.). pp. 87-97 Beijing, Metallurgic Industry Press (1993).

[58] X. M. Zhou, H. B. Zou, J. D. Yang and Y. X. Wang. Sm-Nd isochron age of the Fuchuan Ophiolite Suite in Shexian, Anhui Province and its geological significance (in Chinese with English abstract), *Chin. Sci. Bull.* **34**, 1243-1245 (1989).

Proc. 30th Intern. Geol. Congr. Vol. 1, pp. 111-128
Wang *et al.* (Eds)

Pangaea Cycles, Earth's Rhythms and Possible Earth Expansion

WANG HONGZHEN, LI XIANG, MEI SHILONG and ZHANG SHIHONG
China University of Geosciences, Beijing, 100083, P. R. China

Abstract

Assembling of palaeocontinents into Pangaea in earth history may have occurred periodically at 2500, 1900, 1450, 850 and 250 Ma, with a time interval of 500-600 Ma. These Pangaea occurrences are partly in accord with the important orogenic phases in earth history, and are also comparable with the tectonic megastage boundaries marked by the formation of continental nuclei (2.8 Ga), continental paraplatform (1.8 Ga), continental platform (850 Ma) and Pangaea-250 Ma. Except Pangaea-250, the assembly of earlier pangaeas are all not quite certain. The Piper assembly of Proterozoic continents is used as a reference for Pangaea-1900 and possibly also for Pangaea-2500. Reconstruction has been made of Pangaea-850, which is different from that of Dalziel (1992) in the position of Laurentia.

Rhythms in earth's history are probably a common feature and are notable in many geological processes. Different ranks of rhythms may be discerned in sedimentation, organic evolution, magmatism and geomagnetic reversal. They show a cyclicity and periodicity that are more or less correlatable with each other. They may have reflected a common factor—the different kinds of cosmic cycles. Pangaea cycles of 500-600 Ma may represent the longest cycle known and may be related with the double cosmic year (2700 Ma). A correlation between the different ranks of stratigraphical sequences and those of cosmic cycles is preliminarily assumed.

The increasing length of solar days with time, the recognition of ever more crustal extensional structures on earth, the huge outpour of basalt flow, and the discrepancy in global reconstruction of palaeocontinents in relation to biogeographical interpretation, point to the possibility of an expanding earth, probably limited, asymmetric and punctuated. Following a non-uniformitarian viewpoint we have assumed episodic expansion of the earth after each break-up of the pangaeas with earth radius values of 80 percent after Pangaea-2500, 85 after Pangaea-1900, 90 after Pangaea-1450, 95 after Pangaea-850, and approximate to the present after Pangaea-250. The expansion event probably took place along with the disintegration of each Pangaea, and was followed by a long, comparatively static period. A method of computer reconstruction of palaeocontinents on a smaller globe was designed on the basis of palaeomagnetic data and unchanged area and decreased curvature of the continents. For evaluation of the radius reduction adopted, we have applied the Egyed-Ward model, by use of the available palaeomagnetic data and the method of least diversion of pole values for the same period, to find out the optimum value of the expansion rate. The results for Meso-Neoproterozoic and for the Cambrian are on the whole consistent with data adopted in this paper. Earth expansion is a controversial topic and the present work is preliminary and speculative. The main theme is to investigate the problem with a viewpoint of Neocatastrophism and an idea of punctuated progression in the history of the earth.

Keywords: pangaea cycles, rhythms, bioevents, punctuated progression, earth expansion, palaeocontinent reconstruction.

INTRODUCTION

In the present paper, an attempt is made to present the rhythmic nature of some geological processes in earth's history, the tectonic frames of the earth, the Pangaea cycles and finally the idea of possible earth expansion. As the problems involved are mostly global and to a certain extent speculative, it is probably necessary first to relate the main viewpoints and the main trend of geological thinking adopted here.

The space-time relation in earth's history which forms the main thought in dealing with the problems mentioned above consists of two aspects, the temporary and the spatial, although these two are intimately related to each other. In the temporary aspect, we advocate to the idea of "development by stages", or the idea of "punctuated progression", to borrow a term from palaeontology. The main evidences comprise episodicity, periodicity and irreversibility or progressive development, but not simply cyclicity or recurrence. These comply with the common principle of material development, in which quantitative change always leads to qualitative change. In the spatial aspect, our idea comprises mobilism in global tectonics and heterogeneity of earth materials. The latter is based on the heterogeneous features of the component parts of the lithosphere as well as of the whole earth. Actually the differentiation of geospheres and the lateral inhomogeneity of the geospheres themselves point to the common principle of inhomogeneity, which is the final basis for mobilism. We deem it a tenable approach to combine the idea of punctuated progression in historical development with the idea of mobilism in global tectonics in the research of the earth's history.

In the following we will deal with the topic in three parts: earth's rhythms, pangaea cycles and earth expansion.

RHYTHMS IN EARTH'S HISTORY

The records of earth's history are preserved in sediments, fossils, volcanic and intrusive rocks, and in geological phenomena and deformation of geological bodies. These phenomena display on the whole an apparent recurrence at certain intervals of time in the history of the earth, and may be called the rhythms of the earth. Grabau used the term rhythm in his book of 1940 "The Rhythm of the Ages", and Umbgrove used both rhythm and pulse in 1939 and 1942. It seems that the rhythms in earth's history is probably a universal feature, in the organic and the inorganic world as well. However, there are controversies as to whether they are periodic and synchronous, especially whether there is a world-wide synchroneity in these features. As the periodic and progressive evolution in the organic world may be regarded as established, we will dwell more on the inorganic aspects, notably sedimentation and tectonics.

The rhythms in sedimentation are manifold, but they are best manifested in sequence stratigraphy, the newly established subdiscipline in stratigraphy, which pays special attention to the designation of sequence boundaries that are synchronous, at least within a sedimentary basin but sometimes global in extent. This peculiarity allows for a long distance correlation in stratigraphy with the aid of biostratigraphy and other means, which is difficult by using only traditional methods. Sequence stratigraphy is intimately tied with sea level changes, which, though usually affected by regional tectonic controls, could reflect eustatic changes through careful comparison between cratonic sequences all over the world, at least as far as long term changes are concerned. Thus sequence stratigraphy provides a strong tool to test the world-wide synchroneity in the interaction between lithosphere and hydrosphere.

Since the establishment of Haq's Mesozoic and Cenozoic eustatic sea level change curves of different ranks [7], various authors have tried to improve them and extend them to Palaeozoic and earlier times [29]. In recent years Wang and Shi [40] have, through studies on Phanerozoic sequence stratigraphy of China and a historical review of the problem, established a hierarchy for sequence stratigraphy, and have compared it with other current classifications (Table 1). A possible correlation with various kinds of cosmocycles [28, 32] is suggested. Optimal values of the time ranges for each of the five orders of cycles are given, among which the 2nd order mesosequence with a time range of 30-40 Ma is most distinct [28]. The mesosequence manifests a good accordance with global sea-level changes, biotic extinctions, orogenic episodes, magmatic cycles and palaeomagnetic reversals. These various kinds of geological processes are thought to have been controlled by the cosmic cycle related to crossing of the solar system through the galactic plane. Other cycles include the first order megasequence, probably equivalent to the Sloss sequence and related to the cratonic thermodynamic cycle, and the minor climatic cycles corresponding to the Milankovitch long and short cycles. Of special tectonic and palaeogeographic importance is the megasequence with a time range of 60-120 Ma, which is well manifested on the Sino-Korean (Ordos), and the Tarim continental platforms [41]. The longest rhythm recognized in earth's history is the Pangaea cycle with an average time interval of 550-600 Ma, which is referred to the superrank in cycle order [40]. Although reconstruction of Pangaea on the earth's surface is somewhat uncertain and controversial, the assembly and dissociation of the last Pangaea existent from Late Carboniferous to early Jurassic, generally known as Pangaea-250 Ma, has been studied in some detail.

Table 1. A scheme of hierarchy for sequence stratigraphy (after Wang and Shi, 1996).

Cycle order	Sequences & time ranges	Cosmocycles	Vail *et al.* (1991)	Mitchum *et al.* (1991)	Brett *et al.* (1990)	Cooper (1990)
Super rank	Gigasequence(Gs) 500-600Ma	Double cosmic year				Chelogenic cycle 600-1250Ma
1st	Megasequence(Mg) 60-120Ma	Galactic cycle (Sun crossing spiral arms)	Megasequence >50Ma	Megasequence 200Ma	Megasequence 50-60Ma	Megacycle 250-375Ma Supercycle 70-150Ma
2nd	Mesosequence(Ms) 30-40Ma Orthosequence set 9-12Ma	Solar cycle (Sun crossing Galactic plane)	Supersequence set 27-40Ma Supersequence 9-10Ma	Supersequence set 29-30Ma Supersequence 9-10Ma	Holostrome 10-30Ma	Macrocycle 20-50Ma Mesocycle 5-10Ma
3rd	Orthosequence(Os) 2-5Ma	Oort cycle (Sun approaching to asteroids)	Sequence 0.5-5Ma	Sequence 1-2Ma	Sequence 2-3Ma Subsequence 1-1.5Ma	Cycle 1-3Ma
4th	Subsequence (Ss) 0.1-0.4Ma	Longer Milan-kovitch cycle	Parasequence 0.05-0.5Ma	High-frequency sequence 0.1-0.2Ma	Parasequence set 0.45Ma Parasequence 0.1Ma	Microcycle 0.1Ma
5th	Microsequence(Mc) 0.02-0.04Ma	Shorter Milan-kovitch cycle	Simple sequence 0.01-0.05Ma	5th order sequence 0.01-0.02Ma	Rhythmic bedding 0.02Ma	

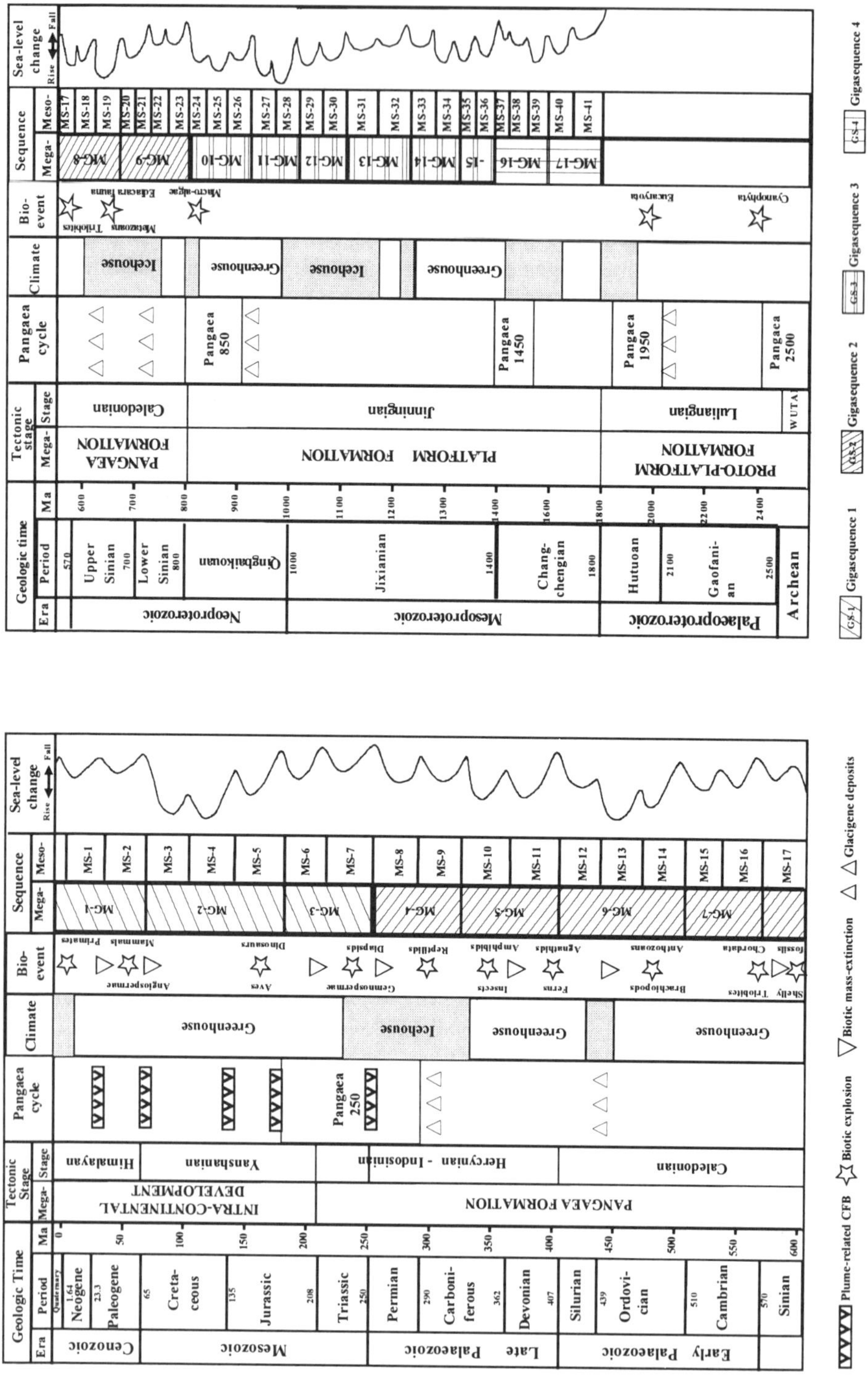

Figure 1. Pangaea cycles, tectonic stages and sequence stratigraphy (after Shi and Wang, 1996, 30th IGC Exhibition)

The evidences of the rhythmic nature of various geological processes are summarized in Figure 1. This is an integrated presentation of the rhythms in earth' history, including Pangaea cycles, tectonic stages, bioevents, sequence stratigraphy and sea level changes. Stratigraphic sequences of different scales, the gigasequence corresponding to Pangaea cycles, the Megasequence or Sloss sequence and the Mesosequence, are subdivided down to 1800 Ma, mainly based on research results recently done in China, mostly from research results of the project on sequence stratigraphy (SSLC) [41]. The sea level change curves are approximate and preliminary. The Phanerozoic part are mainly based on facies and palaeoecological analyses and estimates of depth of sea waters, in addition to marine transgressions and regressions on the palaeocontinents of the time. The main glacial epochs and the presumably plume-related continental flood basalts are shown under the column of Pangaea cycles in juxtaposition against tectonic megastages and stages [37, 39]. Bioevents are shown in terms of biotic explosions and mass extinctions of the successive biota, synthesized from various sources including new results obtained from the Proterozoic of China in recent years. Climatic changes of alternating greenhouse and icehouse effects are tentatively indicated, mainly based on overall biologic and sedimentary evidences. Taking all these into account, it seems that the rhythmic nature in the various lines of records in earth history is manifest and shows a quite good correlation with each other. The Precambrian time scale here used is after Wang and Li 1990 [38], in which the Mesoproterozoic began from 1800 Ma instead of 1600 Ma, as designated in the IUGS 1989 global chart. We regard 1800 Ma as a more reasonable boundary between Messo- and Neoproterozoic both for biological and tectonic reasons. In Figure 1, the Phanerozoic and the Precambian are not in the same scale, and within the Precambrian the scale below 1400 Ma is further reduced.

PANGAEA CYCLES AND THEIR SIGNIFICANCE

The term Pangaea is used to denote the whole assembly of all the palaeocontinents on the earth's surface. The term came with Wegener's hypothesis of continental drift, and was subsequently used for the assembled continent in Permian and Triassic, once called Pangaea A [21], and is called in this paper Pangaea-250 Ma. The term supercontinent is also commonly used, sometimes as synonym of Pangaea. These two may probably better be discriminated from each other. For instance Pangaea-250 consists of two supercontinents, the northern Laurasia and the southern Gondwana or Gondwanaland.

As shown in Figure 1, we have discerned, through preliminary palaeocontinental reconstructions, five Pangaea occurrences since from the beginning of Palaeoproterozoic (2500 Ma), that recurred at a time interval of 550-600 Ma. Taking the optimal age of Pangaea formation as standard, we may call them Pangaea-2500, Pangaea-1900, Pangaea-1450, Pangaea-850 and Pangaea-250 respectively. The time interval between two successive pangaeas may be called a Pangaea cycle, which is more or less comparable with a tectonic megastage [9]. The Pangaea-250 existed for less than 100 Ma (270-180 Ma), and its break-up took probably another 100 Ma. If the average time interval of Pangaea cycles of 550-600 Ma is acceptable, the Pangaea formation is comparatively shorter as compared with the subsequent break-up and stasis periods in the whole cycle. Thus the occurrence of Pangaea probably represents an epoch of special circumstances on the earth, and the study of climatic and tectonic and related conditions at that time might be able to provide valuable data for comparison with global changes to-day.

In regard to the number of pangaeas that had appeared in earth's history, Morel and Irving [21] named Pangaea A to E from 250 Ma to ca 700 Ma. This seems not real. Khain [9] talked of an ancient Pangaea at the end of Archaean, but his second Pangaea came not until 250 Ma. Some authors would suggest 200 Ma for the Wilson cycle, which covers the time interval from the break-up to the re-union of a continent. If we take the Palaeozoic and the modern Atlantic as an example and regard the Late Cambrian as the epoch of widest Palaeozoic Atlantic, the Wilson cycle would be about 500 Ma instead of 200 Ma, and is approximate to our pangaea cycles.

World Tectonic Frame

While making global reconstruction of palaeocontinents, we have found it necessary to determine the tectonic affinities of the separated, interstitial small Precambrian massifs and the fold zones between the main continental platforms or cratons, as palaeomagnetic data are usually lacking in these parts. In the geotectonic studies of China and East Asia, Wang [35] has proposed a tectonic unit of first rank, the tectonic domain. A tectonic domain encloses commonly the continental platform and its surrounding median massifs and fold belts that constitute the complicated continental margin tracts, probably formed of fragmented masses split off the continental platform and later accreted to it again through continent-arc collision. The boundary between the tectonic domains is called convergent crustal consumption zone, which is a main geosuture formed of the final continent-continent collision, while the boundary between tectonic units within a continental margin tract formed of continental-arc or arc-arc collision is called accretional zone of crustal consumption. The second rank of tectonic unit encompasses continental platforms (cratons) on the one hand, and mobile belts composed of Phanerozoic fold zones and Precambrian median massifs on the other. Palaeogeographic and palaeotectonic studies have led us to the belief that it was probably the tectonic domain that constitutes the fundamental unit moving and rotating over the earth's surface in the geologic past.

With this idea, we have made a world map showing the Precambrian platforms and the interstitial median massifs (Fig. 2), largely based on the world tectonic map by Leonov and Khain (1982). Mainly based on the global palaeotectonic analyses in the Neoproterozoic (Jinningian orogenic epoch 850-800 Ma), and in the Late Triassic (Indosinian orogenic epoch), we have recognized five tectonic domains and some thirteen continental platforms in the world (Fig. 2). The tectonic domains seem to have moved on the large as an entity, and the continental platforms have mostly kept their configuration through times. The names of the tectonic domains and their major constituent cratons are given in Figure 2. In this connection we would like to point out that free motion of the major cratons in earth's history may be more constrained than conceived heretofore. Since the publication of Dziewonski *et al.* [5] of the seismic tomography under the Canadian craton, later publications containing seismic tomography have all revealed the craton roots with a depth of over 250 to 300 km [15]. This may be regarded as a further support that the tectonic domains may have behaved more or less as an entity in the history of the earth.

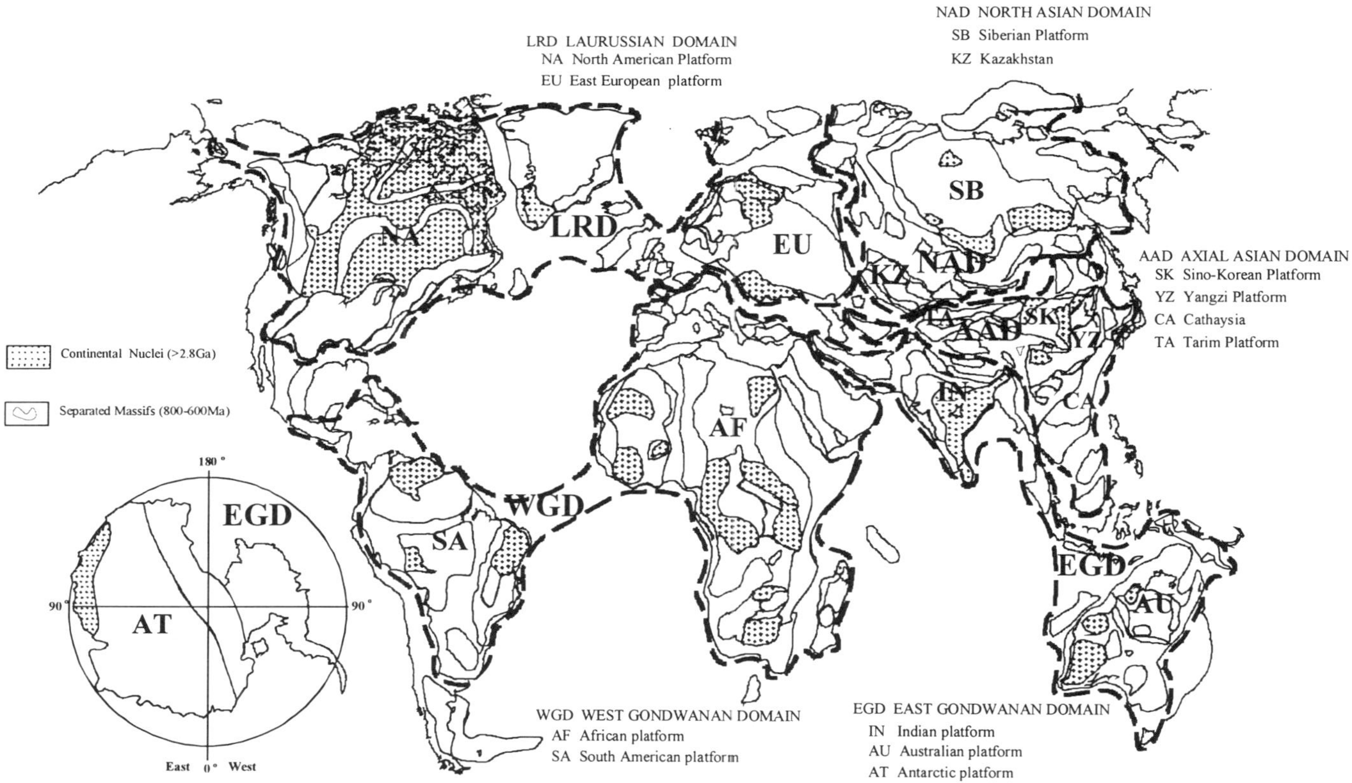

Figure 2. World map showing tectonic domains and the main continental platforms

The Pangaea-2500 Ma is hypothetical, and no reconstruction of it has appeared. Piper's reconstruction [24] of the Proterozoic Pangaea was based on palaeomagnetic data mainly with an age of ca 1900-1800 Ma. He considered that no essential change had occurred during the Proterozoic. His reconstruction may be regarded as representing Pangaea-1900 Ma.

The megastage of Protoplatform formation began at the beginning of Mesoproterozoic (1800 Ma, Figure 1). The tectonic units are believed to have acted since their formation more or less as an entity. Palaeomagnetic data are very meager in the middle part of Mesoproterozoic. A preliminary reconstruction (1450 Ma) is given by Li [14] as an example of the software Gudalu for palaeocontinent reconstruction. The position of the Chinese blocks and of Australia in this map needs more study.

On the whole, palaeomagnetic data of 1400-1500 Ma show a close relationship between Laurentia and Baltica. Siberia was probably independent. Recent data summarized by Zhang *et al.* [43] indicate that the three Chinese cratons, Sino-Korea, Tarim and Yangtze, were not far from each other in the Proterozoic, and the common occurrence of exactly the same fossil *Grypania* in China and western North America suggest close relations between these continents at that time. West Gondwana and East Gondwana were then probably still separate, although it is not certain whether West Gondwana was then completely consolidated.

Pangaea-850 (800-900) Ma.

There are different views on the occurrence and the time interval of a Neoproterozoic Pangaea. The recent work of Hoffman [8] and Dalziel [2] have aroused interest and discussion of the Neoproterozoic supercontinent Rodinia, which has recently been extended almost to a Pangaea. Urug [33] gave a good summary of the Meso- and Neoproterozoic history of Gondwanaland. Two problems deserve notice. The first is the position of Laurentia between East and West Gondwana, as Hoffman and Dalziel suggested. The second is whether the West Gondwanan cratonic blocks were separated from each other, and when did they accrete together and unite with East Gondwana to form Gondwanaland.

First the position of Laurentia in the supercontinent Rodinia. In an open foram organized by Powell [25], one of the negative opinions is its contradiction to the trilobite provincialism in the Early Cambrian. I would like to add another biogeographical point that identical megaalgae fossils were found in East China and western North America. Therefore the joint position between western North America, Australia and East Antarctica as suggested in the map of Rodinia seems not very likely. Li *et al.* [13] tried to put South China between Australia and Laurentia. This is also improbable, as the three palaeocontinents in East China probably collided with each other at their eastern ends (present orientation) in the Jinningian orogeny (850-800 Ma), as is evidenced by collision type of granites along their junctures. This forms the idea of a more or less combined Cathaysiana including Tarim in the Neoproterozxoic.

Concerning the second problem, the Kalahari massif was more related to East Gondwana and probably separated from West Gondwana in the Proterozoic. Li and Powell [12]

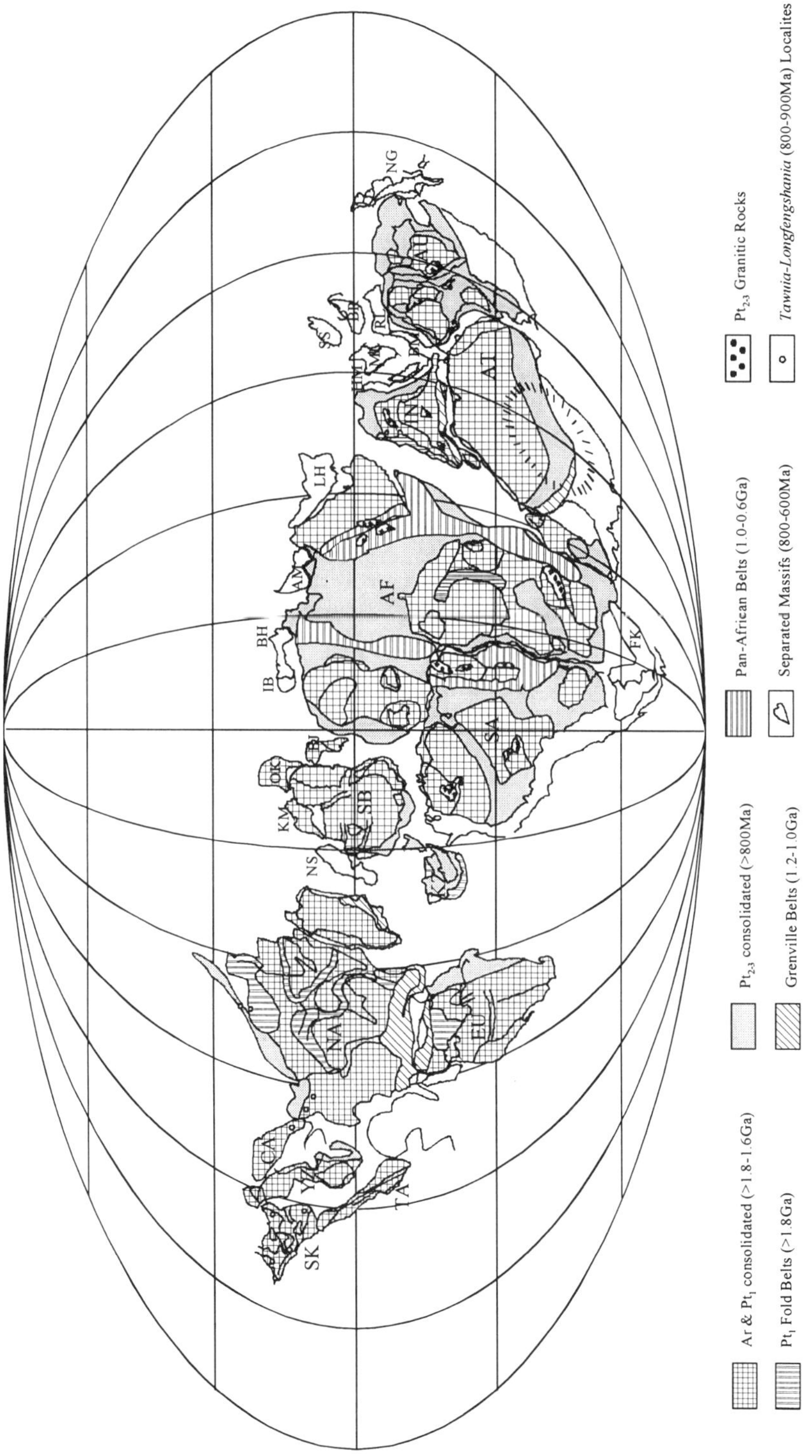

Figure 3. Reconstruction of Pangaea-850 (800-900) Ma

analyzed the palaeomagnetic data and concluded that East Gondwana was formed before 730 Ma and the whole Gondwanaland was formed by 510 Ma at the end of Cambrian. Urug thinks that Mesoproterozoic cratonic blocks existent before 1300 Ma began to be sutured together to form a supercontinent of short duration, which became disintegrated after 870 Ma and reassembled to form Gondwanaland by 500 Ma. We are however inclined to another interpretation that the separate massifs of West Gondwana were coherent parts of our Pangaea-1450 Ma, which started to dissociate after its formation and were reassembled to form our Pangaea-850 Ma. The wide Neoproterozoic orogenic belts within West Gondwana may, except a few with developed ophiolite zones, for the most part be intracratonic aulacogens without much crustal extension.

The reconstruction of Pangaea-850 Ma (Fig. 3) is based on palaeomagnetic data and also on biogeographical, tectonic and climatic evidences. The common occurrence of the profuse *Tawuia-Longfengshania* biota in East China and western North America indicates a close relation between these two continents. The palaeomagnetic data used for reconstruction are meager and are shown in Table 2.

Table 2. Palaeomagnetic data used for reconstruction of Pangaea-850 Ma

Continents	Age (Ma)	Palaeopole position		Palaeo-	References
		Plat	Plon	latitudes	
SK* (115E, 37N)	900	−12.5	170.3	18.0	Zhang, 1991
YZ (108E, 30N)	840	−2.4	33.8	8.0	Zhang, 1991
CA (120E, 30N)	800	29.9	336.6	21.1	Zhang, 1994
NA (95W, 52N)	800-900	18.5	159.0	−0.3	McElh., 1995 (4)
EU (58E, 35N)	800-1000	4.0	244.0	−31.4	McElh., 1995
AF (20E, 10N)	850-1000	71.0	173.0	−6.9	McElh., 1995
SB (110E, 65N)	780-890	−9.5	193.0	5.6	McElh., 1995 (5)

* symbols of continents as in Figure 2.

The Grenville belts (1. 1-0.9 Ga) are considered as a collision record between eastern North America and southern Baltica (Fig.), which might be the Proterozoic predecessor of the Caledonian belt, and the palaeomagnetic data are also consistent. The Pan-African belts beginning from 1.1 Ga are also shown, which may have been resulted from intracratonic aulacogen type of rifting rather than extensive seafloor spreading. West Gondwana may have kept its entity since mid-Mesoproterozoic (1450 Ma), and no large scale extension had occurred between Pangaea-1450 Ma and Pangaea-850 Ma. The position of East Gondwana in the reconstruction map is preliminary, as no good palaeomagnetic data are available. The only palaeomagnetic record bearing an age of 800-900 Ma points to a very high northern latitude for the position of Australia. As profuse Neoproterozoic stromatolites were found in the Amadeus Basin and the Adelaide geosyncline region in Central and South Australia [26, 27, 34], this position for Australia is very unlikely. In this case, we have resorted to tectonic and climatic evidences, and put East Gondwana on the southern low latitudes with the Kalahari massif nearly against East Antarctica. In general, the equatorial and low latitude position of nearly all the palaeocontinents in the map is consistent with the widespread abundant Neoproterozoic stromatolite assemblages which are more or less correlatable in various palaeocontinents.

Pangaea-250 (230-290) Ma

This is the traditional, well recognized Late Palaeozoic Pangaea. The reconstruction map gives more emphasis on phytoprovinces and coral biogeography, which were prominent in the Early Permian. A controversial problem is about the extent and nature of the Palaeotethys. Many palaeontologists would regard the Late Palaeozoic Tethys not as an open oceanic basin, but rather comparatively shallow seas separated by oceanic basins of limited extent [4]. Recently Metcalfe [17] discussed the Late Palaeozoic Southeast Asian terranes and their affinities. Scotese and McKerrow [30] gave a general review of the Palaeozoic biogeography. We have recognized an island chain separating the North and the South Palaeotethys, partly on the basis of coral biogeography [42] and on phytoprovinces. As may be seen on the map (Fig. 4), this island chain extended from Iberia in the west, Lut-Hermand and Qiangtang in the middle and Indochina and Malaya in the east. The island chain divided the Palaeotethys into two parts, and the North Palaeotethys formed a more or less enclosed marine basin with the Yangzi and Cathaysia blocks in the east. The South Palaeotethys may have been deeper and opened in the east to the Protopacific. This is consistent with the pattern of the phytoprovinces, especially the Cathaysian flora which was found all along this island chain from Turkey to Sumatra. Another perplexing problem is the big gap of wide open sea in Northeast Asia on the map. It is well known that the Angaran and the Cathaysian flora began to merge with each other at least in mid-Permian, and Early Permian benthic faunas were dominant in Mongolia and southeastern Siberia. These geological facts are contradictory to the palaeogeographical pattern on the map—a contradiction common to all reconstruction maps so far published.

Lastly a few words about the possible dynamics of the Pangaea cycles. Mantle convection is still the most hopeful to interpret the interaction of the geopheres, although we are not yet clear how mantle convection are generated. Mantle plumes have been intensively studied since the seventies. The Japanese geologists, based mainly on seismic tomographic studies, have regarded the "plume tectonics" as a new paradigm of earth dynamics in interpretation of whole earth tectonics [11, 16]. According to their views, the core-mantle boundary layer (D") above the core-mantle boundary (CMB, 2900 km) is active and is probably responsible for generation of the superplumes with a diameter of over 5000 km. Maruyama and coworkers indicated the presence of another type of "cold plume", which is resulted from subducting plates that pull down the cold lithosphere. At the mantle-boundary (MB, 670 km), the cold lithosphere becomes stagnant and grows in thickness and extent to form a "megalith", until it penetrates the MB and sinks to CMB. Maruyama and coworkers think that the big sinking cold mass will exert a strong influence on the whole mantle convection system after its formation, and will draw all continents toward itself to form finally a supercontinent or Pangaea. The life of a cold sinking "superplume" may be around 400-500 Ma, and the Middle Asia region may serve as an example. This is of course somewhat speculative, but it is interesting to note that the estimated time interval is approximate to our Pangaea cycle and the Wilson cycle.

Mantle convection of lesser scale and shallower depth originate probably from above MB (670 km), the 400 km depth and even shallower. They correspond to the transitional mantle scale and the upper mantle scale convections of Deng *et al.* [3]. We have emphasized the importance of the minor tectonic cycles often met with in the continental margin tract. Continental crustal fragments were often rifted and shifted off the continental

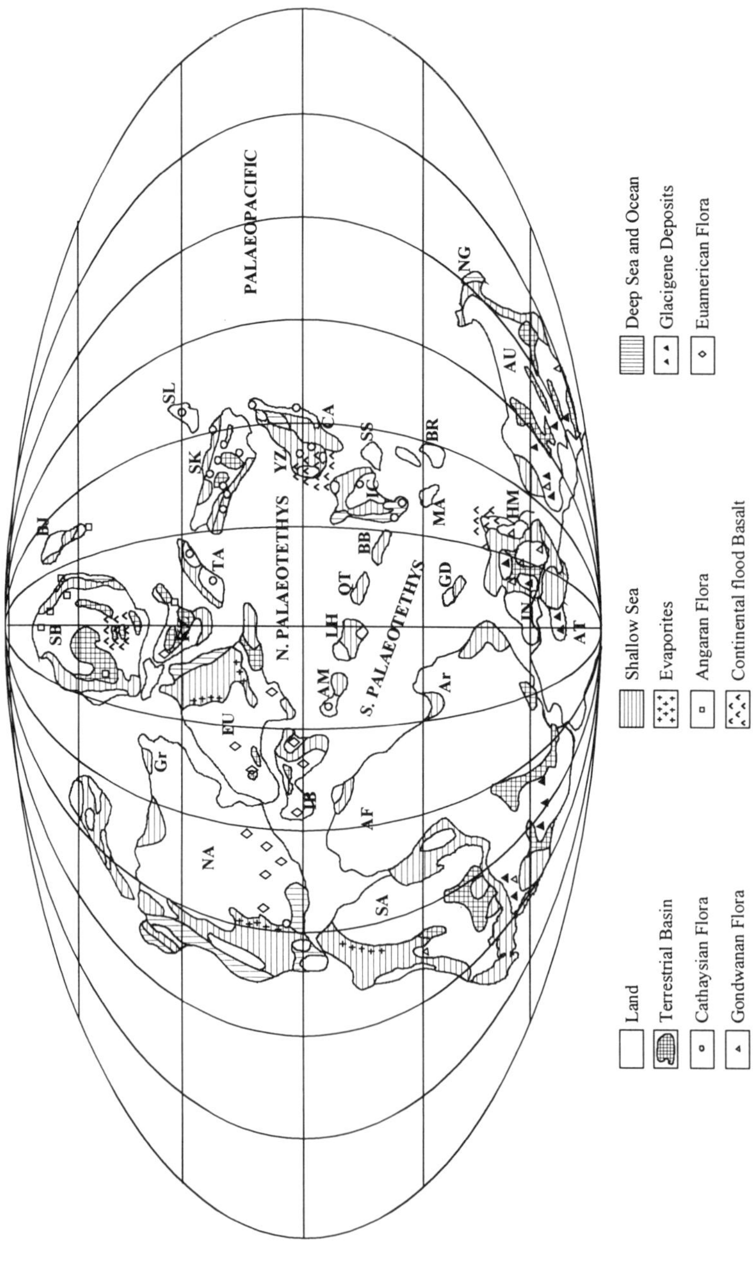

Figure 4. Reconstruction of Pangaea-250 Ma (modified after Wang *et al.*, [42])

margin to form island arcs and marginal seas, which were accreted to the mother continent again through continent-arc collision. Such a scale of tectonic cycle is probably connected with the 670-400 km depth of convection or probably with the still shallower surge tectonics of Meyerhoff [18].

POSSIBLE LIMITED EARTH EXPANSION

The idea of earth expansion has a long history, and was clearly presented by Hilgenberg in 1933. In the late fifties Carey had converted to earth expansion after a long advocation for continental drift. Carey [1] maintains that the earth has undergone since the Mesozoic a radial, fast and accelerated expansion. This idea was shared by Kremp [10]. Owen [22] has pointed out that the fast expansion idea is untenable. He addressed the necessity of earth expansion based on several lines of argumentation, and proposed an increase of 20 percent of the earth's radius since Early Jurassic. Through careful calculation and reconstruction he has shown on the map a shallow sea Tethys instead of a Tethys ocean. However, the entirely inland sea nature of Tethys shown on the map is probably equally not acceptable.

Palaeobiogeographical and palaeogeographical research suggests rather strongly that a reduction of the earth surface area in the Palaeozoic would meditate the contradiction between the reconstruction pattern of the palaeocontinents on the one hand and the geological and palaeobiogeographical evidences on the other. Studies on biologic clocks evidenced by corals, mollusks and stromatolites [23] point to an increase of solar day length which is estimated to have been doubled since the Neoproterozoic. This indicates a slowing down of earth's rotation and is a support to earth expansion. Moreover, recent research in tectonics has revealed much more crustal extension than hitherto recognized. On account of all these reasons, and with a viewpoint of non-uniformitarianism and the idea of punctuated progression in earth's history, Wang [36] suggested an earth expansion of more limited extent than previous authors. If we assume a periodic earth expansion of no more than 5% earth radius shortly after each break-up of the successive pangaeas, then the total increase of earth radius will be no more than 20% since the beginning of Palaeoproterozoic (2500 Ma). The estimated numerical values of expansion are shown in the lower left of Figure 5.

Assuming Earth expansion, reconstruction of the palaeocontinents needs to take care of changes of the curvature of the continental blocks and to define the redistribution of the geographic coordinates of the continents. This can be difficult if the dynamics involved in the curvature changes of the Earth surface is to be considered. Therefore the method we developed here ignores the dynamic aspects and takes care of only geometric calculations, as shown in Figure 5.

Assuming C’ the center of a continent, and A’, B’ two points on the continent, with C’ as the center point of the arc S” connecting A’, B’. Due to earth expanding with radius changing from R’ to R, the arc S” is stretched, becoming arc S’ (dashed), then becomes arc S (arc AB) centered at point C on the outer sphere . Since length S equals length S”, R’P = RQ, or Q = P R’/R, where P is angle <OA’B’ and Q is angle <OAB. A’, B’ and C’ are known, and C = C’, then A, B can be calculated using Q. In reconstructing

the earth that has expanded, C, A, B and C' are known, P will be calculated and so are A' and B'.

There are two assumptions in figure 5. 1) The areas of the drifting continental blocks remain unchanged after expansion, which can be reasonably true for stable platforms where the areal extents at the given time period can be outlined by geological studies. In this case, relative angular distance between any two points within the platform should have been constant had there been no earth expansion. In other words, the angular distance between the two points should have changed if the earth does have expanded; 2) the expansion of the earth causes change of curvature of the continent over its surface in a way as if the center of the continent was pinned, so that each point other than the pinned point within the continent will move geographically toward the pinned point in the case of expansion, and away from the pinned point in the case of shrinking. The later will be the case in reconstructing the palaeocontinents on an earth that had a reduced radius compared to the present.

The estimation of the earth expanding rates has never been made with certainty, as there is no way to test the expanding hypothesis itself, not to mention the rates. Different expansion rates of the earth have been proposed and discussed by various authors, but quantitative evaluation and testing of them are few. The method proposed by Egyed [6] early in the sixties is to calculate the expanding rates by use of the palaeomagnetic data, assuming that the earth's expansion would eventually affect the angular distance between two points occupying the same meridian on the earth. The method was later revised by Eyed, Ward and Van Andel *et al.* to allow the use of group palaeomagnetic data in order to optimize statistically the evaluation result. So far, this method seems to be the most reasonable and prospecting. Nowadays the palaeomagnetic data base over the world has greatly expanded, which allows a re-evaluation in this respect in a more accurate and rapid way. We have developed a forward model to test the validity of this palaeomagnetic method, which simulates the palaeomagnet polar changes stored in rock records of an expanding earth. Using the palaeomagnetic data output by the forward model, we were able to trace down, using the Egyed-Ward method, the expanding rates the model used as the input. This encouraged us to make use of the real world palaeomagnetic data to model the earth expanding rates. The world palaeomagnetic database that we used is widely available in the geological community (McElhinny *et al.*, 1990), a series of reverse calculation with an aim to evaluate the earth's expansion rates in different geological periods have been carried out and some of the results helped working out table 3.

The forward model works on the assumption that palaeomagnetic polar wandering is caused either by continental drifting or by earth expansion. For polar wander caused by continental drift, the scattering pattern of the palaeomagnetic poles will be kept unchanged, whereas in the case caused by earth expanding, the scattering pattern will change. In the model, the area of the major continental platforms remain unchanged and therefore the relative angular distance between points within them should have kept constant if no earth expansion has taken place. If no earth expansion occurred, the pole positions on the same continent in the same period should coincide with each other, or

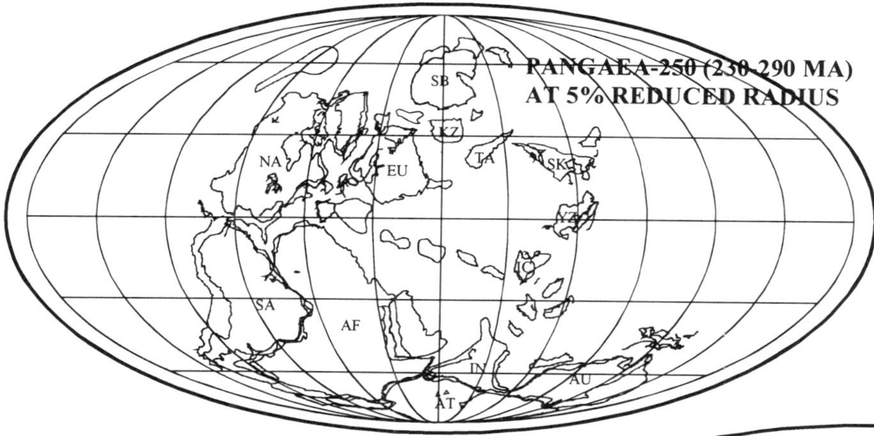

Method of palaeocontinent reconstruction on Earth surface of reduced radii

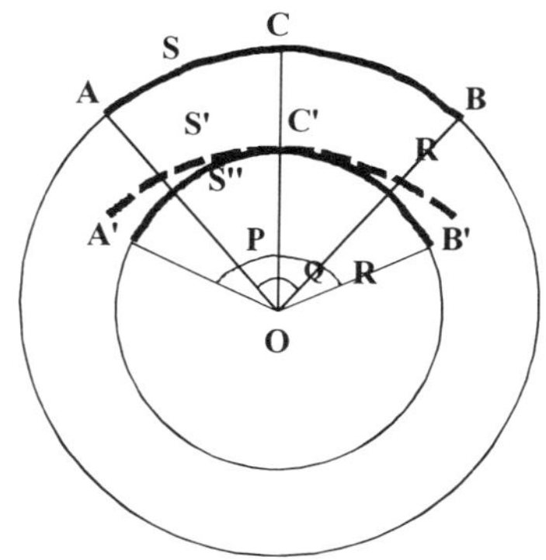

Assuming the Earth undergoing episodic expansion after break-up of each pangaea

After Break-up of	Earth Radii of present
Pangaea-2500 Ma	80%
Pangaea-1950 Ma	85%
Pangaea-1450 Ma	90%
Pangaea-850 Ma	95%
Pangaea-250 Ma	100%

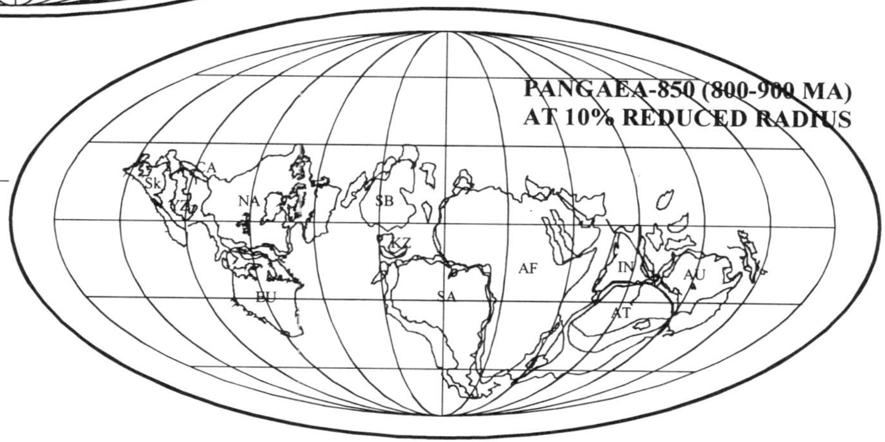

Figure 5. Possible earth expansion and expansion rates.

statistically scattered with the true pole as the expected value. If earth expansion occurred, the pole values would be affected and become diverse. Now we start with the assumption that the palaeomagnetic poles are recorded in rock samples randomly distributed within the stable continent. A series of sites are randomly chosen, and their corresponding pole values in the same period are taken with a given accuracy. We now assume that the earth has expanded at a given rate, and recalculate the pole locations using the method outlined above in figure 4, this will give a new set of data, which may be referred as expansional data. To find out the expansion rate using the expansional data, we use different rates to reduce the earth's radius, re-calculate the pole data, and evaluate the data diversity. Theoretically the expansion rate with the least diverse data of poles will be the correct rate for the period, that is to say, earth expansion will cause the poles to be statistically more diversely distributed, and an earth with reduced radius value that has a minimum dispersion of pole distribution would represent the correct one before expansion. Therefore a trial and error method using different earth expansion rates will eventually find out the right rate of expansion.

Concerning the evaluation of diversity, three methods are available: vector angle, alpha 95 and precision coefficient. Actual calculations have shown that results from Late Mesoproterozoic, Neoproterozoic and Cambrian are to a certain extent consistent with the expansion rates adopted in this paper, as shown in Table 3.

On the whole the concept of earth expansion seems to have found more support in recent years. Any hypothesis based on great mass increase of the earth will meet with difficulties, and earth expansion without essential mass increase would require immense energy resource from the earth's deep interior. In this context, we would suggest that the inner core may have retained ultradense subatomic particles of cosmic origin in the early planetary stage of the earth, which would be capable to provide a source of energy supply required for expansion.

Table 3. Earth expansion rates in the geological past

Geologic age	Percent of present earth radius		
	vector angle	alpha 95	precision coefficient
Ar_3-Pt_1 (2800-1900 Ma)		0.67	0.67
Pt_2 (1400-1000 Ma)	0.87	0.76	0.76
Pt_3 (999-666 Ma)		0.81	0.81
Cm (600-500 Ma)			0.94

A rival hypothesis of earth dynamics to earth expansion is earth pulsation. The idea of a pulsating earth was proposed by W. Bucher, A. W. Grabau and J. M. H. Umbgrove in the thirties, but they did not indicate actual earth expansion and contraction. In the forties A. J. Schneiderov developed a concept of earth pulsation with cataclastic expansions that produce oceans and long slow contractions that cause orogenesis. The pulsation hypothesis was recently advocated and developed by the Russian earth scientists represented by Milanovsky [19, 20], who has especially emphasized the expanding stage since the Mesozoic.

It seems that the essential thing in the study of earth dynamics would be to organize a close collaboration between geologists, geophysicists and astronomers, to investigate whether careful calculation of available parameters in the earth's deep interior will provide valuable results in this basic and interesting field of research.

Acknowledgements

This paper is partly based on the research supported by the National Natural Science Foundation of China (No. 49070139), and on the project Sequence Stratigraphy and Sea-level Changes (SSLÇ), jointly supported by the State Science and Technology Commission and the Ministry of Geology and Mineral Resources, for which we are deeply thankful.

REFERENCES

[1] W. S. Carey. *The Expanding Earth*. Elsevier Scientific Publishing Company. Amsterdam (1976).

[2] I. W. D. Dalziel. Pacific margins of Laurentia and East-Antarctica-Australia as a conjugate rift-pair: Evidence and implications for an Eocambrian supercontinent, *Geology* **19** 598-601 (1991).

[3] Deng Jinfu, Mo Xuanxue and Zhao Hailing. The layered earth material, volcanism and hot spot recycling and global multilayered convection system. In: *Collected papers of a Symposium on Problems in Recent Earth Dynamics*. Ma Zongjin (Ed.). pp. 161-168. Seismology Press, Beijing (1994).

[4] J. M. Dickins, D. R. Choi, and A. N. Yeates. Past distribution of oceans and continents. In: *New Concepts in Global Tectonics*. S. Chattergee and N. Hotton III (Eds.). pp. 193- 202. Texas Tech. Univesity Press, Lubbock (1993)

[5] A. M. Dziewonski and J. H. Woodhouse. Global images in the earth's interior, *Science* **236**, 37-48 (1987).

[6] L. Egyed. *The expanding earth*. New York Acad. Sci. Tr. 23:5, 424- 432 (1961).

[7] B. V. Haq, J. Hardenbol and P. R. Vail. Mesozoic and Cenozoic chronostratigraphy and eustatic cycles. In: *Sea-level Changes: An Integrated Approach*. C. K. Wilgus *et al.* (Eds.). pp. 71-108. SEPM Spec. Publ. 42 (1988).

[8] P. F. Hoffman. Did the break-out of Laurentia turn Gondwanaland inside out? *Science* **252**, 1409-1412 (1991).

[9] V. E. Khain. Tectonic evolution of earth's crust: from early Archaean through late Phanerozoic. *Abstracts, 28th International Geological Congress, Volume 2*, 179-180 (1989).

[10] G. O. W. Kremp. Earth expansion theory *versus* statical earth assumption. In: *New Concrepts in Global Tectonics*. S. Chattergee and N. Hotton III (Eds.). pp. 297-308. Texas Tech. Univesity Press, Lubbock (1992).

[11] M. Kumazawa and S. Maruyama. Whole earth tectonics, *Journ. Jap. Geol. Soc.* **100:1**, 81-102 (1994).

[12] Z. X. Li and C. McA. Powell. Late Proterozoic and Early Palaeozoic palaeomagnetism and the formation of Gondwanaland. In: *Gondwana Eight*. R. H. Findley, R. Urug, M. R. Banks, and J. J. Veevers (Eds.). pp. 9-23. A. A. Balkema Netherlands (1993).

[13] Z. X. Li, Zhang Linghua and C. McA. Powell. South China in Rodinia: part of the missing link between Australia-East Antarcrica and Laurentia? *Geology* **23**, 407-410 (1995).

[14] Li Xiang. Palaeocontinental reconstructions: methodology, computer soft-ware and an example of application, *Jour. Chiina Univ. Geopsciences* **4:1**, 7-13 (1993).

[15] Liu Jianhua *et al.* Three dimensional velocity images of the crust and upper mantle beneath north-south zone in China (in Chinese with English abstract), *Acta Geophysica Sinica* **32:2**, 143-152 (1989)

[16] S. Maruyama. Plume tectonics, *Jour. Geol. Soc. Japan* **100:1**, 24-49 (1994).

[17] I. Metcalfe. Southeast Asian terranes: Gondwanaland origins and evolution. In: *Gondwana Eight*. R. H. Findley, R. Unrug, M. R. Banks, and J. J. Veevers (Eds.). pp. 81-100. A. A. Balkema Netherlands (1993).

[18] A. A. Meyerhoff, I. Taner, A. E. L. Morris, B. D. Martin, W. D. B.Agocs, and H. A. Meyerhoff. Surge tectonics: a new hypothesis of earth dynamics. In: *New Concepts in Global Tectonics*. S. Chattergee and N. Hotton III (Eds.). pp. 151-178. Texas Tech. University Press, Lobbock (1992).

[19] E. E. Milanovsky. Problems in the tectonic development of the earth in the light of the concept of earth pulsation and expansion, *Rev. Geol. Dynam. Geogr. Phys.* **22:1**, 15-29 (1980).

[20] E. E. Milanovsky, The earth's pulsations and recent phase of its prevailing expansion, *Theophrastus' Contributions* **1**, 81-102 (1996).

[21] P. Morel and E. C. Irving. Tentative palaeocontinental maps for the early Phanerozoic and Proterozoic, *Journal of Geology* **86:5**, 535-561 (1978).
[22] H. G. Owen. Has the earth increased in size? In: *New Concepts in Global Tectonics*. S. Chattergee and N. Hotton III (Eds.). pp. 289-296. Texas Tech. Univ. Press, Lubbock (1992).
[23] G. Pannella. Proterozoic sromatolites as palaeontological clocks. *24th Int. Geol. Cong.* Montreal. Sect. I, 50-57 (1972).
[24] I. D. A. Piper. Palaeomagnetic evidence for a Proterozoic supercontinent, *Phil. Trans. Roy. Soc. London* Series **A**., 469- 490 (1976).
[25] C. McA. Powell. Assembly of Gondwanaland-Open Forum. In: *Gondwana Eight*. R. H. Findley, R. Urug, M. R. Banks, and J. J. Veevers (Eds.). pp. 219-237. A. A. Balkema, Netherlands (1993)
[26] W. V. Preiss. The systematics of the South Australian Precambrian and Cambrian stromatolites, Part I, *Trans. Roy. Soc. Australia* **96**, 67-100 (1972).
[27] W. V. Preiss. The systematics of the South Australian Precambrian and Cambrian stromatolites, Part II, *Trans. Roy. Soc. Australia* **97** 91-125 (1973).
[28] M. R. Rampino and R. B. Stothers. Major episodes of geologic change, correlation, time structure and possible causes, *Earth Planet. Sci. Lett.* **114**, 215-217 (1993).
[29] C. A. Ross and J. R. P. Ross. Late Palaeozoic transgressive- regressive deposition. In: *Sea Level Changes: An Integrated Approach*. C,. K. Wilgus *et al.* (Eds.). pp. 227-247. SEMP Spec. Publ. 42 (1988).
[30] C. R. Scotese and W. S. McKerrow. Palaeozoic palaeogeography and biogeography: in introduction to this volume. In: *Palaeozoic Biogeography and Palaeogeography*. W. S. McKerrow and C. R. Scotese (Eds.). pp. 1-24. Geol. Soc. London Memoir **12** (1990).
[31] Shi Xiaoying. 35 Ma: an important natural periodicity in geological history: concept and causes in natural crisis (in Chinese with English abstract), *Earth Science —Jour. China Univ. Geosciences* **21:3**, 235-242 (1996)
[32] S. M. Stothers. Terrestrial mass extinction and galactic plane crossings, *Nature* **313**: 159-160 (1985).
[33] R. Unrug. The Gondwana supercontinent: Middle Proterozoic crustal fragments, Late Proterozoic assembly, and unsolved problems. In: *Gondwana Eight*. R. H. Findley, R. Urug, M. R. Banks, and J. J. Veevers (Eds). pp. 3-8. A. A. Balkema, Netherlands (1993).
[34] M. R. Walter. Stromatolites and the biostratigraphy of the Australian Precambrian and Cambrian, *Palaeont. Asssoc. London, Sp. Pap.* **11** (1972).
[35] Wang Hongzhen. Geotectonic development of China. In: *The Geology of China*. Yang Zunyi, Cheng Yuqi and Wang Hongzhen(Eds.). pp. 235-276. Clarendon Press, Oxford (1986).
[36] Wang Hongzhen. Retrospect on the study of global tectonics (in Chinese with English abstract), *Earth Science Frontiers* **2:1-2**, 37-42 (1995).
[37] Wang Hongzhen. Tectonic evolution and basin development of China. In: *Theoretical and Applied Problems in Geology*. B. A. Sokolov and Zhao Pengda (Eds.). pp. 223-37. Moscow University Press (1996).
[38] Wang Hongzhen and Li Guangcen. *Correlation Table of Stratigraphical Subdivision* (in Chinese and English). Geological Publishing House, Beijing (1990).
[39] Wang Hongzhen and Mo Xuanxue. An outline of the tectonic evolution of China, *Episodes* **18:1-2**, 6-16 (1995).
[40] Wang Hongzhen and Shi Xiaoying. A scheme of the hierarchy for sequence stratigraphy, *Earth Science—Journ. China Univ. Geosciences* **7:1**, 1-12 (1996).
[41] Wang Hongzhen and Shi Xiaoying (Eds.). Sequence Stratigraphy and Sea-Level Changes of China, *Jour. China University Geosciences* **7:1**, 1-137 (1996).
[42] Wang Hongzhen, Wang Xunlian and Chen Jianqiang. Evolutional stages and biogeography of the rugose corals (in Chinese with English abstract). In: *Classification, Evolution and Biogeography of the Palaeozoic corals of China*. Wang Hongzhen *et al.*(Eds.). pp. 175-225. Science Press, Beijing (1989).
[43] Zhang Huimin. On the relation between the continental massifs of East China based on palaeomagnetic data (in Chinese), *Jour. Tianjin Geol. Soc.* **11:1**, 3-12 (1993).

Proc. 30th Intern. Geol. Congr., Vol. 1, pp. 129-142
Wang *et al.* (Eds)

On Crustal Accretion and Mantle Evolution in Qinling Orogenic Belt

ZHANG BENREN, ZHANG HONGFEI, LING WENLI, ZHAO ZHIDAN,
HAN YINWEN, and OUYANG JIANPING
Institute of Geochemistry, China University of Geosciences, Wuhan 430074, P. R. China

Abstract

The Qinling orogen represents the convergence zone between North China and Yangtze cratons. The North and South Qinling, which are separated by the Danfeng-Shangnan megasuture, have been considered to be the active continental margin of North China block and the passive margin of Yangtze block respectively, during the Neoproterozoic and Early Palaeozoic time. Nd model ages of crust-derived rocks indicate that the crust of both the northern and southern Qinling belts were formed during the Proterozoic (2.2-0.8Ga), with the Neoarchaean basement existing only in the South Qinling. The ε Nd (t) *vs.* age plot for mantle-derived rocks reveals that the upper mantle beneath North Qinling was in a stable and strongly depleted state (ε Nd = +7.3 - +6.3) during the entire Proterozoic, whereas the upper mantle beneath South Qinling had developed nearly parallel to the evolutionary line of the depleted mantle from the Palaeo- to Mesoproterozoic. The South Qinling is similar to the Yangtze block in their lithospheric history and property. North Qinling with its own distinctive evolutionary trends of crust and mantle was most likely a separate microcontinent. The identification of crust/mantle recycling in the ancient convergence zone reveals the subduction of oceanic crust and the lateral crustal accretion in North Qinling during Early Neoproterozoic. The vertical crustal accretion accompanied by mafic magma underplating at about 1000 Ma ago in South Qinling was evidenced by the alkaline bimodel nature of the Yaolinghe and Yunxi volcanic rock series, and the source rock of the Indosinian granites.

Keywords: crustal accretion, mantle evolution, crust/mantle recycling, underplating

INTRODUCTION

The East-West trending Qinling orogenic belt represents a complicated convergence zone between the North China and Yangtze cratons or blocks. In recent years, geological and geochemical studies have indicated that the North China and Yangtze blocks should belong to two independent tectono-geochemical provinces differing from each other consistently in Pb isotopic ratios of the crust- and mantle-derived rocks and in Mg Fe, Al, K, Cu, Mo and REE abundance in the crust and upper mantle [34, 29, 25, 26]. Based on Nd model ages and isotopic dating, the crust of North China block has proved to be formed mainly in the Archaean (3.8-2.5Ga), whereas the crust of Yangtze block chiefly in the Palaeo- and Mesoproterozoic (2.1-1.3Ga) [35, 14]. The upper mantle beneath North China block has been revealed to be in a depleted state with ε Nd values vibrating at +3 during the Archaean and Palaeoproterozoic from the Sm-Nd isochronic ages and initial ε Nd values of Precambrian metabasalts [10].

Significant progress has been also made in the interpretation of the Qinling orogen. The Palaeo- and Mesoproterozoic volcanic-sedimentary sequences in both North Qinling and South Qinling were suggested to be formed in a rifting setting [27, 25]. The plate tectonic regime is considered to have been established at the beginning of Neoproterozoic, and the North and South Qinling belts have been identified to be the active continental margin of North China block and the passive continental margin of Yangtze block respectively, during the Neoproterozoic and Early Palaeozoic in these papers. However, the problem of whether the North Qinling was originally a split part from the North China block or it came from elsewhere, remains to be solved. Meanwhile, our knowledge of the crustal accretion and mantle evolution in the Qinling orogen is still scanty.

Based on the rapid accumulation of available geochronological and isotopic data of rocks from the separate lithostratigraphic units in the East Qinling, this paper attempts to explore the crustal accretion, mantle evolution and crust-mantle interaction in various tectonic units of the East Qinling and then to relate the results with the tectonic subdivision and tectonic patterns in the orogen. The aim is to ascertain the tectonic position of North Qinling and to improve our understanding of evolution of the Qinling orogen.

GEOLOGICAL SETTING

The East Qinling includes four tectonic units which are, from north to south, the southern margin of North China Craton (SNC), North Qinling (NQ), South Qinling (SQ) and the northern margin of Yangtze Craton (NYC) (Figure l). The Danfeng-Shangnan fault zone separating the North and South Qinling is commonly regarded as the megasuture between the two cratons. Along the suture is distributed a zone of ophiolite fragments among which the largest and well studied Songshugou ophiolite fragment is composed of metamorphic harzburgite and dunite, dunite cumulate and metatholeiites showing the geochemical characteristics peculiar to both N-MORBs and E-MORBs [33].

The main exposed basement lithostratigraphic units in SNC are the Neoarchaean Taihua high-grade terrain, Dengfeng granite-greenstone terrain, the Palaeoproterozoic Angou Group in Western Henan and coeval Jiangxian and Zhongtiao groups in the Zhongtiao Mountains, and the Mesoproterozoic Xiyanghe Group in the west, and Xionger Group in the east. The main Precambrian sequences in NQ include the Palaeoproterozoic Qinling Group consisting dominantly of paragneisses with minor mafic volcanic and intrusive rocks, the late Mesoproterozoic Kuanping Group, a metavolcanic-sedimentary suite formed in a marginal sea basin north of the Qinling Group [25], and the two Neoproterozoic-Early Palaeozoic metavolcanic-sedimentary series [30, 31], called, respectively the Danfeng Group and Erlangping Group bounding the Qinling Group on the southern and northern sides. In SQ, the main exposed basement lithostratigraphic units are the Palaeoproterozoic Douling Group, Mesoproterozoic Wudang Group and Neoproterozoic Yaolinghe and Yunxi groups in the east, and the Mesoproterozoic

Bikou Group in the west. All consist of metavolcanic-sedimentary sequences. The existence of Neoarchaean Yudongzi Group consisting of paragneisses and metabasalt (amphibolite) in the west makes SQ different from NQ. The Neoarchaean Kongling high-grade terrain in the Yichang area is the oldest known basement of the Yangtze craton. In the eastern part of NYC the Proterozoic includes only the Shennongjia Group. To the west in the Xixiang-Beiba area the exposed basement includes the Palaeoproterozoic Houhe (or Huodiya) Group, Mesoproterozoic Tiechuanshan Group, and Meso- to Neoproterozoic Xixiang Group consisting of Baimianxia Formation in the lower part, Sanhuashi Formation in the upper part, and Sunjiahe Formation at the top. Isotopic ages of these lithostratigraphic units are given in Table 1.

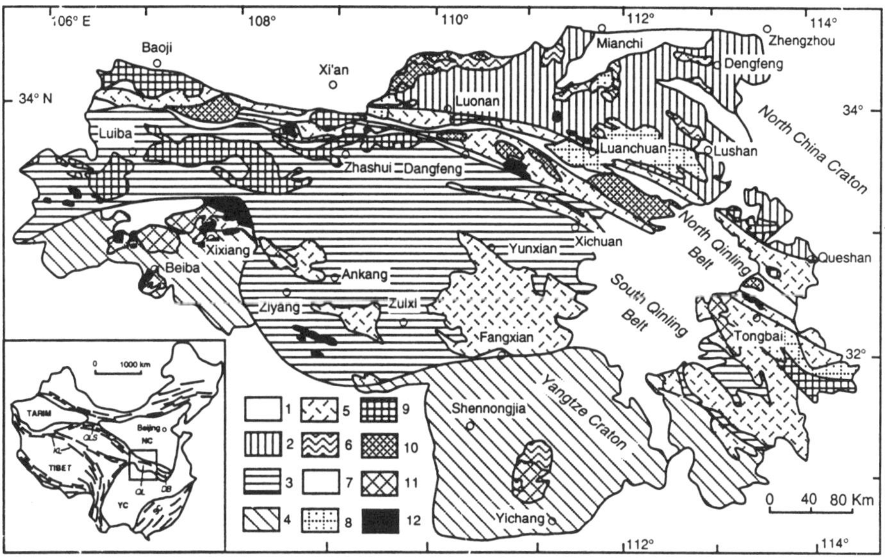

Figure 1. Geological sketch map of East Qinling and its adjacent areas. 1-Quaternary, 2-cover of North China Craton, 3-cover of South Qinling Belt, 4-cover of Yangtze Craton, 5-Proterozoic, 6-Archaean, 7-Neoproterozoic-early Palaeozoic volcanic-sedimentary sequences, 8-late Mesozoic granites, 9-late Palaeozoic-early Mesozoic granites, 10-early Palaeozoic granites, 11-middle-late Proterozoic granites, 12-mafic-ultramafic intrusions. Inset map shows tectonic location of studied region in China. NC-North China Craton, YC-Yangtze Craton, SC-South China Caledonian Belt, KL, QLS and QL denote the Kunlun, the Qilian, and the Qinling Mountains, respectively.

THE CRUSTAL GROWTH

Based on over 350 available Sm-Nd isotopic data of sedimentary and metasedimentary rocks and granitoids compiled from literatures and our study, the Nd model ages of crust-derived rocks relative to the depleted mantle as source of the present MORBs [$(^{143}Nd/^{144}Nd)_{DM}$ = 0.51315, $(^{147}Sm/^{144}Nd)_{DM}$ = 0.2137, [6]] are calculated and histograms of the Nd model age distribution for NC, NQ, SQ and YC are shown in Figure 2.

From the histograms of NQ and SQ, it can be clearly seen that the higher frequences of Nd model age distribution of rocks from NQ and SQ fall almost in the similar range from 2.2 to 0.8 Ga and from 2.2 to 1.0 Ga respectively. This infers that the crust of both was formed basically in the Proterozoic time prior to 0.8 Ga. The Neoarchaean Yudongzi Group in SQ and many samples of metabasaltic rock and gneiss from the Douling Group show Nd model ages older than 2.5 Ga. These facts indicate that the Proterozoic crust in SQ had developed upon an Archaean basement. However, no unambiguous Archaean lithostratigraphic unit has so far been found in NQ and a great quantity of paragneiss and metabasaltic rock samples (18 and 10 respectively) from the Qinling Group exhibit Nd model ages close to the formation age of the group (2000 Ma ±). Moreover, in NQ only a few rock samples with Nd model ages older than 2.5Ga have been reported in literatures.

The histogram of NC indicates that the crust of North China Craton (NC) including its southern part was mainly formed in the Archaean (3.5-2.5 Ga) and subordinately in the Palaeoproterozoic (2.2-1.8 Ga). The Mesoproterozoic volcanics including basalts of Xiyanghe and Xionger groups exhibit negative initial ε Nd values (–4.1-–7.8) and high initial Sr isotopic ratio (0.7072 for mafic rocks, 0.7100 for felsic ones). Based on a detailed analysis of the Nd and Sr isotopic data, the volcanics of Xionger Group were identified to be chiefly the product of the crust-derived magma with minor mantle addition only for the mafic end member [5]. This result suggests that since the Mesoproterozoic the crust of SNC had developed into a new period dominated by substance recycling within the crust with only subordinate new addition from the mantle.

The histogram of YC indicates that the majority of crust of Yangtze Craton (YC) was formed in the Proterozoic (2.4-1.2Ga) but with some crustal component formed in the Neoarchaean as judged from the existence of the Archaean Kongling high-grade terrain. Therefore, NQ differs not only from NC , but also from SQ and YC in the crustal growth history, whereas SQ exhibits its crustal accretion similar to that of YC.

PROPERTY AND EVOLUTION OF THE EARLY UPPER MANTLE

In order to reveal the property and evolution of the early upper mantle beneath the region, the initial ε Nd value *vs.* age plot for Precambrian mantle-derived rocks (metabasaltic rocks) from each of the four tectonic units is made. The data are given in Table 1 and the plot in Figure 3. Most of the metabasaltic rocks, except those from the Xiyanghe and Xionger groups and the Sunjahe Formation, have low initial $^{87}Sr/^{86}Sr$ ratio values (≤0.7046) and high positive initial ε Nd values. Metabasaltic rocks of the Qinling and Kuanping groups even exhibit stable La/Ce ratio which does not vary with the change in K_2O content of samples. These facts indicate that the basaltic rocks had not been significantly contaminated by crustal materials and that the ε Nd (t) and strongly incompatible element ratios of these mantle-derived rocks, particularly those from the Qinling and Kuanping groups, can basically reflect the nature of their mantle sources.

Table 1. Isotopic ages of main lithostratigraphic units and ε Nd (t) of their metabasaltic rocks in the four tectonic units of study region

Lithostratigraphic unit	Isotopic age (Ma)	Method	ε Nd (t)
Southern margin of North China Craton (SNC)			
Taihua high-grade terrain (TH)	2840 [8]	Pb-Pb zircon	+2.9 [*]
Dengfeng greenstone terrain (DEF)	2512 [8]	SHRIMP	+2.2 [9]
Jiangxian Group (JX)	2115 [19]	SHRIMP	+2.3 [19]
Zhongtiao Group (ZT)	2059 [19]	U-Pb zircon	+2.5 [19]
Xiyanghe Group (XY)	1840 [19]	SHRIMP	−4.1 [19]
Xionger Group (XE)	1650 [5]	Sm-Nd isochron	−7.8 [5]
North Qinling Belt (NQ)			
Qinling Group (QL)	2000 ± [31]	U-Pb zircon	
	1987 [31]	Sm-Nd isochron	+7.3 [31]
Kuanping Group (KP)	1124 [31]	Sm-Nd isochron	+6.5–+4.6 [31]
Danfeng Group (DAF)	984 [32]	Sm-Nd isochron	+7.3–+4.9 [30]
Erlangping Group (EL)	788–822 [31]	Sm-Nd isochron	+6.5–+3.7 [31]
Ophiolite (Oph)	1000–1124 [11] [*]	Sm-Nd isochron	+6.8 [11]
South Qinling Belt (SQ)			
Yudongzi Group (YDZ)	2688 [32]	U-Pb zircon	+3.0 TS
Douling Group (DL)	2000 ± [32]	Pb-Pb zircon	+1.3 TS
Bikou Group (BK)	1611 [32]	Sm-Nd isochron	+6.8–+4.0 TS, [32]
Wudang Group (WD)	1263 [*]	Sm-Nd isochron	+5.1 TS
Yaolinghe Group (YLH)	1019 [32]	Sm-Nd isochron	+5.8 TS
Yunxi Group (YX)	1010 [32]	Sm-Nd isochron	+5.8 TS
Northern margin of Yangtze Craton (NYC)			
Kongling high-grade terrain (KL)	2580 [13]	U-Pb zircon	−0.45 [*]
Houhe Group (HH)	2429–2436 TS, [32]	Sm-Nd isochron	+4.1–+4.7 TS
Tiechuanshan Group (TCS)	1669–1699 TS, [32]	Sm-Nd isochron	+7.5–+6.6 TS, [32]
Baimianxia Formation (BMX)	1475 [22]	Sm-Nd isochron	+5.6 TS
Sanhuashi Formation (SHS)	1109 [32]	Sm-Nd isochron	+6.6 [32]
Sunjiahe Formation (SJH)	1013 TS	Sm-Nd isochron	−0.2 TS

Numbers in brackets indicate reference numbers. [*] Zhang, Z.Q. *et al.*, unpublished data. TS denotes data of this study. Labels for lithostratigraphic units and tectonic units are given in parentheses.

The variation trends of initial ε Nd value with age of the Palaeo- to Neoproterozoic metabasaltic rocks from NYC and SQ indicate that the upper mantle beneath NYC developed nearly along the evolutionary line of the depleted mantle (DM) as the sourceof present MORBs from the Palaeo- to Mesoproterozoic and the upper mantle beneath SQ along a line roughly parallel to the depleted mantle evolution line but with lower ε Nd values during the same time (Figure 3). However, the recent geochronological study has revealed that rocks of the Douling, Wudang and Bikou groups have different agesin different localities [34]. If a new ε Nd (t) *vs.* age plot is made on the basis of ε Nd (t) data only of the samples of the three groups having Nd model ages close to the formation ages of these groups listed in Table 1, it can be seen that the evolutionary line for the Palaeo- to Mesoproterozoic upper mantle is also along the depleted mantle evolution line. Therefore, SQ should be similar in the upper mantle evolutionary trend to NYC during the Palaeo- to Mesoproterozoic when the crust in the two tectonic units grew rapidly. The nature of Archaean upper mantle beneath these two tectonic units will not be discussed here.

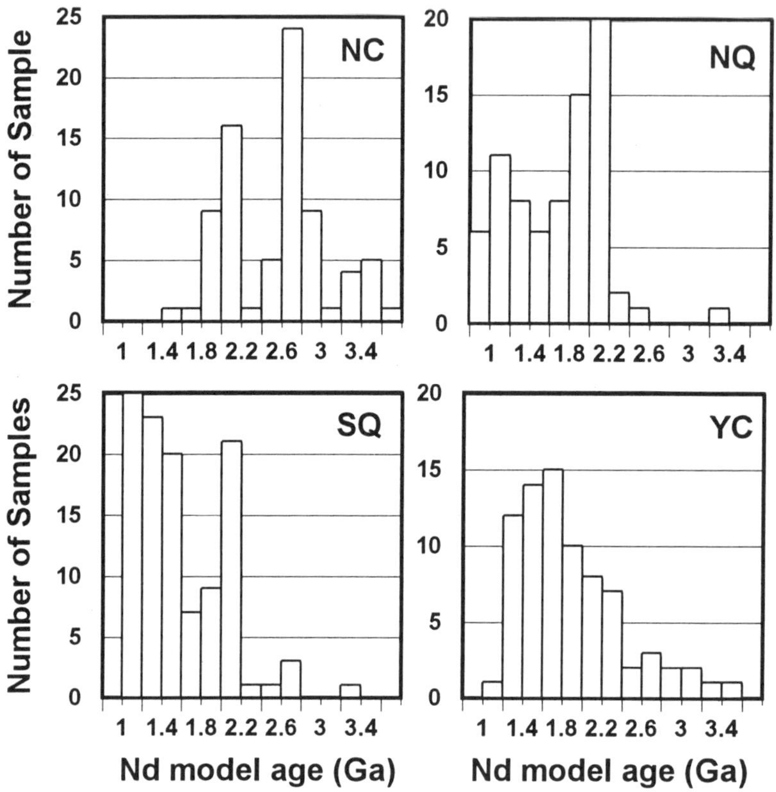

Figure 2. Histograms of Nd model age distribution of clastic and metaclastic sedimentary rocks and granitoids from North China Craton (NC), North Qinling (NQ), South Qinling (SQ) and Yangtze Craton (YC). Data are from this study and from [2, 3, 5, 6, 12, 15, 16, 17, 21, 23, 31] and Ma, D. Q. *et al.*, unpublished report (1991).

The evolutionary trends of NQ and SNC in Figure 3 indicate that the upper mantle beneath the North Qinling was in a stable strongly depleted state with high ε Nd values ranging from +7.3 to +6.3 during the entire Proterozoic, whereas the initial ε Nd values of metabasaltic rocks of the Neoarchaean Taihua and Dengfeng groups and of the Palaeoproterozoic Jiangxian and Zhongtiao groups fall between +2.9 and +2.2, which are close to the initial ε Nd values (about +3) of Archaean metabasaltic rocks widely known in the North China craton [11]. Moreover, the metabasaltic rocks of the Taihua and Dengfeng groups exhibit also low initial $^{87}Sr/^{86}Sr$ ratio values (0.7032-

0.7046 and 0.7027-0.7030, respectively) indicating insignificant contamination of crustal rocks. Therefore, it is reasonable to consider that during the Nearchaean and Palaeoproterozoic the upper mantle beneath SNC was at a stable depleted state with ε Nd values lower than the ε Nd value of the NQ Palaeoproterozoic upper mantle. In addition, the Proterozoic upper mantle beneath NQ does not resemble its coeval counterpart beneath SQ and NYC in the evolutionary trend and in the intensity of depletion.

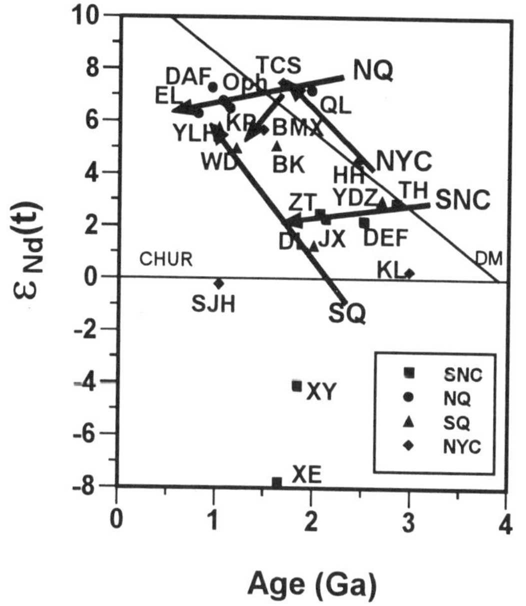

Figure 3. ε Nd (t) *vs.* age plot for Precambrian metabasaltic rocks from the four tectonic units in the study region. This diagram is prepared on the basis of data given in Table 1. Labels for the tectonic and lithostratigraphic units are also given in the table.

Measured ratios between two highly incompatible elements (*e.g.* La/Ce, Nb/La, Th/La etc.) or between two incompatible elements with similar chemical properties (Y/Tb, Zr/Hf, Ta/Nb) are thought not to be significantly affected by known fractionation processes occurring during basalt genesis, but are thought to reflect the ratios of their source [1]. From analysing the data of ratios between these two kinds of elements given in Table 2, it can be deduced that the Palaeo- and Mesoproterozoic upper mantle beneath NQ is characterized by higher Yb/Hf, Nb/La and Th/La ratios compared with the Neoarchaean to Palaeoproterozoic upper mantle beneath SNC, *i.e.* the NQ upper mantle differs from its coeval counterpart of SNC. The similarity of metatholeiites of the Qinling and Kuanping groups to the Neoproterozoic ophiolitic metatholeiite as the relict of oceanic crust not only in Yb/Hf, Nb/La, Th/La and Y/Tb ratios, but also in Pb isotopic ratios infers that these rocks should have come from a common mantle source which was most likely to be the suboceanic mantle. Moreover, the Pb isotopic ratios of all the Proterozoic rocks and feldspar of the Neoproterozoic and Early Palaeozoic granitoids from NQ are usually not only much higher than those of the Precambrian basement rocks and feldspar of the Mesozoic granitoids from SNC, but also higher than those of the Proterozoic basement rocks and feldspar of the Mesozoic granitoids from SQ east of the Mian-Lue area [24]. Synthesizing all these informations, we could suggest that the original North Qinling was most likely a microcontinent formed and developed on the base of an oceanic island.

Table 2. Pb isotopic and trace element ratios of the Precambrian metabasaltic rocks from North Qinling (NQ) and the southern margin of North China Craton (SNC)

Lithostratigraphic unit	Sample No. (a/b)	$^{206}Pb/^{204}Pb$	$^{207}Pb/^{204}Pb$	$^{208}Pb/^{204}Pb$	Yb/Hf	Nb/La	Th/La	Y/Tb
			SNC					
TH (Ar_3)	3/12	17.547	15.490	37.652	0.82	0.79	0.112	34.61
DEF (Ar_3)	1/10	17.267	15.351	37.342	0.52	0.63	0.117	35.85
JX,ZT (Pt_1)	2/10	16.126*	15.265*	37.099*	0.76	0.65	0.121	35.58
XE (Pt_2)	2/14	16.279	15.300	36.047	0.58	0.33	0.063	33.80
			NQ					
QL (Pt_1)	4/11	18.483	15.625	38.359	1.04	1.03	0.181	36.50
KP (Pt_2)	2/49	18.775	15.774	39.105	1.05	1.01	0.148	36.18
Oph (Pt_3)	6/11	18.462	15.579	38.293	1.24	1.32	0.159	35.20
DAF (Pt_3)	10/39	18.366	15.536	38.101	1.15	0.42	0.320	38-28
EL (Pt_3)	3/81	18.233	15.542	38.406	1.11	0.40	0.240	37-27

Majority of Pb isotopic data are from this study and those for the metabasalt of Danfeng Group are from [30]. The data for La, Yb, Nb, Hf and Th of JX, ZT and Oph are from [19] and [33], respectively, and the others are of this study. All the Y and Tb contents analyzed consistently by ICP-AES are from this study. * This is the Pb isotopic composition of metabasalt of the Palaeoproterozoic Angou Group in Western Henan. a and b denote the sample numbers for Pb isotopic and trace element ratios respectively. Labels for lithostratigraphic units are given in Table 1, and geological time labels in parentheses.

CRUST-MANTLE RECYCLING IN THE ANCIENT CONVERGENCE ZONE

In NQ, the coexistence of Neoproterozoic ophiolite suite with rock association and geochemical features characteristic of the oceanic crust, and of the syntectonic metabasalts of Danfeng Group showing depletion in HFSE (Nb, Ta, Zr, Ti), in addition to a northward compositional polarity peculiar to basalts in the arc setting, would provide an opportunity to test the crust-mantle recycling and the subduction of oceanic plate in an old orogen such as the Qinling orogenic belt [25]. The good knowledge of chemical and isotopic characteristics of the early crust and upper mantle in NQ is favourable for this research.

Metabasalts from the ophiolitic fragment, the Danfeng Group and the Erlangping Group representing a back-arc basin sequence are close to each other in the Nd model age (1.1-1.6Ga, 1.0-1.1Ga and 1.3-1.6Ga respectively), the initial ε Nd value and the Pb isotopic ratios (see Table 1 and Table 2). Although the Danfeng and Erlangping metabasalts, like all the arc basalts, are depleted in HFSE, their Yb/Hf and Th/La ratios are still to some extent close to those of the metatholeiites of ophiolite suite and the Qinling and Kuanping groups, The results imply that there should some genetic relation between these mantle-derived rocks.

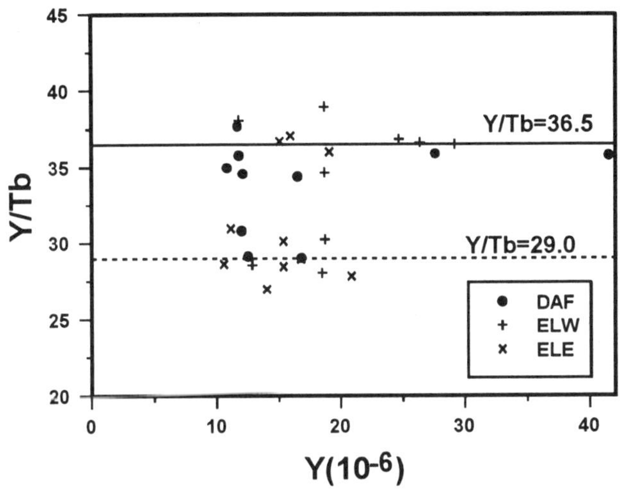

Figure 4. Y/Tb *vs.* Y plot for metabasalts of the Danfeng and Erlangping groups. DAF-Danfeng Group; ELW-Erlangping Group in the area west of the Nanyang basin; ELE-Erlangping Group in the area east of the Nanyang basin

Y and Tb are two incompatible elements exhibiting quite similar geochemical behaviors. Accordingly, their ratio in mantle-derived rocks can be used to indicate the radio of the mantle source [1]. The Y/Tb *vs.* Y plot for the Danfeng and Erlangping basalts reveals distinctly two sources for these rocks with the Y/Tb ratio at about 36.5 and 29 respectively (Figure 4). The source with Y/Tb ratio of 36.5 could be considered as the strongly depleted lithospheric mantle wedge of NQ, which can satisfy the needs for the Danfeng basalt in Y/Tb ratio of 36 ±, high positive initial ε Nd values (>+7.0) and high Pb isotopic ratios. The source with Y/Tb ratio of about 29 could be reasonably interpreted as a subducted oceanic slab together with a small amount of pelagic sediments, as a partial involvement of the deep-sea clay, generally bearing very low Y/Tb ratio (=15 on the average, [20]), is capable to reduce the Y/Tb ratio of the magma substantially. The negative correlation between ε Nd (t) values lower than 7 and Pb isotopic ratios of the Danfeng arc basalt (Table 3) may also support the idea of involvement of pelagic sediments in the origin of Danfeng island-arc basalt, because the addition of pelagic sediments generally enriched in radiogenic lead and much lower ε

Nd values as compared with MORBs [7, 18] may also lead to an increase in Pb isotopic ratios and simultaneously a decrease in ε Nd (t) values in the basalt.

Table 3. Negative correlation between ε Nd (t) and Pb isotopic ratios of the metabasalt of Danfeng Group

Sample No.	ε Nd (t)	$^{206}Pb/^{204}Pb$	$^{207}Pb/^{204}Pb$	$^{208}Pb/^{204}Pb$
2	+7.3	18.3407	15.5577	38.1972
3	+7.1	18.3849	15.5184	38.0407
2	+6.7	18.1042	15.5075	38.0477
1	+6.4	18.2402	15.5458	38.1072
1	+5.4	18.3180	15.5539	38.1434
1	+4.9	19.0527	15.5723	38.1465

Data are from [30]. The errors of measurement for Pb isotope≤1%.

This result supports both the crust-mantle recycling and the subduction of oceanic plate during the ocean-continent interaction in the development of the Qinling orogenic belt, and also confirms the nature of active continental margin of NQ since the beginning of Neoproterozoic. This also leads to the inference that lateral crustal accretion began to occur in NQ in the early Neoproterozoic time.

THE NEOPROTEROZOIC UNDERPLATING

The Indosinian granites (240-189Ma) occurring in both NQ and SQ exhibit the geological and geochemical features peculiar to the late-collision-type granites described by Harris *et al.* in 1986 [4, 26]. The early Neoproterozoic volcanic series of Yaolinghe Group dominated by metabasalt in SQ can be identified as the main source bed for the granites from both SQ and NQ on the basis of the following evidences : (1) Pb isotopic ratios of feldspar of the NQ and SQ granites of this type are similar to each other and close to those of the Yaolinghe metabasalts, but differ clearly from those of both the feldspar of Neoproterozoic and Early Palaeozoic granitoids and various basement rocks in NQ. (2) The Nd model ages of the granites from both SQ and NQ (1.0-1.4 Ga) are well coincident with those of the Yaolinghe metabasalts (1.2-1.4 Ga) and the Yaolinghe mafic volcanics can be a match for the granites in their Nd and Sr isotopic systems (Figure 5). (3)The granites from both SQ and NQ are consistently higher in Nb/Th, Nb/La,Nb/U, Nb/Pb, Ba/Th, Ba/La, Ba/Pb and Ba/U ratios, which are just the characteristics of the SQ mafic basement rocks. The fact that the Indosinian granites in NQ were derived from the source of the SQ basement is a support to the supposition that the SQ basement had underthrusted northwards and imbricated beneath the NQ upper crust [27]. As a special paper concerning this problem was published [28], here we will only discuss the Neoproterozoic underplating.

In the area of SQ west of the Shiquan-Ningshaan line, no exposure of the Yaolinghe Group has ever been found. However, the Indosinian granites of late-collision-type from the area still show almost the same Pb, Nd and Sr isotopic characteristics as the coeval counterparts to the east. This implies that the required main source rock of these granites in the area must have the same Pb-Nd-Sr isotopic systems and Nd model ages as the Yaolinghe metabasalt. Although the Pb isotopic ratios and initial ε Nd value of these granites decrease and their Nd model age (1.2->1.7Ga) increases gradually and slighly as their intrusive position approaches the outcrop of Neoarchaean Yudongzi Group in the west (see Figure 5), these variation trends can be easily ascribed to the addition of more and more materials of the Yudongzi Group (T_{DM}=2.77-3.19Ga, ε Nd (200Ma) =–16--25; $^{206}Pb/^{204}Pb$=16.401, $^{207}Pb/^{204}Pb$ =15.231, $^{208}Pb/^{204}Pb$=36.219) into the magma source.

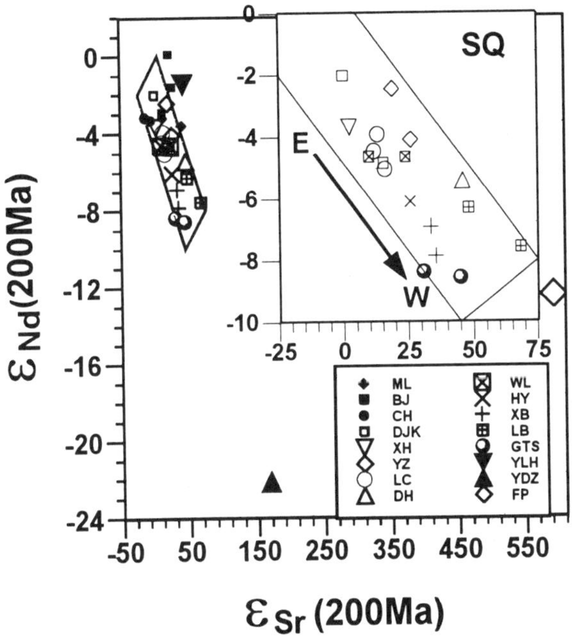

Figure 5. ε Nd (200 Ma) *vs.* ε Sr (200 Ma) plot for the Indosinian late-collision-type granites from both North and South Qinling belts and related basement rocks. 200 Ma is the average age of granitoids. Labels for the pluton names: plutons in the North Qinling belt: ML-Mangling, BJ-Baoji, CH-Cuihuashan; plutons in the South Qinling belt from east to west: DJK-Dongjiangkou, XH-Xiaohekou, YZ-Yanzhiba, LC-Laocheng, DH-Donghetaizi, WL-Wulong, HY-Huayang, XB-Xiba, LB-Liuba, GTS-Guangtoushan. YLH-the average of metabasalt of Yaolinghe Group, YDZ-the average of rocks (paragneiss, amphibolite) of Yudongzi Group, FP-gneiss of the Fuping Group. Inset rectangle denotes the variation in ε Nd and ε Sr values of granitoids with their intrusive position from east to west.

Near Fuping in a dome structure was found to compose mainly of the Palaeoproterozoic Fuping crystalline complex (1900-1800Ma, Zhang, Z.-Q. , unpub. data) which is overlain directly by the Sinian-Devonian strata for lacking of the Middle and Upper Proterozoic. A large Indosinian granite pluton, the Wulong granite, intruded into the core of the dome and formed a clear metamorphic aureole in the surrounding rocks. In this case, the magma must have been produced at depth *via* partial melting of a rock which is younger than the overlying Fuping Complex and completely resembles the Yaolinghe metabasalt in geochemical characteristics. In addition, the Yaolinghe and

Yunxi volcanics yield the same Sm-Nd isochronous age of about 1000 Ma and the same initial ε Nd value of +5.8 [32]. The Yaolinghe volcanic rock series is dominated by alkaline mafic rocks, whereas the Yunxi volcanic rock series by alkaline felsic ones. Both constitute a synchronous alkaline bimodel suite indicaing a rifting setting generally related to the anomalous heat flow of mantle. Therefore, this example may well demonstrate that the mafic magma underplating had widely occurred in SQ during the early Neoproteroic with the Yaolinghe volcanics representing the eruptive products of the deep mafic magma in the east.

CONCLUSIONS

From all results presented above, the following conclusions may be drawn:

1. The continental crust in East Qinling was formed mainly during the Proterozoic prior to 0.8 Ga, the Neoarchaean basement existing only in South Qinling.

2. North Qinling was most likely a microcontinent originated from a segment of the crust newly formed over the strongly depleted mantle in the Palaeoproterozoic, whereas South Qinling exhibits its early lithospheric history similar to that of the Yangtze craton.

3. The Palaeoproterozoic and Mesoproterozoic crustal growth in the East Qinling generally took the form of vertical accretion accompanied by some lateral displacement in response to rifting processes. The plate tectonic regime began to dominate over the Qinling orogenic belt at the beginning of Neoproterozoic. The early Neoproterozoic crustal accretion was in a lateral form related to subduction of the Qinling oceanic plate in the ancient convergence zone on the south of the North Qinling active continental margin, while it was still in the vertical form related to the rifting pattern in the South Qinling passive margin. The Neoproterozoic (1.0 Ga ±) underplating might have occurred widely in South Qinling.

Acknowledgements

This research is supported by the National Natural Science Foundation of China (Grant No. 49290102). We are grateful to Zhang Zongqing, You Zhendong, Ma Daquan and Zhang Zhufu for permission to use their unpublished isotopic data in this paper.

REFERENCES

[1] H. Bougault, J. B. Joron and M. Treuil. The primordial chrondrite nature and large scale heterogeneities in the mantle: evidence from high and low partition coefficient elements in oceanic basalts, *Philos. Trans. R. Soc. London*, Ser. A **297**, 203-213 (1980).

[2] J. F. Chen, T. X. Zhou, F. M. Xing, X. Xu and K. A. Foland. Nd isotopic composition and detrical source of epimetamorphic sedimentary and sedimentary rocks from the Southern Anhui, *Chinese Sci. Bull.* **34:20**, 1572-1574 (1989).

[3] J. F. Chen, T. X. Zhou, X. M. Li, K. A. Foland, C. Y. Huang and W. Lu. Sr and Nd isotopic constraints

on source of the Yanshanian intermediate-acid rocks from the Southern Anhui (in Chinese with English abstract), *Geochemica* **3**, 263-268 (1993).

[4] N. B. W. Harris, J. A. Pearce and A. G. Tindle. Geochemical characteristics of collision-zone magmatism. In: *Collision Tectonics*. Coward, M.P. and Ries,A.C. (Eds.). pp. 67-81. Geol. Soc. Spec. Publication No. **19** (1986).

[5] X. Huang and L. R. Wu. Nd-Sr isotopes of granitoids from Shaanxi Province and their significance for tectonic evolution (in Chinese with English abstract), *Acta Petrologica Sinica* **6:2**, 1-11 (1990).

[6] B. M. Jahn, B. Auvray, Q. H. Shen, D. Y. Liu, Z. Q. Zhang *et al.* Archaean crustal evolution in China: the Taishan complex, and evidence for juvenile crustal addition from long-term depleted mantle, *Precamb. Res.* **38**, 381-403 (1988).

[7] R. W. Kay, S. S. Sun and C. N. Lee-hu. Pb and Sr isotopes in volcanic rocks from the Aleutian Islands and Pribilof Islands, Alaska, *Geochim. Cosmochim. Acta* **42**, 263-273 (1978).

[8] A. Kroner, W. Compston, G. W. Zhang, A. L. Guo and W. Todt. Age and tectonic setting of Late Archaean greenstone-gneiss terrain in Henan Province, China, as revealed by single-grain zircon dating, *Geology* **16**, 211-215 (1988).

[9] S. G. Li, S. R. Hart, A. L. Guo, G. W. Zhang. Whole rock Sm-Nd isotopic age of the Dengfeng Group in the Central Henan and its tectonic implication, *Chinese Sci. Bull.* **22**, 1728-1731 (1987).

[10] S. G. Li and Z. Q. Zhang. Nd isotopic composition and evolution of the Archaean upper mantle in North China and their constraints on heterogeneity of the lithospheric mantle in the region (in Chinese with English abstract), *Geochemica* **4**, 277-285 (1990).

[11] S. G. Li, Y. Z. Chen, G. W. Zhang, Z. Q. Zhang. An alpinotype peridotite block emplaced at 1000Ma ago: evidence for the late Proterozoic plate tectonics regime in North Qinling (in Chinese with English abstract), *Geological Review* **37:3**, 235-241 (1991).

[12] H. F. Ling, W. Z. Shen, B. T. Zhang *et al.* Nd isotopic composition and material source of sedimentary rocks deposited before and after the Sinian in the Xiushui area, Jiangxi (in Chinese with English abstract), *Acta Petrologica Sinica* **8:2**, 190-193 (1992).

[13] G. L. Liu. New progress in the chronological study of Kongling Group (in Chinese with English abstract), *Chinese Regional Geology* **1**, 93 (1987).

[14] S. N. Lu. Precambrian Mantle- Crust Evolution in Eastern China. *Abstracts, 30th International Geological Congress,* Vol. **1 of 3**, 11, Beijing, China, 4-14 August 1996.

[15] C. X. Ma and X. K. Xiang. Preliminary study of Nd model ages of the Precambrian metamorphic strata in Northeastern Jiangxi (in Chinese with English abstract), *Scientia Geologica Sinica* **28:2**, 145-150 (1993).

[16] G. S. Qiao, M. G. Zhai and Y. H. Yan. Sm-Nd isotopic dating of the early Archaean in Eastern Hebei (in Chinese with English abstract), *Scientia Geologica Sinica* **1**, 51-54 (1987).

[17] G. S. Qiao, M. G. Zhai and Y. H. Yan. Isotopic geochronological study of Archaean rocks from the Anshan area (in Chinese with English abstract), *Scientia Geologica Sinica* **2**, 159-166 (1990).

[18] P. H. Reynolds, E. Dasch. Lead isotopes in marine manganese nodules and the ore lead growth curve, *J. Geophys. Res.* **76**, 5124-5129 (1971).

[19] Sun Dazhong and Hu Weixing. *Precambrian Chronotectonic Framework and Chronocrustal Structure in Zhongtiao Mountains* (in Chinese). Geological Publishing House, Beijing (1993).

[20] K. K. Turekian and K. H. Wedepohl. Distribution of the elements in major units of the Earth crust, *Geol. Soc. Amer. Bull.* **72**, 172-192 (1961).

[21] Wang Shougiong. The Proterozoic in the Wudang area (in Chinese), *Hubei Geology* **7**, 1-9 (1993).

[22] L. Q. Xia, Z. C. Xia and X. Y. Xu. Properties of middle-late Proterozoic volcanic rocks in South Qinling and Precambrian continental break-up, *Science in China (Series D)* **39:3**, 256-265 (1996).

[23] Xing Fengming and Xu Xiang. Nd,Sr and Pb isotopic characteristics of Mesozoic granitoids from Southern Anhui (in Chinese), *Anhui Geology* **3:1**, 35-41 (1993).

[24] J. F. Xu, B. R. Zhang, Y. W. Han. High-radiogenic lead of the Proterozoic basement rocks from North Qinling and its implications, *Chinese Science Bulletin* **41:19**, 1771-1774 (1996).

[25] B. R. Zhang, T. C. Luo, S. Gao, J. P. Ouyang, D. X. Chen *et al.* Geochemical Study of The Lithosphere, Tectonics And Metallogenesis In: *The Qinling-Dabashan Region* (in Chinese with English abstract), pp. 446. Press of China University of Geosciences, Wuhan (1994).

[26] B. R. Zhang, T. C. Luo, S. Gao, J. P. Ouyang, Y. W. Han and C. L. Gao. Geochemical constraints on

the evolution of North China and Yangtze blocks, *Journal of Southeast Asian Earth Sciences* **9:4**, 405-416 (1994).

[27] G. W. Zhang, Q. R. Meng, Z. P. Yu, Y. Sun *et al.* Orogenesis and dynamics of the Qinling Orogen, *Science in China (Series D)* **39:3**, 225-234 (1996).

[28] H. F. Zhang, B. R. Zhang, Z. D. Zhao and T. C. Luo. Continental crust subduction and collision along Shangdan Tectonic Belt of East Qinling, China—Evidence from Pb,Nd and Sr isotopes of granitoids, *Science in China (Series D)* **39: 3**, 273-282 (1996).

[29] Zhang Ligang, Wang Kefa, Chen Zhensheng, Liu Jinxiu and Li Zhitong. Lead isotope composition of feldspar of Mesozoic granites and subdivision of lead isotopic provinces in Eastern China (in Chinese), *Chinese Sci. Bull.* **38: 3**, 254-257 (1993).

[30] Q. Zhang, Z. Q. Zhang, Y. Sun and S. Han. Trace element and isotopic geochemistry of metabasalts from Danfeng Group in Shangxian-Danfeng area, Shaanxi Province (in Chinese with English abstract), *Acta Petrologica Sinica* **11: 1**, 43-54 (1995).

[31] Zhang Zhongqing, Liu Dunyi and Fu Guoming. *Chronological Study of Metamorphic Strata in North Qinling* (in Chinese). Geological Publishing House, Beijing (1994).

[32] Z. Q. Zhang, G. W. Zhang, G. M. Fu *et al.* Geochronology of metamorphic strata in the Qinling Mountains and its tectonic implications, *Science in China (Series D)* **39: 3**, 283-292 (1996).

[33] D. W. Zhou, Z. J. Zhang, Y. P. Dong and L. Liu. Geological and geochemical characteristics of Proterozoic Songshugou ophiolite piece from Shangnan County, Qinling (in Chinese with English abstract), *Acta Petrologica Sinica* **11**, 154-164 (1995).

[34] B. Q. Zhu. Tri-dimension spacial topological diagrams of ore lead isotopes and their application to the division of geochemical provinces and mineralizations (in Chinese with English abstract), *Geochimica* **3,** 209-215 (1993).

[35] B. Q. Zhu, X. L.Tu, R. Zou and H. C Liu. Geochemical division and affinity of the blocks in East Asia and its implications for tectonic evolution. *Abstracts*, *30th International Geological Congress,* Vol. **1 of 3**, 210, Beijing, China, 4-14 August (1996).

Proc. 30th Intern. Geol. Congr., Vol. 1, pp. 143-152
Wang *et al.* (Eds)

Episodic Growth of the Continental Crust in SE China: Nd Isotopic Approach

ZHOU XINHUA, HU SHILING, S. L. GOLDSTEIN*, LI JILIANG and HAO JIE
Institute of Geology, Chinese Academy of Sciences, Beijing, 100029, P. R. China
**Lamont-Doherty Geol. Obs., Columbia University, Palisades, NY 10964, USA*

Abstract

The growth pattern of continental crust is a fundamental issue in earth sciences and has been a long-existing hot debate in geological community. Geochemical evolution of sediments and granitoids has been proposed by several authors in late 1980's, which is quite effective to recognize juvenile addition, crustal forming event and recycling mechanism. The growth pattern and rate variation can also be studied. Based on a systematic sampling of major crust rocks, mainly granitoids, sedimentary cover and metasediments in South China, combined with the published data, geochronology study, survey of T_{GC}, generalized crust residence age and Nd isotope approach have been employed to investigate growth pattern of the continent crust in South China. Analysis of age histogram and spectrum of T_{GC} distribution as well as the Sm-Nd isotope system (each rock formation for a given geological age is calculated to its initial Nd isotope ratio, $\varepsilon_{Nd}(T)$) provides strong arguments to support the episodic growth model. Nd isotope data of crustal rocks in the region do not show gradual changes with time, but the evolution curve suggests that episodic crustal growth occurred in *ca.* 2.5 Ga, 1.8-2.0 Ga, 1.4 Ga, 1.0-1.2 Ga in Precambrian, and 0.4 Ga and *ca.* 0.2 Ga in Phanerozoic. The $\varepsilon_{Nd}(T)$ oscillate widely in between, especially in Phanerozoic. These imply a high frequency of mixing between distinct sources. All the above implications would place severe constraints on the interpretation of tectonic evolution of South China.

Keywords: growth, continental crust, Nd isotope, SE China

INTRODUCTION

The growth pattern and rate of the continental crust is one of the fundamental issues in earth sciences and a hot debate in geological community. A variety of models have been proposed since mid-1960's, in which two extremes are the continuing and even accelerating one [17], and a steady-state or recycling one [3, 4]. Lately, most authors, dominated by geochemists and sedimentologists, preferred models in between [18, 25, 1, 30, 31], which agree that the majority of crust was formed by the end of the Archean. Although people have tried different approaches to test their models, the details of change of growth rate with time, either over a large area or a fine structure in a certain region remain uncertain. Since mid-80's, Nd isotopic compositions of clastic sediments have yielded much of information of growth history of the continental crust as they reflect average crustal residence age [23, 28, 29, 11, 9, 5, 19, 20]. The aims of this study are to test Nd isotope approach for sediment evolution over a large region with quite complicated geological history, like South China, to recognize the growth pattern and the change of growth rate, if possible, and to make a comparative study to fine structures in a single passive margin, such as in North China [34].

GEOLOGICAL BACKGROUND

The tectonics of southern China has been a great controversy in the geology of the country. This region is located to the south of the Proterozoic Yangzi Block and is traditionally called Cathayia, or the Phanerozoic South China Fold Belt. The extensive Mesozoic and Cenozoic magmatism in the region is related to a west-dipping Jurassic and Cretaceous subduction complex. Part of South China underlain by the Precambrian and Palaeozoic terranes. For the regional tectonics and sampling area, the readers are referred to a relevant sketch map [19, Fig. 1], which is a sister paper and part of the data source of this study. It has been interpreted as para-platform, the Cathayia landmass, Caledonian geosyncline, arc-trench-basin system and collision tectonic either in late Precambrian or in Indosinian time [16, 12, 13, 14, 21]. However, most acceptable view is that it is a complex collage of sutured-bounded tectono-stratigraphic terranes, which were most likely assembled mainly during Mesozoic time [19, 21].

In this paper, the authors try to use the present available database of Nd isotopes of most crustal rocks, mainly granitoids and sedimentary rocks, to explore the growth pattern of continental crust in South China.

ND ISOTOPE APPROACH AND DATA PRESENTATION

Sm-Nd isotope systematics of the fine-grained clastic sediments deposited over some 3.8 Ga of the Earth history have proved to be exceptionally valuable for investigation of problems, such as sedimentary provenance and more general questions of evaluating models of continental growth and evolution. McCulloch and Wasserburg [23] proposed the Sm-Nd Chondritic Uniform Reservoir model age (T_{CHUR}), which provides a new way to estimate the average age of continental crust. Depaolo [6, 7] made a significant modification to this conception, and postulated on the depleted Mantle model age (T_{DM}), so that the calculated ages would better fit the observed geological data. A remarked contribution in applying this approach to continental growth was made by O'Nions and his group [26], who proposed the idea of crustal residence age (T_{CR}), the period of time for which the rare earth component of the sediments have resided within the continental crust. The basis of the approach is to utilize radiogenic Nd isotope to monitor the changing rare earth budget of the continental crust. It is well established that typical continental crust has an uniform Sm/Nd ratio which is 40% lower than the estimated bulk-Earth value, and further that clastic sediments are generated and recycled with minimal perturbation of this ratio [10, 11]. This approach has been successfully applied to a variety of case studies [*e.g.* 2, 24, 9, 34].

However, a few points should be stressed here. First, the crustal residence age calculated for a given region may be an average value reflecting contributions from several different sources. In general, therefore it can not be regarded as dating a specific crustal-forming event, but may provide an estimation for exposed continental crust from which the sediments were eroded. Secondly, all calculations of T_{CHUR}, T_{DM} and T_{CR}

are based on single stage evolution, which do not applicable to quite many geological cases, as pointed out by several authors [6, 22]. A two-stage model age calculation has been well accepted, which makes most results more geologically meaningful [8]. Thirdly, the application of T_{CR} is severely restricted in sedimentary system by its definition, although on the other hand, T_{DM} has been widely applied to granitoids. There is indeed no any difference in calculating T_{DM} and T_{CR}. To improve this situation and make things clearer and simpler, the authors attempted to extend T_{CR} to a broader meaning, that is to apply this conception to most crustal rocks, such as granitoids, metasediments and ortho-metamorphic rocks, named as generalized crustal residence age, T_{GC} instead of T_{DM} [33]. The prerequisites for T_{GC} are that the rock system can be approximately regarded as a two-stage system, and the Sm/Nd ratio of the first stage can be reasonably estimated empirically or by relevant model.

To make a broader geological coverage, the data collected by this study are taken from published literatures over an area of ten provinces [31], and some are cited from the authors' unpublished results. All Nd isotopic compositions are corrected to the initial ratio, as the epsilon notation. The available geochronological information for each sets of the data are used for this correction. The Nd isotope ratios are normalized to $^{146}Nd/^{144}Nd=0.7219$. The present day $^{147}Sm/^{144}Nd$ and $^{143}Nd/^{144}Nd$ ratios of Chondritic Uniform Reservoir (CHUR) and Depleted Mantle Reservoir (DM) are 0.1967, 0.512638 and 0.2136, 0.513150, respectively. For the calculation of T_{GC} of granitoids and metasediments, the $^{147}Sm/^{144}Nd$ ratio of first stage is taken as 0.1200, based on a statistics of more than 270 samples from South China.

GEOCHRONOLOGICAL RECORDS

Since the main purpose of this part of study is to find a way to express or to designate the general gross features of age spectrum in SE China instead of an original and local chronological study, it should be noted first that, (1) the database used to construct Fig. 1 has been done for filtering based on the data quality, such as complete information of sampling, analytical quality and self-consistence, (2) there is no need to distinguish the real geological meaning for each data set, such as emplacement, metamorphism or cooling ages, as they can be regarded in a broad sense as a record of thermal events. The histogram of age dating is plotted in Figure 1a, for the data < 900 Ma, that is in Phanerozoic and Sinian; and 1b, for the data in Palaeoproterozoic, respectively. The remarked peaks in 120-150 Ma, 190-210 Ma, 390-410 Ma and 850-900 Ma in Fig. 1a have been usually interpreted as the chronological records of Yanshanian, Indosinian, Caledonian and Neoproterzoic tectonic events in South China. As mentioned in the previous section, the Phanerozoic tectonics is commonly regarded as "intracrustal" in nature by many authors, whereas much of geological and geochemical arguments suggest a major collision between two blocks, Yangzi and South China in Neoproterozoic. The thermal events recorded in age dating in Neoproterozoic seem to reflect a "major growth" of continental crust in the region.

The histogram of age dating in Palaeoproterozoic and Neoarchaean is shown in Figure 1b, which indicates two major events, 1800-2000 Ma, and 2200-2300 Ma in

Palaeoproterozoic and one event, 2500 Ma in Neoarchean. The preservation of ancient and Precambrian crust in South China have been only confirmed in the 1980's [*e.g.* 15]. It is commonly believed that this was the cratonization of South China Block. The above observation is further supported by the zircon dating record as shown in Figure 1c. In addition, it reveals that an event occurred in Mesopropterozoic, around 1200-1400 Ma, which has been well documented and was interpreted as continental rifting and generation of new crust material [35].

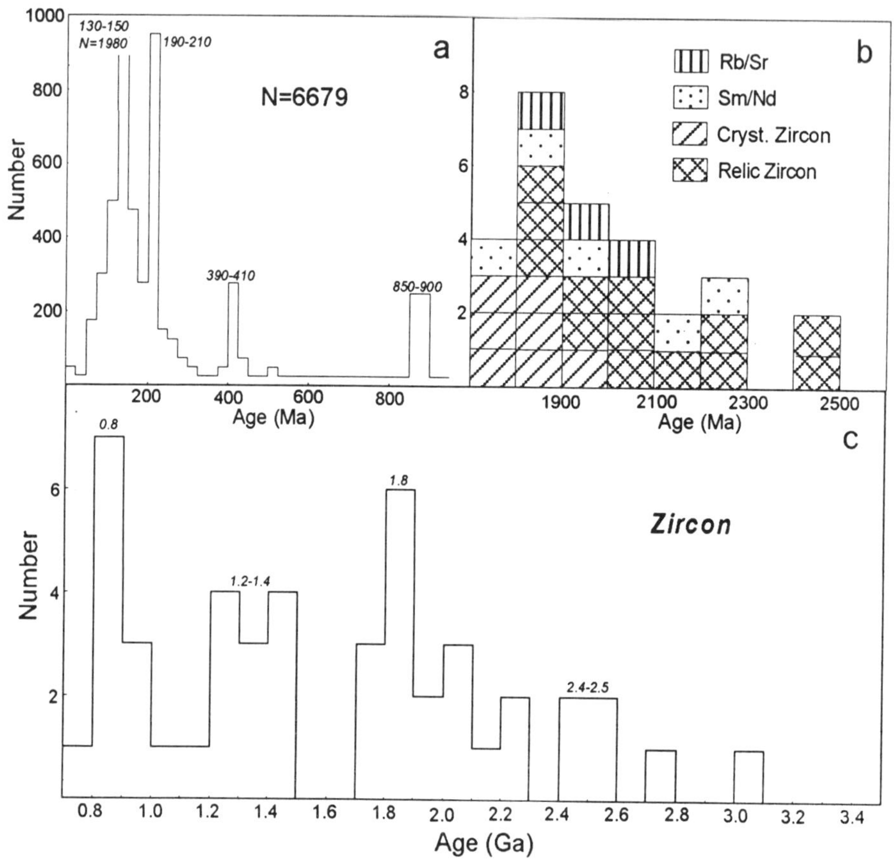

Figure 1. Histogram of isotopic dating in South China. a) data < 900 Ma, b) data 1700-2500 Ma, c) zircon data.

It is not a safe way to identify addition of new crust material, *i.e.* accretion or growth event of crust, based merely on geochronological record. Geological and isotope geochemical study would provide more creditable criteria to distinguish these events from intracrustal records.

ND ISOTOPE MODEL AGES

A general survey of Nd crust residence age T_{GC} has been performed for the available data of granitoid and sediments in South China, as shown in Fig. 2a, 2b, and 2c. Remarked peaks in 1.1-1.4 Ga, 1.7-1.8 Ga and 2.0-2.1 Ga highlight this histogram. Although the calculation of T_{GC} is based on two-stage model, as defined in the previous section, it is necessary to know whether they represent crust growth events. The period of 1.7-1.8 Ga was a world-wide geological record, as was the timing of cratonization in North China Block. It was also reflected as a major peak in age dating (Fig. 1b). However, the peak of T_{GC} in 2.0-2.1 Ga seems to be a mixture phenomenon with different components, as it is neither shown in any other records, nor supported by any geological and geochemical studies in SE China, although it may be important in some other area, such as the Siberian Craton. We have already indicated that the peak in 1.1-1.4 Ga represents a crust forming event, as it is recorded by both age datings of a variety of mafic rocks [35] and Nd model ages. An additional evidence comes from lack of this peak in the spectrum of sediments T_{GC}.

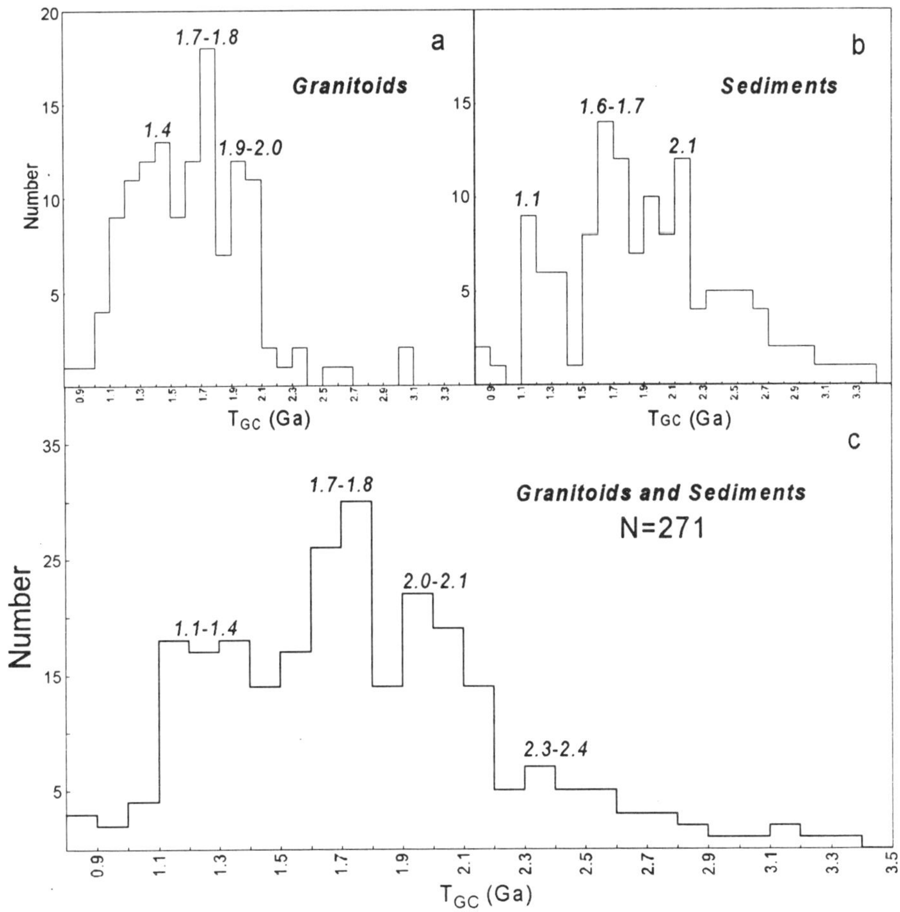

Figure 2. Histogram of T_{GC}, the generalized crust residence age in South China. a) granitoids, b) sediments, c) granitoids and sediments.

ND ISOTOPE SYSTEM AND EVOLUTION OF GRANITOIDS AND SEDIMENTS

In this part, we would like to emphasize and to examine the nature of crustal growth component, which should be differentiated for the first time from the mantle. Then, we would on this basis attempt to provide further evidences for the episodic signature of crustal growth in South China as recorded in Sm-Nd systems of granitoids and sediments, the major components of upper crust in the region.

Figure 3 is the Sm-Nd isotope diagram of granitoids in South China, with crystallization ages (T), from Mesozoic to Neoproterozoic. Figure 3a is a plot of $^{143}Nd/^{144}Nd$ *vs.* $^{147}Sm/^{144}Nd$, the so-called isochron type plot, which suggests that the major source material of granitoids is in Proterozoic, 1.0-2.0 Ga, with a mean age of 1.5-1.6 Ga, by their single-stage T_{DM}. A few of scattered data are caused by high fractionated Sm/Nd ratio, an indicator of later stage differentiation. When the same data set are plotted onto $\varepsilon_{Nd}(T)$ *vs.* T, Figure 3b, the characteristics of Nd isotope evolution is clearly shown. The domination of mantle-derived origin of Precambrian granitoids, though few, highlights its early stage evolution. This is a firm evidence showing juvenile addition to crust growth in that period. In Phanerozoic, the source materials for granitoids are dominated by old and recycled crust, as indicated by the negative $\varepsilon_{Nd}(T)$ values (down to −14). Three reference lines of Nd isotope evolution are drawn for the single-stage systems starting from a primordial mantle {$\varepsilon_{Nd}(T)$=1-2} in 1.4 Ga, 1.7 Ga and 2.0 Ga, with an average Sm/Nd=0.1200, respectively in which most of data are enclosed. Some fine structures of Phanerozoic type is also shown in this diagram. The two least negative $\varepsilon_{Nd}(T)$ values are observed in *ca.* 400 Ma and *ca.* 200 Ma. One datum point of 200 Ma showing positive value indicates that the minor new addition occurred in limited space and time, and was masked by mixing with voluminous old crust. Several geological evidences related to Caledonian and Indosinian tectonics in South China provide further support to this interpretation [21, 27].

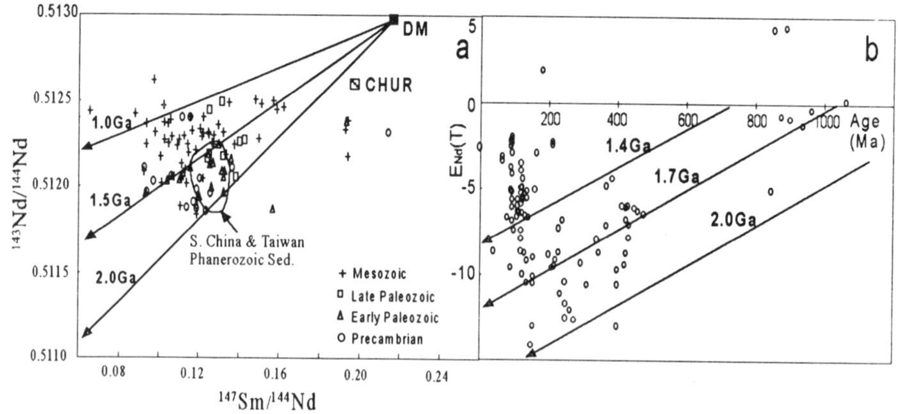

Figure 3. Sm-Nd system of granitoids in South China. a) Sm-Nd isochron plot, b) $\varepsilon_{Nd}(T)$ *vs.* age

Similar patterns are shown in Figure 4a and 4b for sediments in South China. The Sm-

Nd isochron plot suggests that older sources have been involved in their provenances as indicated by T_{DM} up to 2.5-3.0 Ga, which represent the data of Precambrian metasediments in Zhejiang and Fujian Provinces. Figure 4b provides a strong evidence for the episodic crustal growth in the region. No matter where and how old the sampling localities or model ages are, most of the Precambrian data, up to 2.5 Ga, show clear affinity to mantle derivation, in which several peaks of 1.0-1.2 Ga, 1.8 Ga and 2.5 Ga comprise data sets with fairly positive $\varepsilon_{Nd}(T)$ values. In Phanerozoic, the date set is a mimic to the pattern of granitoids, implying the similar source and evolutionary history.

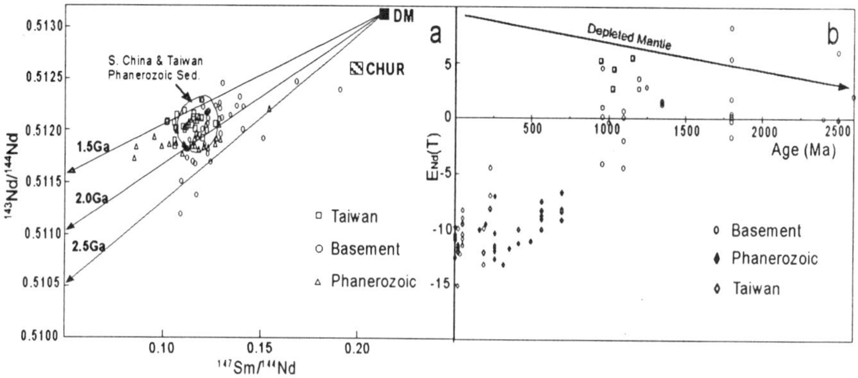

Figure 4. Sm Nd system of sediments in South China. a) Sm-Nd isochron plot, b) $\varepsilon_{Nd}(T)$ *vs.* age.

EPISODIC CRUST GROWTH IN SOUTH CHINA AND ITS GLOBAL COMPARISON

A combination of Figure 3b and 4b may give us a complete picture of Nd isotope evolution path of the upper crust in South China, as shown in Figure 5. The evolution curve, composed of all the available data, does show an episodic feature, instead a continuous or gradual one. The major crust-forming events occurred in 2.5 Ga, 1.8 Ga, 1.4 Ga and 1.0-1.2 Ga, respectively, although minor growth in Phanerozoic, such as in 400 Ma and 200 Ma, also played a significant role in crustal evolution. Three reference lines are plotted onto Figure 5 to provide constraints on source characteristics. The most depleted data points are close to the depleted mantle path [10], and the most enriched data, mostly with negative $\varepsilon_{Nd}(T)$ value, are close to the reference line of 2.5 Ga old crust for the most pre-Sinian data, and to the line of 1.8 Ga old crust for most Sinian and Phanerozoic data. This observation implies that depleted mantle and the 1.8 Ga and 2.5 Ga old crust have been the three major components involved in crustal growth and accretion in South China since from 2.5-3.0 Ga.

For a global comparison, the South China data are plotted onto a diagram of T_{GC} *vs.* geological age, as designed for T_{strat}. The shaded area represents a world-wide trend for global sediments. The South China data well fit this pattern, which suggests, once more, that the major crustal growth was in the Archaean and Palaeoproterozoic, and the recycled crust component only played a significant role later on, by mixing with the episodic addition of minor mantle-derived juvenile material.

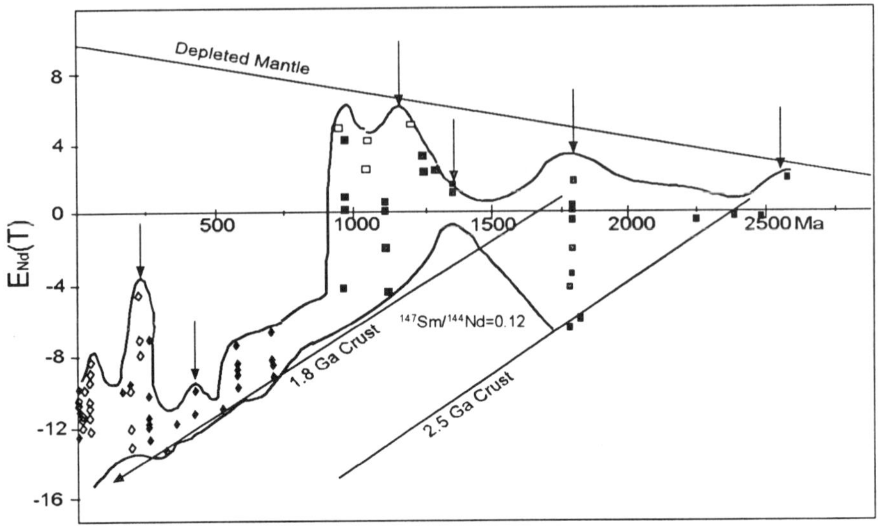

Figure 5. Nd isotope evolution of upper crust in South China.

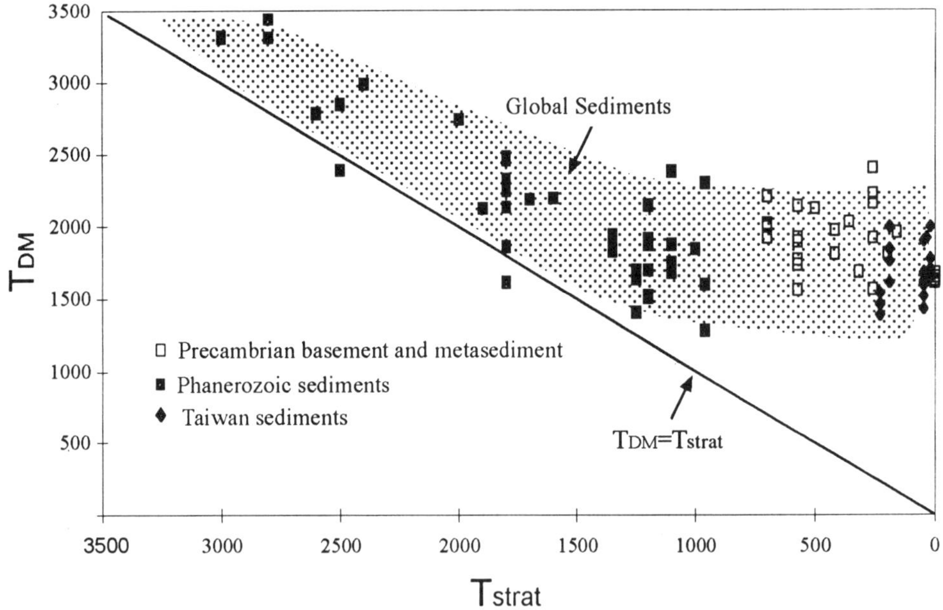

Figure 6. T_{DM} *vs.* T_{strat} plot, with comparison with global trend.

CONCLUDING REMARKS

(1) Episodic continental growth is unequivocally observed in the crust records for the region studied in South China, through the approaches of geochronology, Nd isotope model ages and Sm-Nd isotope evolution.

(2) The major crustal growth was in Archaean and Palaeoproterozoic, whereas the new juvenile additions were masked by mixing with voluminous old crust in Phanerozoic.

(3) Recycled crustal material is the major component throughout the geological record, especially in post Archaean times.

(4) Episodic growth occurred at 2.5 Ga, 1.8 Ga, 1.4 Ga, 1.0-1.2 Ga, *ca.* 0.4 and *ca.* 0.2 Ga in South China, in comparison with mainly Neoarchaean and Palaeoproterozoic in North China.

(5) High Frequency oscillation during growth episodes and multiple component mixing is characteristic of the Phanerozoic crust evolution.

(6) The sediment examples studied in South China may be typical in the growth pattern of a global nature.

Acknowledgements

The authors are indebted to Professors Sun Shu and A.W. Hofmann for constructive discussions and comments during the senior author (XHZ)'s academic visit in Mainz and in the course of this project. We are also grateful to Professor Borming Jahn for fruitful discussions. This work is supported by NNSF of China and Max-Planck-Institute fur Chemie.

REFERENCES

[1] C. J. Allegre. Chemical Geodynamics, *Tectonophysics* **81**, 109 (1982).

[2] C. J. Allegre and D. Rousseau. The growth of the continent through geological time studied by Nd isotope analysis of shales, *Ear. Planet. Sci. Lett.* **67**, 19-34 (1984).

[3] R. L. Armstrong. A model for Sr and Pb isotope evolution in a dynamic Earth, *Rev. Geophysics* **6**, 175-199 (1968).

[4] R. L. Armstrong. Radiogenic isotopes: the case for crustal recycling on a near-steady-state no-continental growth Earth, *Philos. Trans. R. Soc. London* **A301**, 443-472 (1981).

[5] C. H. Chen, B. M. Jahn, T. Lee, C. H. Chen and J. Cornichet. Sm-Nd isotopic geochemistry of sediments from Taiwan and implications for the tectonic evolution of southeast China, *Chem. Geol.* **88**, 317-332 (1990).

[6] D. J. Depaolo. Nd isotope in the Colorado Front Range and crust-mantle evolution in the Proterozoic, *Nature* **291**, 193-196 (1981).

[7] D. J. Depaolo. The mean life of continents: Estimates of continent recycling rates from Nd and Hf isotopic data and implications for mantle structure, *Geophys. Res. Lett.* **8**, 705-708 (1983).

[8] D. J. DePaolo, A. M. Linn and G. Schbert. The continental crustal age distribution: Methods of determining mantle separation ages from Sm-Nd isotopic and application to the Southeastern United States. *J. Geophys. Rew.* **96**, B, 2071-2088 (1991).

[9] C. D. Frost and D. Winston. Nd isotopic systematics of coarse- and fine-grained sediments: Examples from the Middle Proterozoic Belts-Purcell Supergroup, *J. Geol.* **95**, 309-327 (1987).

[10] S. L. Goldstein and R. K. O'Nions. Nd and Sr isotopic relationships in pelagic clays and ferromanganese deposits, *Nature* **292**, 324-327 (1981).

[11] S. L. Goldstein, R. K. O'Nions and P. J. Hamilton. A Sm-Nd isotopic study of atmospheric dusts and particulates from major river system, *Ear. Planet. Sci. Lett.* **70**, 221-236 (1984).

[12] L. Z. Guo, Y. S. Shi and R. S. Ma. Tectonic framework and evolution of South China, In: *Proceedings of International Academic Exchange for Geological Papers* **(1)**. Geological Press, Beijing (1980).

[13] K. J. Hsu, S. Sun, J. L. Li and H. H. Chen. Huanan Alps, not South China Platform, *Sci. Sinica* **B**, 1107-1115 (1987).

[14] K. J. Hsu, J. L. Li, H. H. Chen, Q. C. Wang, S. Sun and A. M. C. Sengor. Tectonics of South China: Key to understanding West Pacific geology, *Tectonophysics* **183**, 9-39 (1990).

[15] X. J. Hu, J. K. Xu, C. X. Tong and T. H. Chen. *Precambrian Geology of Southeast Zhejiang*, Geological Press, Beijing (1991).

[16] J. Q. Huang. The tectonic framework of China, *Acta Geologica Sinica* **51**, 117-135 (1977).

[17] P. M. Hurley. Absolute abundance and distribution of Rb, K and Sr in the earth, *Geochim. Cosmochim. Acta* **32**, 273 (1968).

[18] S. B. Jacobeson and G. J. Wasserburg. Transport model for crust and mantle evolution, *Tectonophysics* **75**, 163 (1981).

[19] B. M. Jahn, X. H. Zhou and J. L. Li. Formation and tectonic evolution of Southeastern China and Taiwan: Isotopic and geochemical constraints, *Tectonophysics* **183**, 145-160 (1990).

[20] C. Y. Lan, T. Lee, X. H. Zhou and K. C. Kim. Nd isotope study of Precambrian basement of South Korea: Evidence for early Archean crust, *Geology* **23**, 249-252 (1995).

[21] J. L. Li. On the tectonics of Southeast China, In: *Structure and Evolution of Southeast China Lithosphere*. J. L. Li (Ed.). pp. 1-24. Publishers of Science and Technology of China, Beijing (1992).

[22] T. C. Liew and M. Y. McCulloch. Genesis of granitoid batholiths of peninsular Malaysia and implications for models of crustal evolution: Evidence from a Nd-Sr isotopic and U-Pb zircon study, *Geochim. Cosmochim. Acta* **49**, 587-600 (1985).

[23] M. T. McCulloch and G. J. Wasserburg. Sm-Nd and Rb-Sr chronology of continental crust formation, *Science* **200**, 1003-1011 (1978).

[24] A. Michard, P. Gurrier, M. Soudant and F. Albarede. Nd isotopes in French Phanerozoic shales: External *vs.* internal aspects of crustal evolution, *Geochim. Cosmochim. Acta* **49**, 601-610 (1985).

[25] R. K. O'Nions and P. J. Hamilton. Isotope and trace element models of crustal evolution, *Philos. Trans. R. Soc. London* **A301**, 473 (1981).

[26] R. K. O'Nions P. J. Hamilton and P. J. Hooker. A Nd-isotope investigation of sediments related to crustal development in the British Isles, *Ear. Planet. Sci. Lett.* **63**, 229-240 (1983).

[27] S. L. Ren, J. L. Li, X. H. Zhou, J. Hao, J. B. Zhang and X. J. Hu. Xiongshan diabase swarm of North Fujian: Petrochemistry and tectonic implications, In: *Annual Report of the Lab of Lithosphere Tectonic Evolution, IGCAS*, **1993-1994**, pp. 89-94 (1994).

[28] S. R. Taylor and S. M. McLennan. The composition and evolution of the continental crust: rare earth element evidence from sedimentary rocks, *Philos. Trans. R. Soc. London* **A301**, 381-399 (1981).

[29] S. R. Taylor and S. M. McLennan. *The continental crust: its composition and evolution*. Blackwell Scientific Publications, Oxford (1985).

[30] J. Veizer and S. L. Jansen. Basement and sedimentary recycling and continental evolution, *J. Geol.* **87**, 341-370 (1979).

[31] J. Veizer and S. L. Jansen. Basement and sedimentary recycling 2: Time dimension to global tectonics, *J. Geol.* **93**, 625-643 (1985).

[32] X. H. Zhou. Isotope geochemical constraints on tectonic evolution of SE China, In: *Isotope Geochemical Researches in China*, G.Z. Tu (Ed.). pp. 203-245. Science Press, Beijing (1996).

[33] X. H. Zhou, H. Cheng and Z. H. Zhong. The source province of metamorphic rocks from southwest Zhejiang — a case study, In: *Memoir of Lithosphere Tectonic Evolution Research* **(1)**, pp.114-120 (1992).

[34] X. H. Zhou and S. L. Goldstein. Continental growth and sediment evolution: Nd isotope profile of the Sinian Formation, North China, *Contribution to ICOG, Australian Geological Society, Abstracts* **27**, 116 (1990).

[35] X. H. Zhou, S. L. Hu, X. J. Hu, S. W. Dong, R. H. Jiang, B. T. Xu and J. B. Zhang. Mesoproterozoic rifting event in South China: Inferred from isotope system, In: *Annual Report of the Lab of Lithosphere Tectonic Evolution, IGCAS*, **1993-1994**, pp.179-182 (1994).

Proc. 30th Intern. Geol. Congr. Vol. 1, pp. 153-158
Wang *et al.* (Eds)

On the Way from Plate Tectonics to Global Geodynamics

VICTOR E. KHAIN
Institute of the Lithosphere, Russian Academy of Sciences, 22 Staromonetny per., 109180 Moscow, Russia

Abstract

The plate tectonics theory had a profound and positive impact on the progress of Earth sciences. Its main principles were confirmed by later observations. But at the same time it has become obvious that these principles were primarily oversimplified, that the theory underestimated some substantial phenomena of Earth structure and evolution (intraplate tectonics and magmatism, in particular), and finally that it could be applied only to the uppermost part of the solid Earth, to a depth of *ca.* 400 km, and to the latest interval of its history, *ca.* 1000 Ma. All these necessitate the replacement of the classical plate tectonics by a genuine global geodynamic model, comprising plate tectonics as one of its elements. Main requirements to such a future model are formulated.

Keywords: plate tectonics, global geodynamics

INTRODUCTION

The advent of plate tectonics nearly 30 years ago had a profound revolutionary and positive impact on the development of the Earth sciences. Main principles of this theory, formulated in 1967-68, were confirmed in the course of further exploration by deep-sea drilling, observations from submersibles, seismic tomography and, finally, by cosmic geodesy, including the recent GPS measurements. But at the same time the enormous amount of new information showed that the natural phenomena were much more complicated than their description in the original theory, that this theory had not taken into account all the aspects of tectonic and magmatic activities, especially its intraplate manifestations, that the explanation of this activity by mantle convection needs serious elaboration and, finally that the sphere of application of plate tectonics is restricted in space and time. In space it is limited to the uppermost part of the solid Earth (tectosphere), up to 400 km depth, and in time to the latest interval of Earth history, from 1.0 Ga. onwards. All these necessitate the revision of the main plate tectonic principles, their refinement and, ultimately, the replacement of this theory by a more wide, genuine global geodynamic model, which will comprise plate tectonics as one of its important elements.

PLATE TECTONICS: ORIGINAL PRINCIPLES IN THE LIGHT OF FURTHER EXPERIENCE

Nearly all the principles formulated in 1967-68 need now some modification [3]. Let us examine them one after another.

1. Lithosphere and asthenosphere

The first prerequisite of plate tectonics is the subdivision of the upper solid Earth into two layers with very different rheology—lithosphere and asthenosphere. This distinction remains valid up to now, but it is necessary to recognize that the properties of both layers are very variable laterally and the lithosphere is not homogenous in its vertical section as was formerly imagined. At first it was admitted that the thickness of the lithosphere under the oceans was of the order of 50-60 km and 100-120 km under the continents. Now we know that this thickness could be as small as 3-4 km under the spreading ridges, *e.g.* the East pacific rise [5], and could be more than 200 km, even reaching 400 km (South Africa), under the old cratons [7]. The viscosity of the asthenosphere, depending mainly on heat flow, changes substantially in a lateral direction, and the viscosity contrast between the two layers under the cratons is so small that their boundary is difficult to detect. Some researchers even expressed doubts about the existence of the asthenosphere in such regions. But it is just here that isostatic conditions are fulfilled most obviously (in regions of recent or former glaciation), which is a convincing proof of the presence of asthenosphere.

The rheological vertical profile of the lithosphere shows a pronounced strain contrast between the upper crust and lithospheric mantle on the one hand, which are brittle, and the lower crust on the other hand, which is much weaker and ductile. This contrast is at the place of crustal delamination, now established in many regions, especially in those that have experienced a continent-continent collision (*e.g.* Himalaya and Tibet).

2. Subdivision of the lithosphere into plates

In the original model of modern distribution of lithosphere plates seven major and seven medium-sized plates were recognized. This pattern does not take into account the existence of a wide belt of diffuse seismicity in the Mediterranean region and Central Asia, and the analogous but smaller belts in Alaska and the USA Cordillera. These facts in the first case led to recognition of the existence of an ensemble of about twenty microplates between Eurasia and the Gondwanan continents Africa, Arabia and India [8]. Palaeotectonic analysis of ancient mobile belts have also discerned numerous microcontinents, representing separate microplates, and even miniplates, nowadays often called terranes. Such microplates are now established in the oceans, *e.g.* Easter or Azores microplates. So the real pattern of distribution of lithospheric plates appears much more complex than it was suggested primarily.

3. Types of relative motions of the plates

Three types of such motions were established: divergence, manifested by spreading; convergence, represented by subduction and collision; and motions along transform faults. Each of these types must be considered now as much more complicated in its apparent expression.

Spreading is ordinarily preceded by continental rifting and the major rift zones divide the continental plates into subplates, *e.g.* the East African rift system separates the Somalia subplate from the main African plate. Spreading is not always strictly symmetrical and, for example, the ocean floor relief is therefore not quite similar on both sides of the Mid-Atlantic ridge. The spreading axes are often subdivided into overlapping segments, which manifest jumping of smaller or larger scale, as in the Greenland-Norwegian Sea, beginning with propagation of one of the axes beyond a transform fault.

Convergence of lithospheric plates is manifested not only by subduction and collision, but also by obduction, eduction (reverse, ascending movement with exhumation of subducted material, which experienced high and ultra-high pressure) and large strike-slip faulting. Subduction is not always accompanied by accretion of sedimentary wedges, but often by tectonic erosion instead of accretion [2].

Motion along transform faults is ordinarily not merely strike-slip, but is in many cases transtensive, creating transform trenches with manifestations of serpentinite diapirism and magmatic activity, and in other cases transpressive, with thrusting of one wall over another. The character of motion could change along the transform strike, as is the case with the Azores-Gibraltar transform fault, which is transtensive in the west and transpressive in the east. Even slight changes in the position of the poles of rotation will produce a very complex pattern of structure and relief of the transform zones, as was detected in the Romanche zone in the equatorial Atlantic [1].

4. The principle of plate motions

The obedience of plate motions to the Euler's theorem remains the only principle of the classic plate tectonics which does not need any modification.

5. The Earth's lithospheric extension, expressed by spreading, is automatically compensated by subduction and collision and the volume and radius of the Earth consequently remain constant

This postulate of the theory is confirmed in general form by successive reconstructions, concerning in particular the evolution of triple functions, but could not be taken as absolute. The distribution in time of fold thrust deformations, granite plutonism and regional metamorphism shows a certain periodicity of the activation of compression, but the manifestation of ophiolite formation and LT/HP metamorphism may mark the same relation to extension [4]. These data must be regarded as proofs of the alternation of epochs of dominant extension with those of dominant compression, and consequently of some pulsation of the Earth's volume change, but probably not exceeding a few percent. The fact of heat flow diminution in time leads to the conclusion that a general tendency of slow contraction of the Earth's volume is very probable.

6. Mechanism of plate motions

Plate motions are due to the drag exerted on the lower surface of the plates by the asthenosphere flow, directed from the spreading axes to the subduction zones. This drag is not considered now as a unique mechanism of plate motions and two other forces are recognized—ridge-push and slab-pull. The role of ridge-push is confirmed by the study of the stress fields in the plates interior [9].

7. Ultimate cause of the kinematics of plate tectonics

The ultimate cause of plate tectonics kinematics is seen in the mantle convection. This convection was regarded primarily as whole-mantle and of thermal origin. Nowadays many researchers recognize the possibility of separate convection cells in the upper and lower mantle, and the convection itself as more probably thermo-chemical, not purely thermal.

MAIN DEFICIENCIES OF THE PLATE TECTONICS CONCEPT

The plate tectonics theory successfully explained the most important elements of Earth's lithosphere–tectonic and magmatic activity connected with plate boundaries. But it did not provide such explanation for intraplate phenomena: deformations and magmatism. Concerning deformations, this gap was fulfilled by recent study of the stress distribution in the continental crust. It was discovered that these stresses are mainly compressive and oriented normal or to spreading axes in the oceans or to the collision sutures on the continents [9]. It is quite natural that under the action of these stress deformations are produced in the first place in zones of preexisting weakness–rifts and ancient sutures. But intraplate magmatism could not find its explanation in the plate tectonics frame. This circumstance was early recognized and already in the mid-sixties a special hypothesis was proposed, *i.e.* the widely known Wilson-Morgan's concept of hot spots-mantle plumes. In spite of its recognition, this concept contains some points which permit of a different interpretation. One of them concerns the depth of origination of mantle plumes in the asthenosphere, on the bottom of upper mantle or at the mantle/core interface. Of even more importance is the question of interaction between plumes and convective cells in the mantle. It is one of the most important problems of geodynamics.

Another phenomenon not considered in classical plate tectonics was the periodicity in the change of intensity of tectonic and magmatic activity and the alternating of epochs of dominant extension and compression already mentioned. Obviously this flaw was due to the fact that plate tectonics was primarily concerned only with the last 160 Ma. of Earth's history. The most natural explanation of this periodicity is the uneven and cyclical change in the amount of heat flow discharged on the surface through the mantle and the crust.

Classical plate tectonics disregarded also the very probable existence of a general evolutionary trend in the manifestation of endogenic activity and the change in style of this activity, produced in the first place by the secular decrease of heat and fluid flow from the deep interior of the planet.

Plate tectonics ignored also the fact of the existence of a distinctive global net of fractures and lineaments, confirmed by satellite imagery (Sounder's rhegmatic net). The most probable origin of this net is its connection with periodical changes of the speed of rotation of our planet and the corresponding changes of its ellipticity.

It is furthermore important that plate tectonics action has clear limitations in space and time. Seismic tomography has revealed that the distribution of hot and cold regions in the mantle corresponds rather well to surficial features, mid-oceanic ridges, ancient cratons and their shields, only to a certain depth. We must admit that plate tectonics is operative only to the depth of the boundary between upper mantle *s. str.* and the transition zone, *i.e.* the depth of 400 km. Regarding the question of its temporal frame, it must be recognized that all the plate tectonics indicators, ophiolites, LT/HP metamorphics, calc-alkaline volcano-plutonic belts, are known mostly from the last billion of years and very few before that date. So it could be ascertained that plate tectonics represents the mode of behavior of endogenic activity of the outermost shell of the solid Earth during the last billion years of its history. It does not mean that plate tectonics appeared only since the Neoproterozoic. It certainly appeared much earlier, but had not found its full expression until in Neoproterozoic. In the earliest stages of Earth's history before 3.0 or 3.5 Ga, the endogenic activity of the Earth was manifested probably in the form of plume tectonics, as is seen now on Venus, which experienced recently a transition from plume to plate tectonics.

It must be clear from the aforesaid, that we have now arrived at the turning point, when the need for the plate tectonics model of Earth's evolution to be replaced by a wider and more comprehensive global geodynamic model has become quite obvious. The work in this direction has begun already in different countries, including Japan and Russia. Below I will try to formulate the main requirements which should be addressed to such a model.

MAIN PRINCIPLES OF A GENUINE GLOBAL GEODYNAMIC MODEL

1. First of all this model must take into account that the solid Earth is a multilayered body with each layer of geosphere envelope representing a quasi-autonomous system with specific character not only of mineralogical, phase composition and rheological properties, but also of the mode of heat-mass transfer.

2. At the same time, all the envelopes are constantly in dynamic interaction, and the type of this interaction changes in time, depending on the effectiveness of heat transfer to the Earth's surface and the relative changes of the Rayleigh number of the mantle at different levels.

3. Two main types of heat-mass transfer coexist in the solid Earth convection and advection, the latter in the form of mantle plumes, and forming respectively two main geodynamic styles, plate tectonics and plume tectonics. These two operate simultaneously and interdependently, the main plumes being situated at the intersection of the sides of Rayleigh Benard connective cells, at present and in the past.

4. The relative importance of plate tectonics and plume tectonics probably changes in depth and time, with increasing role of plume tectonics in depth and with decreasing role in geologic time. The interrelations between plate tectonics and plume tectonics may be expressed by the following statement: plume tectonics manifestations' in the lithosphere are controlled by plate tectonics, but plate tectonics itself is directed from the depth by plume tectonics.

5. Regarding the influence of the Earth's axial rotation on mantle flow, on the origin of certain regularities in the structural plan of the lithosphere, and on the formation of the rhegmatic net of planetary faults, lineaments and fractures must be taken into account.

6. The same concerns the effect of the Earth's motion along its galactic orbit, with periodical crossing of concentrations of cosmic dust, which could explain the periodicity of changes in endogenic activity of the order of the galactic year (*ca.* 215 Ma). The explanation of the periodicity(cyclicity) of higher ranks could lie also in the resonance phenomena between the impulse of astronomical factors and the processes in the Earth's interior (heat flow, phase changes), and the shortest cyclicity corresponds to changes in the Earth's rotational regime invoked by the Milankovitch theory.

The new global geodynamic model proposed in 1994 by a group of Japanese geoscientists (M. Maruyama, S. Kumazawa *et al.* [6]) satisfies most of the requirements formulated above, except points 5 and 6. The interaction between plumetectonics and platetectonics needs further elaboration, including the origin of the periodicity of changes in endogenic activity, reversals of the magnetic field, transgressions and regressions, changes of the Earth's rotation speed and the volume and shape of the Earth.

REFERENCES

[1] E. Bonatti. The Ocean Floor Observations, Theory and imagination. *Intern. Symp. Abstracts. Acad. Nat. Lincei. Roma.* p. 5 (1994).

[2] R. E. von. Huene and D. W. Scholl. Observations at convergent margins concerning sediment subduction, erosion and the growth of continental crust, *Rev. Geoph.* **29**, 279-316 (1991).

[3] V. E. Khain and M. G. Lomize. *Geotectonics with elements of geodynamics* (in Russian). Moscow Univ. Press, Moscow (1995).

[4] V. E. Khain and K. B. Seslavinsky. Global rhythms in the Phanerozoic endogenic activity of the Earth (in Russian), *Strat. Geol. Correl.* **6**, 40-63 (1994).

[5] K.C. MacDonald and P. J. Fox. The mid-ocean ridge, *Scientific American* **6**, 72-79 (1990).

[6] S. Maruyama, M. Kumazava and S. Kawakami. Towards a new paradigm on the Earth's dynamics, *Jour. Geol. Soc. Japan* **100**, 1-3 (1994).

[7] L. Polet and D. L. Anderson. Depth extent of cratons inferred from tomography studies, *Geology* **23**, 205-208 (1995).

[8] S. A. Ushakov, O. P. Ivanov and Yu. I. Prozorov. Small plates of the Alpine-Himalayan belt. In: *Global tectonic and dynamic natural processes.* pp.3-14. Moscow (1984).

[9] M. L. Zoback *et al.* Global pattern of tectonic stress, *Nature* **341**, 291-298 (1989).

Proc. 30th Intern. Geol. Congr., Vol. 1, pp. 159-175
Wang *et al.* (Eds)

Oxygen Isotope Fractionation in $MgSiO_3$ and Mg_2SiO_4 Polymorphs: Implications for the Chemical Structure of the Mantle and Isotopic Nonequilibuium in Mantle Assemblages

ZHENG YONGFEI
Department of Earth and Space Sciences, University of Sciences and Technology of China, Hefei 230026, P. R. China

Abstract

The modified increment method has been applied to the calculation of oxygen isotope fractionation factors for the polymorphic phases of MgSiO3 and Mg_2SiO_4. The results suggest the following sequence of ^{18}O-enrichment in the different structures of the mantle silicates: pyroxene $Mg_2Si_2O_6$ > olivine Mg_2SiO_4 > spinel-structured Mg_2SiO_4 > ilmenite-structured $MgSiO_3$ > perovskite-structured $MgSiO_3$. Assuming isotopic equilibrium between the various phases of the mantle, the chemical structure of the mantle can be described by the following sequence of ^{18}O-enrichment: upper mantle > transition zone > lower mantle. Such an oxygen isotope layering results from differences in the chemical composition and crystal structure of mineral phases at different mantle depths.

Oxygen isotope fractionations between pyroxene and olivine in mantle assemblages can be negative under closed-system conditions, provided that pyroxene has a derivation of the lower mantle from the polymorphic transition of the perovskite-structured $MgSiO_3$ without isotopic reequilibration. It can thus have inherited the oxygen isotope composition of the lower mantle minerals without isotopic reequilibration despite the change in crystal structure. Oxygen isotope inheritance in mineral formation by polymorphic transformations is very common in nature and in laboratory experiments. Therefore, the occurrence of the lower mantle materials can be discovered by studies on the oxygen isotope fractionation relationship in the mineral assemblages of the mantle-derived rocks. Moreover, spinel or corundum formed by the decomposition of garnet in the mantle could be out of isotopic equilibrium as normally defined with coexisting pyroxene and olivine due to the oxygen isotope inheritance in the Mg-O-Al of Al-O structural units of garnet.

Keywords: oxygen isotopes, mineral polymorphs, mantle assemblages, chemical layering, equilibrium partitioning, nonequilibrium fractionation

INTRODUCTION

Numerous experimental studies on the mineralogy of the Earth's mantle have suggested that the perovskite-structured $MgSiO_3$ is the dominant silicate phase in the lower mantle [15, 22, 27, 28, 32], whereas the olivine-structured Mg_2SiO_4 is the most abundant mineral phase in the upper mantle. In the transition zone (*i.e.* at depths between 400 and 670 km), the spinel-structured Mg_2SiO_4 is the dominant mineral [3, 17, 27, 37]. Increasing pressure and temperature with depth drives various phase changes and chemical reactions within the mantle. Near 400 km depth a phase transition occurs, with $MgSiO_4$ olivine converting to the high-pressure modified (β) spinel structure. At depths about 670 to 700 km, the spinel-structured $MgSiO_4$ transforms to perovskite-structured $MgSiO_3$ and MgO. Pryoxenes and garnets will also transform to the perovskite structure.

Phases from the higher pressures and temperatures of the lower mantle are unstable in the lower pressure-temperature environments of the upper mantle and crust. Thus polymorphic transition and mineral reaction are inevitable during the upward transportation of materials from the lower mantle to the upper mantle and finally to the crust. From a theoretical geochemistry perspective, the Earth consists of well oxidized mantle and crust, and a metallic core. The ratio of oxygen atoms to all other atoms combined is between 4/3 and 3/2 in the mantle, and significantly higher in the crust. Oxygen is also the largest of the abundant ions. Therefore, information on the equilibrium oxygen isotope properties of minerals within the mantle is essential for understanding the chemical structure of the mantle.

The crystal structure of minerals can influence their isotopic properties depending on the nature of the interatomic interaction in the various structural forms [33]. Because mineral polymorphs have different bond strengths and hence different vibrational frequencies, they may be expected to exhibit different behaviors with respect to oxygen isotope fractionations. This has been theoretically illustrated for the polymorphs of SiO_2, TiO_2 and Al_2SiO_5 [18, 43, 51, 53]. Because the lower mantle has significantly different mineral phases from the upper mantle, oxygen isotope fractionation can be expected to exist between them.

The polymorphic phases of $MgSiO_3$ and Mg_2SiO_4 are predominant phases in the mantle at different depths. Oxygen isotope fractionation in enstatite and forsterite which comprise the upper mantle minerals have been experimentally determined [9, 38]. However, data for phases constituting the lower mantle and the transition zone are not available. This paper presents a theoretical calculation on oxygen isotope fractionation in all the polymorphic phases of $MgSiO_3$ and Mg_2SiO_4 by means of the modified increment method [42, 47]. The results have been discussed in order to gain an insight into nonequilibrium isotope fractionations among the mantle-derived minerals.

CALCULATION METHOD AND RESULTS

The increment method was primarily developed on the basis of the principle of mineral crystal-chemistry [42]. The approach was brought to public attention by comparison of the calculated fractionations with the experimentally determined data [35]. Zheng [47, 50] has modified the increment method to calculation of thermodynamic oxygen isotope factors for metal oxides, wolframates, anhydrous and hydroxyl-bearing silicates [47-50, 52]. The modified increment method has proven to be a valid approach for calculating oxygen isotope fractionation factors in solid minerals as a function of statistical mechanical and crystal structural affects.

The method of calculating I-^{18}O indices for solid minerals has described by Zheng [47, 50] in detail and is not repeated here. In principle, the degree of ^{18}O-enrichment in a given mineral can be quantitatively represented by the size of its oxygen isotope index I-^{18}O . Quartz is taken as the reference mineral and thus its I-^{18}O index is defined as

1.0000. The greater the I-^{18}O index of a mineral, the more ^{18}O- enriched it is. The I-^{18}O value of the mineral is calculated by summing the normalized ^{18}O- increment (i'_{ct-o}) for different cation-oxygen bonds in its crystal structure. The ^{18}O-increment is determined by the effects of cation-oxygen bond strength (C_{ct-o}) and cation mass on isotopic substitution (W_{ct-o}). The cation-oxygen bond strength is defined as a function of cation oxidation state (V), coordination number (CN_{ct}) and corresponding ionic radii ($r_{ct}+r_o$). Thus, the I-^{18}O index of a mineral results from a marriage of crystal chemistry with the relationship between vibrational frequency and reduced mass.

Table 1. Calculation of the normalized ^{18}O-increments in the cation-oxygen bonds of mantle minerals

Bond	CN_{ct}	CN_o	$r_{ct}+r_o$(A)	M_{ct}	W_{ct-o}	C_{ct-o}	i_{ct-o}	i'_{ct-o}
Si-O	4	2	1.61	28.09	1.03748	0.62112	0.02285	1.0000
Si-O	4	4	1.64	28.09	1.03748	0.60976	0.02244	0.9821
Si-O	6	4	1.78	28.09	1.03748	0.37453	0.01378	0.6031
Si-O	6	6	1.80	28.09	1.03748	0.37037	0.01363	0.5964
AL-O	6	4	1.91	26.98	1.03690	0.26178	0.00949	0.4151
AL-O	4	4	1.77	26.98	1.03690	0.42373	0.01535	0.6718
Mg-O	6	4	2.10	24.31	1.03537	0.15873	0.00552	0.2415
Mg-O	12	6	2.54	24.31	1.03537	0.06562	0.00228	0.0998

Table 2. The thermodynamic oxygen isotope factors for $MgSiO_3$ and Mg_2SiO_4 polymorphs in the mantle ($10^3\ln\beta = Ax + Bx^2 + Cx^3$, where $x=10^6/T^2$)

Chem. Form.	Structure	I-^{18}O	A	B	C
$CaTiO_3$	perovskite	0.4450	5.392	−0.165	0.0055
$FeTiO_3$	ilmenite	0.5132	6.218	−0.190	0.0063
$MgAl_2O_4$	normal spinel	0.4024	4.875	−0.149	0.0049
	inverse spinel	0.4612	5.588	−0.171	0.0057
	special spinel	0.5931	7.186	−0.219	0.0073
Al_2O_3	normal corundum	0.4106	4.975	−0.152	0.0051
	special corundum	0.7380	8.942	−0.273	0.0091
$MgSiO_3$	perovskite	0.4267	5.170	−0.158	0.0052
	ilmenite	0.4780	5.791	−0.177	0.0059
	enstatite	0.7791	9.440	−0.288	0.0096
	special	0.7131	8.640	−0.264	0.0088
Mg_2SiO_4	spinel	0.6034	7.311	−0.223	0.0074
	forsterite	0.6717	8.138	−0.249	0.0083
$Mg_3Al_2[SiO_4]_3$	pyrope	0.7194	8.716	−0.266	0.0088
	oxide	0.6309	7.644	−0.233	0.0078
SiO_2	quartz	1.0000	12.116	−0.370	0.0123
	coesite	0.9896	11.990	−0.366	0.0122
	stishovite	0.8837	10.707	−0.327	0.0108
$CaCO_3$	calcite		11.781	−0.420	0.0158

The present calculations deal with the following structures of $MgSiO_3$ and Mg_2SiO_4

polymorphs: (1) the perovskite-structured $MgSiO_3$; (2) the ilmenite-structured $MgSiO_3$; and (3) the spinel-structured Mg_2SiO_4. The oxygen isotope fractionations in pyroxene Mg_2SiO_6 and olivine Mg_2SiO_4 have been calculated by Zheng [50]; those in ilmenite $FeTiO_3$ and spinel $MgAl_2O_4$ and perovskite $CaTiO_3$ have been calculated by Zheng [47, 52]. The I-^{18}O indice for the mantle silicates of the oxide structures are calculated in the same way as that for the metal oxides [47]. The normalized ^{18}O-increments of the cation-oxygen bonds in the minerals are calculated and presented in Table 1, together with the Crystal-chemical parameters used. The coordination numbers and interatomic distances are after Muller and Roy [31]. The I-^{18}O indices obtained are listed in Table 2.

The reduced partition function ratios (equivalent to the thermodynamic isotope factors in the increment method) for quartz were calculated by Clayton and Kieffer [8]. This was done by applying a small " adjustment factor " to the theoretical data of Kieffer [21] in order to bring them into the best agreement with the experimental data for the calcite-quartz system [6]. The combined theoretical-experimental data for the quartz $10^3 \ln \beta$ factors were represented for temperatures greater than 400 K by the equation:

$$10^3 \ln \beta_{quartz} = 12.116x - 0.370x^2 + 0.0123x^3 \quad (1)$$

Here $x = 10^6/T^2$ (T in Kelvin).

Eq. (1) has a zero-intercept and thus it can accurately describe the high-temperature behavior of oxygen isotope partitioning in quartz. However, its application to low-temperature fractionations becomes invalid because the 1/T term is not included. For the high-temperature mantle minerals, nevertheless, Eq. (1) can be well used as the $10^3 \ln \beta$ equation of the reference mineral and therefore applied to calculation of their thermodynamic oxygen isotope factors. Furthermore, the low-temperature correction term (D) introduced by Zheng [47] can be removed in dealing with the high-temperature systems. Consequently, the $10^3 \ln \beta$ factors for the mantle minerals are calculated by the relation:

$$10^3 \ln \beta_{mineral} = I\text{-}{}^{18}O_{mineral} \bullet 10^3 \ln \beta_{quartz} \quad (2)$$

The results are algebraically represented in Table 2 by the polynominal equation like Eq. (1):

$$10^3 \ln \beta = Ax + Bx^2 + Cx^3 \quad (3)$$

Figure 1 depicts the temperature dependence of thermodynamic oxygen isotope factors for the polymorphs of $MgSiO_3$ and Mg_2SiO_4. Oxygen isotope fractionation factors among major phases in the mantle are thus small (mostly < 0.5‰ in $\delta^{18}O$), but they must be taken into account in dealing with mantle-derived mineral assemblages.

According to the thermodynamic oxygen isotope factors in Table 2, the fractionation factor equations can be obtained for mineral-mineral systems by the use of the relation:

$$10^3 \ln \alpha x - y = 10^3 \ln \beta x - 10^3 \ln \beta y \quad (4)$$

The theoretical calibrations are valid only at the temperatures above *ca.* 300°C if the low-temperature correction [47] is not applied. The uncertainty in the fractionation factors $10^3 \ln \alpha$ derived from the modified increment method is within ±5% of the factor values [47, 50], This indicates that the calculated fractionations are in the error of 10±0.5‰ or 1±0.05‰. Such an uncertainty is small enough to resolve the small $\delta^{18}O$

differences found in the mantle minerals (about 0.5‰).

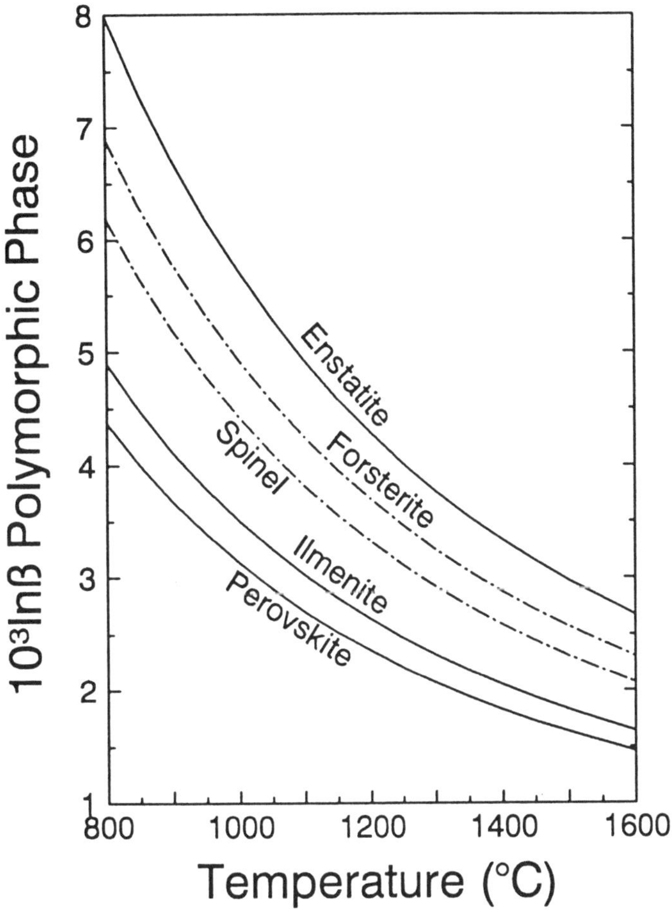

Figure 1. Temperature dependence of thermodynamic oxygen isotope factors for the polymorphic phases of $MgSiO_3$ (*solid curve*) and Mg_2SiO_4 (*dot-and-dash curve*) in different crystal structures.

Comparing with the previous treatments on the common rock-forming minerals [47, 49, 50], the application of Eq. (1) to the present calculations eliminates the constant term (non-zero intercept) in the resultant $10^3 \ln \beta$ and $10^3 \ln \alpha$ equations. It appears that the oxygen isotope fractionations between the high-temperature mantle minerals are accurately represented by a polynomial form like Eq. (3). Substantially, the regressed $10^3 \ln \beta$ equation by Zheng [50] for the reference mineral quartz is still valid in the temperature range from 1200 to 2000°C despite the presence of the constant term. Therefore, the differences in the mineral $10^3 \ln \beta$ factors due to the fit in 1/T-space and to the non-zero intercept will be offset then calculating the fractionation factors $10^3 \ln \alpha$ between phases by Eq. (4). This is also favored by the increment method itself, because the theoretical fractionation curves are internally consistent between mineral pairs [42,

47, 49, 50].

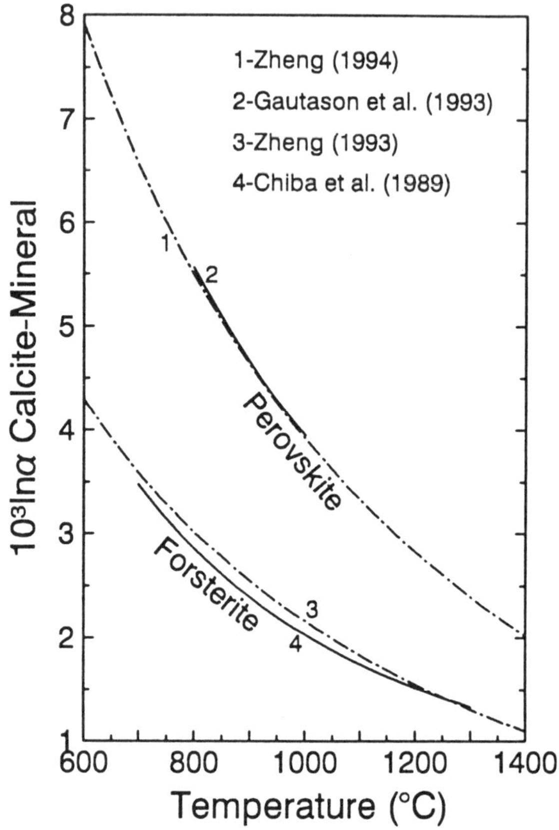

Figure 2. Comparison of the theoretical calculations with the experimental determinations on oxygen isotope fractionations for the systems calcite-perovskite (Zheng [52] *vs*. Gautason *et al.* [11] and calcite-forsterite (Chiba *et al.* [9] *vs*. Zheng [50]).

DISCUSSION

Oxygen isotope fractionation between calcite and perovskite $CaTiO_3$ have been experimentally determined using the partial exchange technique at 800 and 1000°C, respectively [11]. Zheng [52] has theoretically calculated oxygen isotope fractionation in perovskite. As shown in Figure 2, the experimental data [11] are in excellent agreement with the theoretical calibrations [52]. Figure 2 also illustrates the agreement for the calcite-forsterite system between the experimental determination [9] and the theoretical calculation [50]. The corroboration of the modified increment method for the common rock-forming minerals suggests that all extension of the methodology to the mantle minerals is potentially valid.

Like perovskite $CaTiO_3$, the I-^{18}O index of the perovskite-structured $MgSiO_3$ are very small (0.4276), as shown in Table 2. They are less than those of the ilmenite-structured counterparts (0.4780). According to the principle of the increment method [42, 47], the size of mineral I-^{18}O indices is a quantitative indicator of the degree of I^{18}O-enrichment. Apparently, the perovskite-structured silicates are depleted in ^{18}O relative to the ilmenite-structured counterparts at isotopic equilibrium. Pyroxene $Mg_2Si_2O_6$ has a very great I-^{18}O index of 0.7791 [50]. Thus it is most enriched in ^{18}O with respect to the other polymorphs of $MgSiO_3$ (Fig. 1).

For the polymorphs of Mg_2SiO_4, the I-^{18}O index of the spinel-structured Mg_2SiO_4 is 0.6034, which is less than that of the forsterite Mg_2SiO_4 (0.6717 [50]). Hence the spinel-structured Mg_2SiO_4 is considerably depleted in ^{18}O relative to the forsterite (Fig. 1). Pyrope $Mg_3Al_2[SiO_4]_3$ has an I-^{18}O index of 0.7194 [50]). and thus it is depleted in ^{18}O relative to pyroxene but enriched in ^{18}O relative to olivine.

However, effect of pressure on stable isotope partitioning can become significant under mantle conditions. Polyakov and Kharlashina [34] have reckoned that pressures of tens of kbars may produce a measurable effect on isotopic fractionation and even change the sign of the isotopic shift. At the pressures of the deeper mantle, for instance, the $Mg_3Al_2[SiO_4]_3$ may isotopically behave like a metal oxide (metallization) and therefore assume an I-^{18}O index of 0.6309. In this case, it would be depleted in ^{18}O relative to forsterite Mg_2SiO_4 but enriched in ^{18}O with respect to the spinel-structured Mg_2SiO_4. On the other hand, if $MgSiO_3$ in the mantle could take the garnet-structure, it would behave isotopically like garnets and thus be depleted in ^{18}O with respect to the normal pyroxene.

IMPLICATIONS FOR THE CHEMICAL STRUCTURE OF THE MANTLE

By comparing the I-^{18}O indices of the $MgSiO_3$ and Mg_2SiO_4 polymorphs and the similar structure of metal oxides in Table 2, the sequence of ^{18}O-enrichment can be obtained as follows: pyroxene $Mg_2Si_2O_6$ > olivine Mg_2SiO_4 > spinel Mg_2SiO_4 > ilmenite $FeTiO_3$ > ilmenite $MgSiO_3$ > perovskite $CaTiO_3$ > perovskite $MgSiO_3$ > spinel $MgAl_2O_4$. Apparently, the perovskite-structured silicates in the lower mantle and the spinel-structured silicates in the transition zone would fractionate differently with each other than pyroxenes and olivines in the upper mantle. If there would be complete isotopic equilibration in the mantle, the spinel-structured silicates in the lower mantle but depleted in ^{18}O relative to the olivine in the upper mantle.

Figure 3 plots the I-^{18}O indices for the mantle minerals versus their $\delta^{18}O$ values at 1200 and 1600°C, respectively. An arbitrary $\delta^{18}O$ value of perovskite $MgSiO_3$is taken as the initial point to illustrate the relative sequence of ^{18}O-enrichment in the mantle minerals. It can be expected that the oxygen isotope fractionation between forsterite in the upper mantle and the spinel-structured $MgSiO_4$ in the transition zone is 0.37‰ at 1200°C and the fractionation between the spinel-structured $MgSiO_4$ in the transition zone and the perovskite-structured $MgSiO_3$ in the lower mantle is 0.59‰ at 1600°C. The higher the temperatures at boundaries between the mineralogical layers of the mantle, the smaller the fractionations between them. At 2000°C the fractionation between the spinel-structured $MgSiO_4$ and the perovskite-structured $MgSiO_3$is only 0.40‰. Assuming isotopic equilibrium on a whole earth scale. the chemical structure of the mantle can be represented by the following sequence of ^{18}O-enrichment: upper mantle > transition zone > lower mantle.

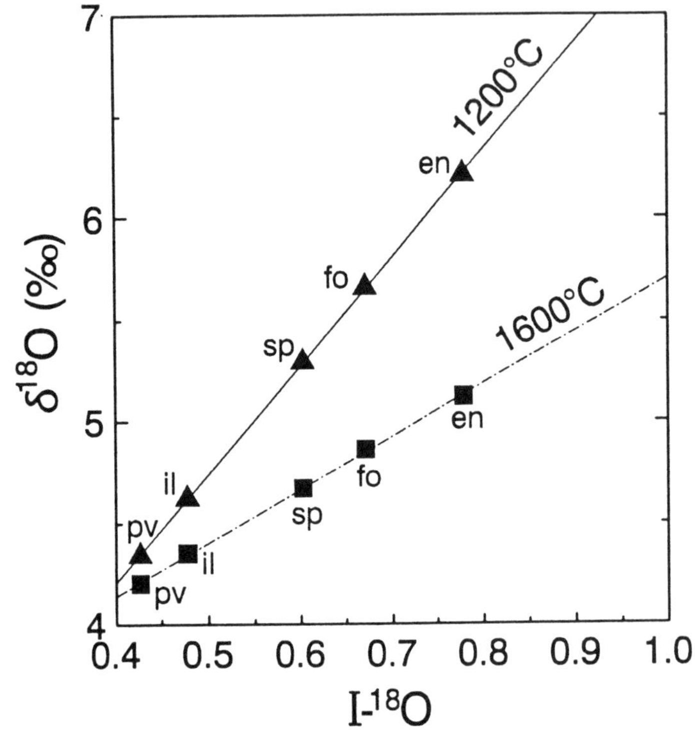

Figure 3. The relative $\delta^{18}O$ values of the mantle minerals plotted versus the I-^{18}O indices for perovskite $MgSiO_3$ (pv), ilmenite $MgSiO_3$ (il), spinel $MgSiO_4$ (sp), forsterite Mg_2SiO_4 (fo) and enstatite $MgSiO_3$ (en). Temperature is assumed to be 1200°C and 1600°C, respectively. The arbitrary $\delta^{18}O$-enrichment in the mantle minerals.

Essentially, the oxygen isotope layering of the mantle is determined by differences in the chemical composition and crystal structure of mineral phases at different mantle depths. The physico-chemical differentiation of the Earth in its forming stage may be the mechanism by which the Earth as a whole would become layered in chemical and oxygen isotope compositions. This becomes plausible if the mantle would consist of discrete reservoirs that were mutually in isotopic equilibrium. It is possible that the upper and lower mantles would be in a complete oxygen isotope communication by convection. These reservoirs would interact on such a scale as to develop oxygen isotope equilibrium between the perovskite-structured and low $\delta^{18}O$ lower mantle and the olivine-structured and higher $\delta^{18}O$ upper mantle. The recycling of crustal materials into mantle depths by pate subduction would only result in local mixing effects with respect to the isotopic zonation in the mantle on a whole earth scale.

The upper mantle may be primarily made up of magnesium-rich olivine (forsterite), pyroxene (enstatite) and garnet (pyrope). The peridotite mantle model is dominated by olivine along with pyroxene [36]; the "pyrolite" model is a nearly 1:1 mixture of olivine and pyroxene [37]; and the eclogite model is a mixture of pyroxene and garnet [1]. According to the theoretical calculations [50] and the experimental determinations [9, 38, 39], garnet is enriched in ^{18}O relative to olivine but depleted in ^{18}O relative to pyroxene. At 1200°C, the oxygen isotope fractionation between enstatite and pyrope is 0.3‰ and that between enstatite and forsterite is 0.6‰ [50]. The three minerals are considerably depleted in ^{18}O relative to water (H_2O) and carbon dioxide (CO_2) even at mantle temperatures (-4 to -2‰ [50]). Because H_2O may probably be the most abundant fluid species in the Earth's interior [4, 19, 44, 45], mantle degassing can cause the mantle-derived mineral assemblages to be considerably depleted in ^{18}O with respect to the primary mantle.

The lower mantle is the largest single region of the Earth's interior, accounting for 55% of its volume. In contrast to the case for the upper mantle, from which we have samples, we cannot determine the isotopic composition of the lower mantle by direct observation, yet the isotopic geochemistry of its major elements has important implications for the evolution of the Earth and its current thermal and chemical state. The possibility of oxygen isotope stratification—distinct upper and lower mantle compositions—implies that the Earth's mantle would convect in at least two layers and that the Earth would have evolved slowly and retained much of its primordial heat and large-scale geochemical heterogeneities.

IMPLICATIONS FOR ISOTOPIC NONEQUILIBRIUM IN MANTLE ASSEMBLAGES

There is an increasing interest in the oxygen isotope geochemistry of the mantle. A number of oxygen isotope studies have been carried out on the mineral assemblages of the mantle [10, 13, 14, 16, 20, 23-25, 30, 41]. The fractionations between the mantle-derived minerals are observed to be often out of the experimentally determined equilibrium values under the upper mantle conditions [20, 24, 41]. This was interpreted as the result of open-system interactions with fluids possessing exotic oxygen isotope compositions derived from subducted oceanic crust [12, 20]. However, because oxygen is by far the most abundant element in the mantle (about 50% in weight), the addition of substantial amounts of fluid with distinctly different $\delta^{18}O$ values is required to produce even small shifts in the oxygen isotope composition of the mantle [26, 40]. Therefore, other mechanisms have to be searched in order to resolve the dilemma.

An empirical olivine-pyroxene geothermometer was calibrated by Kyser *et al.* [24] on the basis of analyses of terrestrial and lunar volcanic rocks and mantle nodules. The authors observed that oxygen isotope fractionations between olivine and pyroxene change from positive to negative (cross-over) in the mantle nodules at 1150°C. Specifically, the negative isotopic fractionations occur in high-pressure spinel lherzolites from Massif Central (France) and San Quintin (Baja California). Gregory and Taylor [12] argued against the validity of this cross-over and attributed it to the result of a nonequilibrium assemblage involving fluid metasomatism under the open-system conditions in the mantle nodules. The experimental determinations on the carbonate-silicate systems [9, 38] and the theoretical calculations on anhydrous silicates [50] indicate that there is no cross-over in isotopic fractionation between olivine Mg_2SiO_4 and pyroxene $Mg_2Si_2O_6$ at high temperatures.

As pointed out at the beginning of the introduction, the mantle is dominantly composed of olivines in the upper part, of the perovskite-structured silicates in the lower part, and of the spinel-structured silicates in the transition zone, There may exist considerable amounts of pyroxenes in the upper mantle [13, 46]. It is possible that the presently observed pyroxenes in the mantle-derived rocks may partly be the products of polymorphic transformation from the perovskite-structured $MgSiO_3$ in the lower mantle. This process may take place during the upward transportation of the mantle materials from the lower mantle to the upper mantle and finally to the crust. Similarly, the olivines occurred in the crust could partly be the products of polymorphic transformation from the spinel-structured Mg_2SiO_4 in the transition zone of the mantle. In this context. The mantle-derived mineral assemblages are not cogenetic in petrology and thus are in geochemical nonequilibrium.

The present calculation of oxygen isotope fractionation in the mantle minerals can provide a resolution to the controversies involving nonequilidrium fractionations among the mantle-derived mineral assemblages. Figure 4 depicts the temperature dependence of oxygen isotope fractionations between the $MgSiO_3$ and Mg_2SiO_4 polymorphs. The oxygen isotope fractionations between perovskite-structured $MgSiO_3$ and forsterite Mg_2SiO_4 and between perovskite-structured $MgSiO_3$ and spinel-structured Mg_2SiO_4 are -1.35‰ and-0.9‰, respectively, at 1200^0C. This suggests that oxygen isotope fractionations between olivine and pyroxene can be negative in mantle assemblages, provided that pyroxene and olivine have inherited the oxygen isotope compositions of the precursors perovskite $MgSiO_3$ in the lower mantle and spinel Mg_2SiO_4 in the transition zone, respectively. In this regard, oxygen isotope reequilibration between olivine and pyroxene has not taken place during and after the polymorphic transition. Although the rates of isotopic exchange by oxygen diffusion could be large enough to reequilibrate isotopes during the upward transportation of the mantle materials, the temperature dependence of oxygen isotope fractionation in the high-pressure phases would be conveyed to their low-pressure polymorphs due to oxygen isotope inheritance in polymorphic transition. The olivine and pyroxene in the mantle assemblages can thus preserve the oxygen isotope feature inherited from their polymorphs at the mantle depth. The primary feature of oxygen isotopes would be not altered despite the change in the crystal structure of the minerals.

It is interesting to note that the theoretically expected isotopic fractionations in mineral polymorphs have been not measured in laboratory investigations. No measurable effect was observed on oxygen isotope fractionation between $CaCO_3$ and H_2O at 500^0C due to the calcite-aragonite phase change which occurs between 11 and 13 kbar experiments [5]. Negligible isotope fractionation also occurs in the experiments involving polymorphic transition from β-quartz to α-quartz [7, 29]. However, the theoretical calculations have predicted significant oxygen isotope fractionations in the polymorphs of SiO_2 and $CaCO_3$ [18, 43, 51]. The inconsistency with the experimental data [5, 7, 29] implies that the oxygen isotope signature of polymorphic phases can be inherited in the structural change of $CaCO_3$ and SiO_2 without isotopic reequilibration with ambient materials. According to a comprehensive study on oxygen isotope fractionation in magnetites, Zheng [54]suggests that the magnetite of an inverse spinel-structure has often inherited the oxygen isotope feature of a spinel-structural magnetite. The oxygen isotope inheritance in mineral formation by polymorphic transition is very common in nature, thus it can be used to discover the exposition of the unstable mantle minerals under the crustal conditions.

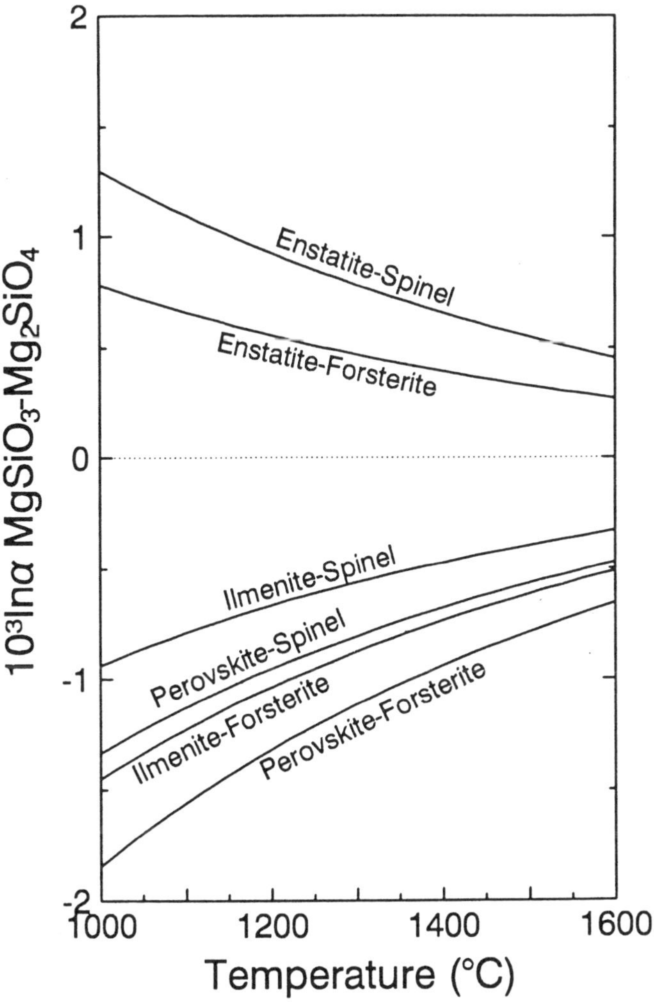

Figure 4. Temperature dependence of oxygen isotope fractionations between the polymorphs of $MgSiO_3$ and Mg_2SiO_4 in different crystal structures.

In terms of the above discussion, the oxygen isotope fractionation between pyroxene and olivine can be reversed if they would have primarily derived from the lower mantle and the transition zone, respectively, and thus formed via phase change without isotopic reequilibration. The isotopic fractionations out of the closed-system equilibrium exchange can result from oxygen isotope inheritance in the high-pressure minerals during polymorphic transformation or in a given structural unit of a mineral in its decomposition. In this regard, the negative fractionations between pyroxene and olivine, if they do exist in nature [24, 30, 41], could imply a primary source of the lower mantle for the pyroxene. The change in isotopic fractionation from negative to positive may indicate a kinetic process for isotopic reequilibration during the upward transportation of the mantle materials from the lower part to the upper part.

Kyser *et al.* [23-25] observed that, in some xenoliths of the mantle origin, garnets are sometimes rimmed by spinel. This is not a polymorphic transformation, however. The transition from garnet to spinel must involve the participation of pyroxene and thus imply the following mineral reaction in the mantle:

$$Mg_3Al_2[SiO_4]_3 + Mg_2SiO_4 = MgAl_2O_4 + 2Mg_2Si_2O_6 \qquad (5)$$

As a result, the spinel derived from such a decomposition of garnet could inherit the oxygen isotope composition of Mg-O-A1 structural units in the precursor garnet. This may proceed in the following two steps of minera1 reaction:

$$\underset{0.7194}{Mg_3^{VIII}Al_2^{VI}[Si^{IV}O_4^{IV}]_3} \rightarrow \underset{0.5931}{Mg^{VIII}AL_2^{VI}O_2^{VI}O_2^{IV}} + \underset{0.7131}{Mg_2^{VIII}Si_2^{IV}O_6^{IV}} + \underset{0.9896}{Si^{IV}O_2^{II}} \qquad (6)$$

$$\underset{0.6717}{Mg_2^{VI}Si^{IV}O_4^{IV}} + \underset{0.9896}{Si^{IV}O_2^{II}} \rightarrow \underset{0.7791}{Mg_2^{VI}Si_2^{IV}O_2^{IV}O_4^{III}} \qquad (7)$$

where the number below the chemical formula denotes its I-^{18}O index.

According to the principle of the increment method [42, 47, 50], it can be calculated that the I-^{18}O index for the special spinel $Mg^{VIII}Al_2^{VI}O_2^{VI}O_2^{IV}$ is 0.5931. This value is significantly different either from the I-^{18}O index of 0.4024 for the normal spinel $Mg^{IV}Al_2^{VI}O_4^{IV}$ or from that of 0.4612 for the inverse spinel $Mg^{VI}Al^{VI}Al^{IV}O_4^{IV}$. Likewise, the I-$^{18}O$ index for $Mg_2^{VIII}Si_2^{IV}O_6^{IV}$ is 0.7131, which differs from that of 0.7791 for the enstatite $Mg_2^{VI}Si_2^{IV}O_2^{IV}O_4^{III}$. For this reason, the special spine1 formed by the decomposition of garnet could be considerably enriched in ^{18}O relative to the normal spinel. And the resultant pyroxene could have a given $\delta^{18}O$ variation, whose lower limit may be like the oxygen isotope composition of the precursor garnet (I-^{18}O =0.7194 for pyrope [50]). Figure 5 illustrates the fractionation relationships between enstatite, forsterite and corundum as well as their chemical counterparts produced by the mineral reactions. In this context, the spinel derived from the decomposition of garnet in the mantle could be not in oxygen isotope equilibrium as usually defined with the coexisting pyroxene.

On the other hand, pyrope could be decomposed to form corundum and enstatite under the mantle conditions:

$$Mg_3^{VIII}Al_2^{VI}[Si^{IV}O_4^{IV}]_3 \rightarrow Al_2^{VI}O_3^{IV} + 3Mg^{VIII}Si^{IV}O_3^{IV} \tag{8}$$

0.7194 0.7380 0.7131

In this case, the oxygen isotope composition of Al-O structural units in pyrope can be conveyed to the decomposed product Al_2O_3. As a result, I-^{18}O value for the Al_2O_3 is 0.7380 (instead of 0.4106 for normal corundum) and that for $MgSiO_3$ 0.7131(instead of 0.7791 for enstatite). Therefore, the corundum formed by the decomposition of garnet in the mantle could even be slightly enriched in ^{18}O with respect to both enstatite and forsterite (Fig. 6). In other words, the corundum and enstatite in the mantle could be not in isotopic equilibrium as normally defined with each other due to the oxygen isotope inheritance in Al_2O_3 from the Al-O polymers of garnet. Such an oxygen isotope inheritance in the formation of "secondary" mineral has been well documented for magnates [54].

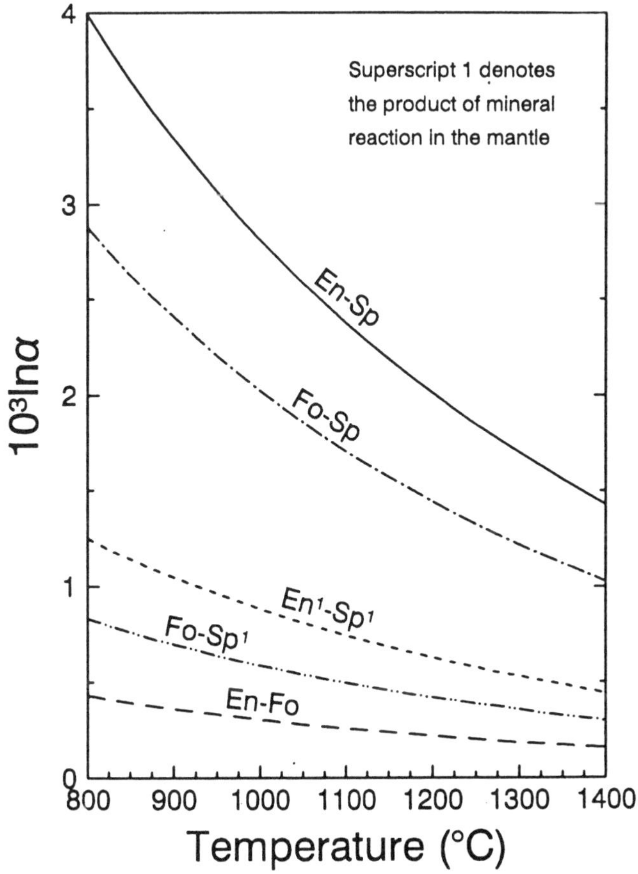

Figure 5. Temperature dependence of oxygen isotope fractionations between enstatite (En), forsterite (Fo) and spinel (Sp) as well as the products of pyrope decomposition and mineral reaction in the mantle.

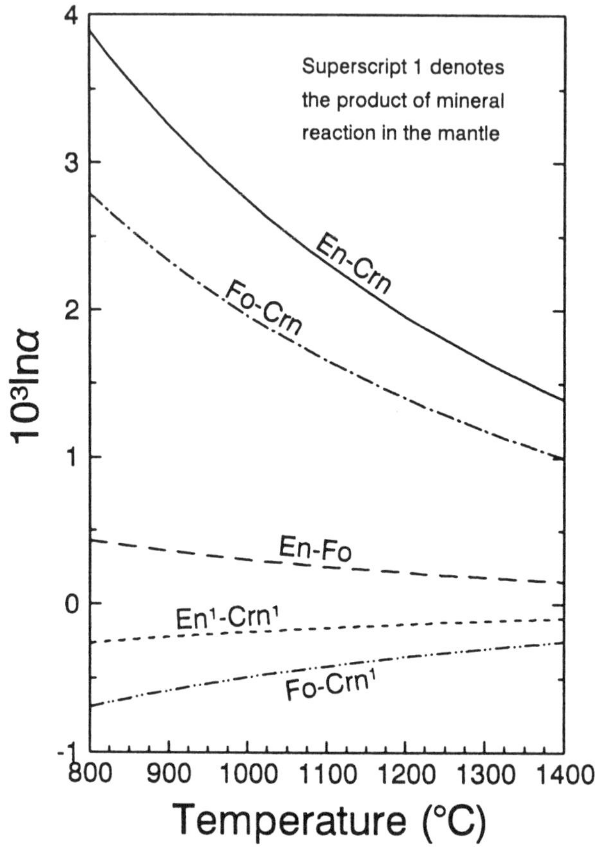

Figure 6. Temperature dependence of oxygen isotope fractionations between enstatite (En), forsterite (Fo)and corundum (Crn) as well as the products of pyrope decomposition in the mantle.

CONCLUSIONS

The oxygen isotope studies can have an insight into the mass balance of geochemical processes in the Earth's crust and mantle. Calculation of oxygen isotope fractionation in the polymorphic phases of $MgSiO_3$ and Mg_2SiO_4 suggests that the perovskite-structured silicates in the lower mantle are significantly depleted in ^{18}O relative to the spinal-structured silicates in the transition zone. They are considerably depleted in ^{18}O relative to olivine in the upper mantle at isotopic equilibrium. The oxygen isotope layering of the mantle is compatible with differences in the chemical composition and crystal structure of mineral phases at different mantle depths. Oxygen isotope fractionations among major phases in the mantle are small (mostly $< 0.5‰$ in $\delta^{18}O$), but not negligible. Although the most direct approach to determining the isotopic composition of the lower mantle has been to difficulties in sampling, it is possible to study it by means of oxygen isotope perspective. The polymorphic transition from the unstable phases to the stable phases can lead the laters to assume the oxygen isotope composition of the formers. This is a likely cause for the nonequilibrium fractionations between the mantle-derived minerals in nature.

If pyroxene in the mantle-derived assemblages is the product of polymorphic transformation from the perovskite $MgSiO_3$, the present calculations can provide a resolution to the controversy involving the negative oxygen isotope fractionations between olivine and pyroxene in the mantle assemblages. If such negative fractionations do exist in nature, it can be interpreted to indicate that the mantle assemblages were primarily derived from the lower mantle and have inherited the oxygen isotope compositions of their polymorphs at the lower mantle depth. In this regard, oxygen isotope study can be used to discover the occurrence of the lower mantle materials in the crust. This provides data for understanding the chemical geodynamics of the mantle.

Moreover, the decomposition of garnet in the mantle can result in the abnormal behavior of oxygen isotope partitioning in product spinel or corundum due to oxygen isotope inheritance in the Mg-O-Al or Al-O structural units of garnet. Specifically, the spinel could be out of isotopic equilibrium with the coexisting pyroxene, while corundum could even be slightly enriched in ^{18}O relative to the pyroxene and olivine.

Acknowledgements

This study has been supported by the funds from in the Natural Science Foundation of China and the Chinese Academy of Sciences within the framework of the project "Stable Isotope Geochemistry of the Earth's Crust and Mantle". Thanks are due to Drs. J. Hoefs, T. K. Kyser, D. Mattey, Z. D. Sharp and Y. G. Xu for their helpful comments on an early version of the manuscript.

REFERENCES

[1] D. L. Anderson. The chemical composition and evolution of the mantle: Advances in earth and planetary sciences. In: *High-Pressure Research in Geophysics*, S. Akimoto and M. H. Manghnani (eds.). pp. 301-318. K. Reidel. Hingham, Mass. (1982)

[2] D. L. Anderson. Chemical composition of the mantle, *J. Geophys. Res.* **88**, 41-52 (1983).

[3] D. L. Anderson and J. D. Bass. The transition region of the Earth's upper mantle, *Nature* **320**, 321-328 (1986).

[4] D. R. Bell and G. R. Rossman. Water in the Earth's mantle: The role of nominally angydrous minerals, *Science* **255**, 1931-1397 (1992).

[5] R. N. Clayton, J. R. Goldsmith, K. J. Karel, T. K. Mayeda and R. C. Newton. Limits on the effect of pressure on isotopic fractionation, *Geochim. Cosmochim. Acta* **39**, 1197-1201 (1975).

[6] R. N. Clayton, J. R. Goldsmith and T. K. Mayeda. Oxygen isotope isotope fractionation in quartz, albite, anorthite and calcite, *Geochim. Cosmochim. Acta* **53**, 725-733 (1989).

[7] R. N. Clayton, J. R. O' Neil and T. K. Mayeda. Oxygen isotope exchange betweem quartz and water, *J. Geophys. Res.* **77**, 3057-3067 (1972).

[8] R. N. Clayton and S. W. Kieffer. Oxygen isotopic thermometer calibrations. In: *Stable Isotope Geochemistry: A Tribute to Samuel Epstein,* H. P. Taylor, Jr., J. R. O' Neil and I. R. Kaplan (Eds.). pp. 3-10. Geochemical Society Special Publication No. **3** (1991).

[9] H. Chiba, T. Chacko, R. N Clayton and J. R Goldsmith. Oxygen isotope fractionations involving diopside, forsterite, magnetite and calcite: Application to geothermometry, *Geochim. Cosmochim. Acta* **53**, 2985-2995 (1989).

[10] G. D. Garlick, I. D. MacGregor and D. E. vogel. Oxygen isotope ratios in eclogites from kimberlites,

Science **172**, 1025-1027 (1971).

[11] B. Gautason, T. Chacko and K. Muehlenbachs. Oxygen isotope partitioning among perovskite ($CaTiO_3$), cassiterite (SnO_2) and calcite ($CaCO_3$). *Program and Abstracts of the jointed Annual Meeting of GAC and MAC*. Edmonton, A **34** (1993).

[12] R. T. Gregory and H. P. Taylor Jr. Possible non-equilibriun $^{18}O/^{16}O$ effects in mantle nodules. an altenative to the Kyser-O' Neil-Carmichael $^{18}O/^{16}O$ geotyermometer, *Contrib. Mineral. Petrol.* **93**, 114-119 (1986).

[13] R. S. Harmon, J. Hoefs and K. H. wedepohl. Stable isotope (O, S, H) relationships in Tertiary basalts and their mantle xenoliths from the Northern Hessian Depression, w. Gcrmany, *Contrib. Mineral. Pctrol.* **95**, 350-369 (1987).

[14] R. S. Harmon, P. D. Kempton, H. G. Stosch, J. Hoefs and D. A. Ionov. $^{18}O/^{18}O$ ratios in anhydrous spinel iherzolite xenoliths from the Shavaryn-Tsaram volcano, Mongolia, *Earth Planet. Sci. Lett.* **81**, 193-202 (1986/1987).

[15] R. J. Hemley and R. E. Cohen. Silicate perovskite, *Ann. Rev. Earth Planet. Sci.* **20**, 553-600 (1992)

[16] D. A. Ionov, R. S. Harmon, C. France-Lanord, P. B. Greenwood and I. V. Ashchepkov. Oxygen isotope cimposition of garnet and spinel periditites in the continental mantle: Evidence from the Vitim xenolith suite, southern siberia, *Geochim, Cosmochim. Acta* **58**, 1463-1470 (1994).

[17] E. Ito and E. Takahashi. Postspinel transformation in the system Mg_2SiO_4-$_F$e2S$_i$O4 and some geophysical implications, *J. Geophys. Res.* **94**, 10637-10646 (1989).

[18] I. Kawabe. Calculation of oxygen isotope fractionation in quartz-water system with special reference to the low temperature fractionation, *Geochim. Cosmochim. Acta* **42**, 613-621 (1978).

[19] T. Kawamoto, R. L. Hervig and J. R. Holloway. Experimental evidence for a hydrlus transition zone in the early Earth's mantle, *Earth Planet. Sci. Lett.* **142**, 587-592 (1996).

[20] P. D. Kempton, R. S. Harmon, H. G. Stosch, J. Hoefs and C. J. Hawkesworth. Open-system O-isotope behavior and trace element enrichment in the sub-Eifel mantle, *Earth Planet. Sci. Lett.* **89**, 273-287 (1988).

[21] S. W. Kieffer. Thermodynamics and lattice vibration of minerals: 5. Application to phase equilibria, isotopic fractionation, and high pressure thermodynamic proporties, Rev. *Geophys. space Phys.* **20**, 827-849 (1982).

[22] E. Knittle and R. Jeanloz. Synthesis and equation of state of (Mg, Fe)SiO_3 perovskite to over 100 gigapascals, *Sciencc* **235**, 668-670 (1987).

[23] T. K. Kyser. Stable isotopes in the continental lithospheric mantle. In: *Continental Mantle*, M. Menzies (Ed.). pp. 127-156. Clarendon Press (1990).

[24] T. K. Kyser, J. R. O' Neil and I. S. E. Carmichael. Oxygen isotope thermometry of basic lavas and mantle nodule, *Contrib. Mineral. Petrol.* 77, 11-23 (1981).

[25] T.K. Kyser, J. R. O, Neil and I. S. E. Carmichael. Genetic relations among basic lavas and ultramafic nodules: evidence from oxygen isotope compositions, *Contrib. Mineral. Petrol.* **81**, 88-102 (1982).

[26] T. K. Kyser, J. R. O' Neil and I. S. E. Carmichael. Reply to " Possible non-equilibrium oxygen isotope effects in mantle nodules. an alternative to the Kyser-P' Neil-Carmichael $^{18}O/^{16}O$ geothermometer " , *Contrib. Mineral. Petrol.* **93**, 120-123 (1986).

[27] L. G. Liu. Orthorhombic perovskite phases observed in olivine, pyroxene and garnet at high pressures and temperatures, *Phys. Earth Planet. Inter.* **11**, 289-298 (1976).

[28] H. K. Mao, T. Yagi and P. M. Bell. Mineralogy of the Earth's deep mantle: quenching experiments on mineral compositions at high pressures and temperatures, *Yearbook Carnegie Inst. Washington* **76**, 502-504 (1977)

[29] Y. Matsuhisa, J. R. Goldsmith and R. N. Clayton. Oxygen isotope fractionation in the systen quartz-albite-anorthite-water, *Geochim. Cosmochim. Acta* **43**. 1131-1140 (1979).

[30] D. Mattey, D. Lowry and C. Macpherson. Oxygen isotope composition of mantle peridotite, *Earth Planet. Sci. Lett.* **128**, 231-241 (1994).

[31] O. Muller and R. Roy. *The Major Ternary Structural Families.* Springer-Verlag, Berlin Heidelberg New York (1974).

[32] A. Navrotsky and D. J. Wiedner(Eds.). *Perouskite: A Structure of Great Interest to Gephysics and Material Science,* Washington DC, Am. Geophys. Union (1989)

[33] J. R. O' Neil. Theoretical and experimental aspects of isotopic fractionation. In: *Stable Isotopes in High Temperature Geological Processes,* J. W. Valley, H. P. Taylor Jr. and J. R. O' Neil (Eds.). pp. 1-40. Rev. Mineral. **16** (1986).

[34] V. B. Polyakov and N. N. Kharlashina. Effect of pressure on equilibrium isotopic fraction, *Geochim. Cosmochim. Acta* **58**, 4739-4750 (1994).

[35] R. Richter and S. Hoernes. The application of the increment method in comparison with expermentally derived and calculated O-isotope fractionations, *Chem. Erde* **48**, 1-18 (1988).

[36] R. Richter and D. P. McKenzie. on some consquences and possible causes of layered mantle convection, *J. Geophys. Res.* **86**, 6133-6142 (1981).

[37] A. E. Ringwood. *Composition and Petrology of the Earth's Mantle.* Mc-Graw-Hill, New York, 618pp (1975).

[38] J. M. Rosenbaum, T. K. Kyser and D. Walker. High temperature oxygen isotope fractionation in the enstatite-olivine-$BaCO_3$ system, *Geochim. Cosmochim. Acta* **58**, 2653-2660 (1994).

[39] J. M. Rosenbaum and D. Mattey. Equilibrium garnet-calcite oxygen isotope fractionation, *Geochim. Cosmochim. Acta* **59**, 2839-2842 (1995).

[40] J. M. Rosenbaum, D. Walker and T. K. Kyser. Oxygen isotope fractionation in the mantle, *Geochim. Cosmochim. Acta* **58**, 4767-4777 (1994).

[41] J. M. Rosenbaum, A. Zindler and J. L. Rubenstone. Mantle flrids: Evidence from fluid inclusions, *Geochim. Cosmochim. Acta* **60**, 3229-3252 (1996).

[42] H. Schiitze. Der Isotopenindex-eine Inkrementmethode zur naherungsweisen Berechnung von Isotopenaustauschgleichgewichten zwischen kristallinen Substanzen, *Chem. Erde* **39**, 321-334 (1980).

[43] Y. Shiro and H. Sakai. Calculation of the reduced partition function ratios of α-, β-quartzs and calcite, *Bull, Chem. Soc. Japan* **45**, 2355-2359 (1972).

[44] J. R. Smyth. A crystallographic model for hydrous wadsleyite (β-Mg_2SiO_4): An ocean in the Earth's interior? *Am. Mineral.* **79**, 1021-1024 (1994).

[45] A. B. Thompson. Water in the Earth's upper mantle, *Nature* **358**, 295-302 (1992).

[46] P. J. Wyllie. Magma genesis. plate tectonics, and chemical differentiation of the Earth, *Rev. Geophys.* **26**, 370-404 (1988).

[47] Y. F. Zheng. Calculation of oxygen isotope fractionation in metal oxides, *Geochim. Cosmochim. Acta* **55**, 2299-2307 (1991).

[48] Y. F. Zheng. Oxygen isotope fractionation ix wolframite, *Eur. J. Mineral.* **4**, 1331-1335 (1992).

[49] Y. F. Zheng. Calculation of oxygen isotope fractionation in hydroxyl-bearing silicates, *Earth Planet. Sci. Lett.* **120**, 247-263 (1993).

[50] Y. F. Zheng. Calculation of oxygen isotope fractionation in anhydrous silicate minerals, *Geochim. Cosmochim. Acta* **57**, 1079-1091 (1993).

[51] Y. F. Zheng. Oxygen isotope fractionation in SiO_2 and Al_2SiO_5 polymorphs: effect of crystal structure, *Eur. J. Mineral.* **5**, 651-658(1993).

[52] Y. F. Zheng. Oxygen isotope fractionation in perovskite, *Chinese Science Bulletin* **39**, 1989-1993 (1994).

[53] Y. F. Zheng. Oxygen isotope fractionation in TiO_2 Polymorphs and application to geothermometry of eclogites, *Chinese J. Geochem.* **14**, 1-12 (1995).

[54] Y. F. Zheng. Oxygen isotope fractionation in magnetites; structural effect and oxygen inheritance, *Chem. Geol.* **121**, 309-316 (1995).

Proc. 30th Intern. Geol. Congr. Vol. 1, pp.177-185
Wang *et al.* (Eds)

Mesoproterozoic Microbiotas of the Northern Hemisphere and the Meso-Neoproterozoic Transition

V. N. SERGEEV
Geological Institute of RAS, 109017, Pyzhevskii per., 7, Moscow, Russia.

Abstract

The silicified peritidal carbonates of the Mesoproterozoic (Lower to Middle Riphean) Kotuikan and Yusmastakh formations of the Billyakh Group, Anabar Uplift, Siberia, contain diverse microfossil assemblages dominated by fossilized akinetes of the Genus *Archaeoellipsoides* and associated short trichomes and enthophysalidacean cyanobacteria. All these fossils bear close counterparts with living cyanobacteria. However, similar *Archaeoellipsoides*-dominated silicified microbiotas were found only in Mesoproterozoic formations in Canada, China and elsewhere, but are missing in the Neoproterozoic deposits. This biostratigraphic and evolutionary paradox seems to reflect ecological effects of the newly evolved eukaryotes and the changing chemical composition of sea water and sea floor substrate. The morphologically complex unicellular eukaryote microorganisms are missing in the Mesoproterozoic microbiotas.

The principal biological event occuring at the Meso-Neoproterozoic transition was the explosive radiation of eukaryote microorganisms and the consequent appearance of morphologically complex microfossils in the Neoproterozoic. However, this transition was not instantaneous, and many taxa widely distributed in the Neoproterozoic are known in the latest Mesoproterozoic. Nonetheless, the composition of conservative communities in the restricted peritidal environments has also changed. This perhaps reflects the "hidden" expansion of the eukaryote microorganisms and the changing substrate conditions.

Keywords: microfossils, Proterozoic, Siberia, Billyakh Group, cyanobacteria, eukaryotes.

INTRODUCTION

During the last 30 years significant progress has been made in the understanding of the nature and stratigraphic significance of Proterozoic microorganisms. Numerous discovery of organic-walled and silicified microbiotas have revealed the high diversity and complexity of Precambrian microorganisms. Specially impressive results have been obtained in the last decade, and the discovery of remnants of morphologically complex eukaryote unicellular organisms in Neoproterozoic rocks has provided a good means of biostratigraphic correlation. However, much less is known about the composition of Mesoproterozoic (Early-Middle Riphean) microbiotas when compared with the Neo- and even the Palaeoproterozoic fossils. Mesoproterozoic microbiotas were found in the early studies of Precambrian fossils. It was only in recent years that numerous well preserved Mesoproterozoic microbiotas were found from Siberia and elsewhere in the northern Hemisphere, which reveal the high diversity of the Mesoproterozoic microbial life. Perhaps the most typical Mesoproterozoic assemblages are those found from silicified peritidal carbonates. They are dominated by the ellipsoidal Genus *Archaeoellipsoides*—fossilized akinetes of nostocalean cyanobacteria, and the diverse, predominantly short trichomes interpreted as cyanobacterial hormogonia, along with the coccoidal chroococcacean and entophysalidacean cyanobacteria.

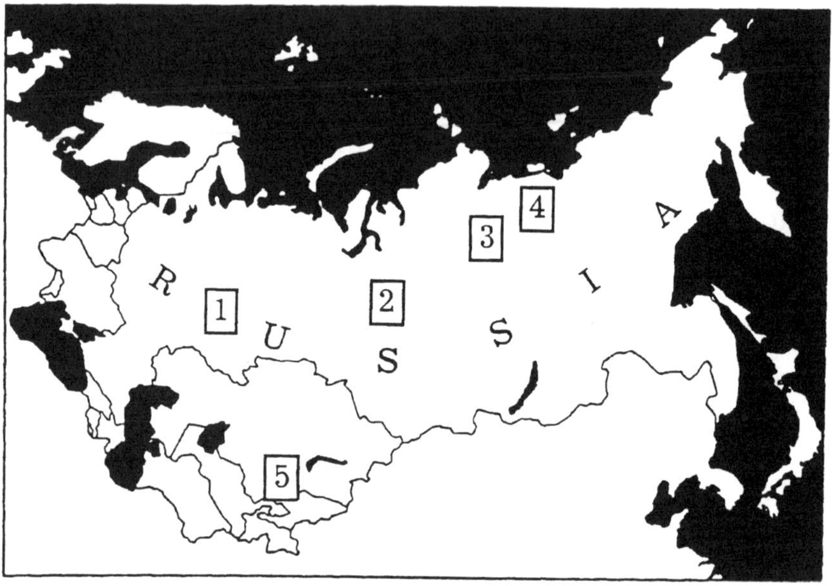

Figure 1. Locality map of the main Meso-Neoproterozoic microbiotas of Russia and Kazakhstan. Rectangles show the main fossiliferous localities studied. 1. Southern Ural Mountains. 2. Turukhansk Uplift. 3. Anabar Uplift. 4. Olenek Uplift. 5. Lesser Karatau Range.

COMPOSITION OF THE MESOPROTEROZOIC MICROBIOTAS

The most representative and diverse Mesoproterozoic (Lower-Middle Riphean) assemblages of silicified microfossils came from Siberia, China, Northern Canada and other parts in northern Hemisphere. One of the best microbiota containing exceptionally well preserved microfossils came from silicified peritidal carbonates of the Mesoproterozoic Kotuikan and Yusmastakh formations in Anabar Uplift, northern Siberia [1, 20, 23, 27]. The Anabar Uplift is located in the northern part of the Siberian Platform (Fig. 1). Here, unmetamorphosed Mesoproterozoic deposits of the Mukun and Billyakh groups rest with angular unconformity on the Archaean and Palaeoproterozoic crystalline complex exposed in the core of the structure and are disconformably overlain by the Vendian (Neoproterozoic III) and Cambrian deposits. The Billiakh Group is subdivided into three formations (in ascending order): the siliciclastic Ust'-Ilya, the predominantly carbonate Kotuikan, and the Usmastakh, the latter formations again each subdivided into two members. Lenses and nodules of gray to black cherts, some fossiliferous, are common in the carbonates. Carbon isotope profile throughout the Kotuikan Formation is consistent with the stromatolite sequences and the radiometric data that indicate an early Mesoproterozoic (Early Riphean) age[10]. The distinctive precipitation fabrics in the upper Kotuikan carbonates are also well known in Mesoproterozoic peritidal successions, but are rare or absent in comparable Neoproterozoic rocks [5, 12]. K-Ar and Rb-Sr age data obtained from glauconite in the basal part of the Ust'-Ilya Formation yields ages of 1459±10 and 1483±5 Ma respectively [3]. Carbon isotopes for Yusmastakh carbonates differ little from those of the Kotuikan Formation. The silicified microfossil assemblages are also quite similar. These data as well as the stromatolites strongly suggest a Mesoproterozoic age at least for the fossiliferous lower member of the Yusmastakh Formation.

The Kotuikan and Yusmastakh microfossil assemblages contain 33 species belonging to 17 genera [23]. Differences between the assemblages from the two formations are minor. Therefore, all silicified fossils are considered as one unit. This assemblage is dominated by large and distinctive ellipsoidal akinetes of nostocalean cyanobacteria (Genus *Archaeoellipsoides*) and by problematic spheroidal unicellular forms (*Myxococcoides grandis*); both are allochthonous and presumably planktonic. The assemblage also includes distinctive mat-forming scytonematacean (Genus *Circumvaginalis*) and entophysalidalean cyanobacteria, diverse short trichomes interpreted as cyanobacterial hormogonia or germinated akinetes, rarely with longer trichomes, and several types of colonial unicellular forms. These populations are scattered more or less randomly in originally thin micritic event beds and in sea-floor cements that abound in silicified Kotuikan horizons. Many taxa in the Kotuikan-Yusmastakh assemblage are long-ranged. They are abundant in the Palaeoproterozoic (Lower-Middle Riphean) successions and continue upwards into the Neoproterozoic (Late Riphean). Nevertheless, the overall character of the assemblage is distinctly Mesoproterozoic, and its major features are shared by the coeval forms from Canada, China, and India. Similar microbiotas occur in the Dismal Lake Group of Canada [8], the Gaoyuzhuang and Wumishan formations of China [30, 31], and elsewhere. Similar assemblages of silicified microfossils with *Archaeoellipsoides*, but not dominating, occur in the Middle Riphean Debengda Formation of the Olenek Uplift, northern Siberia [21], and the Kheinjua Formation of Central India. These assemblages are dominated by enthophysalidacean cyanobacteria and tufted mats of empty sheaths of oscillatoriacean cyanobacteria assigned to the Genus *Siphonophycus*.

The Billyakh and other Mesoproterozoic silicified prokaryote assemblages present an apparent biostratigraphic and evolutionary paradox. These peritidal assemblages contain abundant and morphologically distinctive microfossils that bear close counterparts with living cyanobacteria. But many of these forms are rare or absent in cherts from Neoproterozoic tidal flats. The Mesoproterozoic/Neoproterozoic (Middle-Upper Riphean) differences in permineralized prokaryotes may be partially a reflection of the evolution of eukaryotes. For example, freshened peritidal pools inferred to be the habitat of *Archaeoellipsoides*-producing cyanobacteria are today dominated by eukaryote algae. It is reasonable to suppose that the radiating Neoproterozoic eukaryotes displaced the previously dominating nostocalean cyanobacteria. If so, one may expect that the displacement was gradual and statistical rather than sudden and absolute. Indeed, *Archaeoellipsoides* does occur locally in Neoproterozoic cherts in Chichkan Formation, Lesser Karatau Range, southern Kazakhstan [18, 19]. The Genus *Brevitrichoides*, an organic-walled ellipsoid form found in some Neoproterozoic shales [25] may also be an *Archaeoellipsoides*-like akinete. Probably, a high level of CaCO3 supersaturation, evidenced by the abundance of precipitation structures in many

Mesoproterozoic carbonates, triggered the full transformation of *Anabaena*-like filaments into chains of akinetes. This may explain the abundance of *Archaeoellipsoides* akinetes in the Mesoproterozoic deposits. Chains of akinetes are present in the Kotuikan (Fig. 14.1-14.3, and 14. 16) and Wumishan formations ([31], Fig. 8.B) and probably in the Dismal Lakes Group ([8], figs. 13.B, 13.E, 13.F).

Other silicified Mesoproterozoic assemblages differ in composition from those discussed above. They are represented predominantly by empty sheaths of hormogonian cyanobacteria (Genus *Siphonophycus*), remnants of chroococcacean and entophysalidacean cyanobacteria and sometimes by microfossils of other types. Microbiotas of this general composition have been described from the Satka and Avzyan formations of the Southern Ural Mountains, the Sukhaya Tunguska Formation of the Turukhansk Uplift, Northern Siberia, and the Bylot Supergroup of Canada [7, 14, 16, 19, 21]. These microbiotas are similar in composition to many others of Palaeo- and Neoproterozoic age from restricted peritidal environments. Existence of two types of Mesoproterozoic microbiotas from the peritidal carbonates can be related to specific environments, probably chemical composition of sea water and composition of sea floor substrate. Assemblages of organic-walled microfossils in Mesoproterozoic shales often contain large, simple spheroid and coiled spiral forms that are probably remnants of eukaryote organisms; *e.g.* in the Ust'-Ilya and Kotuikan formations of the Anabar Uplift, in the Belt Supergroup of North America, and in the Gaoyuzhuang Formation of North China.

Billyakh fossils are among the most diverse and best preserved assemblages yet described from Mesoproterozoic rocks. They constitute a true Lagerstaetten that reveals a relatively wide and clear view of the Mesoproterozoic life. Perhaps the most striking aspect of the view they afford is the absence of morphologically complex eukaryotes. There are no cellularly preserved green, red, or chromophyte algae; no vase-shaped protists, and no acanthomorphic acritarchs, although a quality of preservation and palaeoenvironmental sampling range will yield invariably such fossils in Neoproterozoic rocks. This strongly suggests that taphonomic and/or sampling bias cannot explain the absence of morphologically complex eukaryote fossils in rocks older than 1000-1200 Ma. Thus the Billyakh biota supports the idea that the major groups of "higher" eukaryotes began their explosive diversification near the Mesoproterozoic/ Neoproterozoic boundary [9, 19]. This of course does not indicate that nucleated cells were absent before this time. After all, molecular phylogenies imply that the "big bang" of eukaryote evolution occurred relatively late in the history of the group [24]. Billyakh cherts contain numerous problematic spheroids (*Myxococcoides grandis*, *Phanerosphaerops magnicellularis*) that could be eukaryote, while associated shales contain abundant relatively large leiosphaeroid acritarchs that are probably eukaryotes. Actritarchs of probable eukaryote origin and macroscopic impressions and compressed remains that are surely eukaryotes occur in late Palaeoproterozoic and Mesoproterozoic all around the world [17]. However, eukaryote diversity is relatively low in these assemblages, and no morphological features seem to link any of these fossils to extant algal or protozoan phyla. As suggested by Knoll [9], at least some of these fossils may have belonged to extinct groups of early eukaryotes.

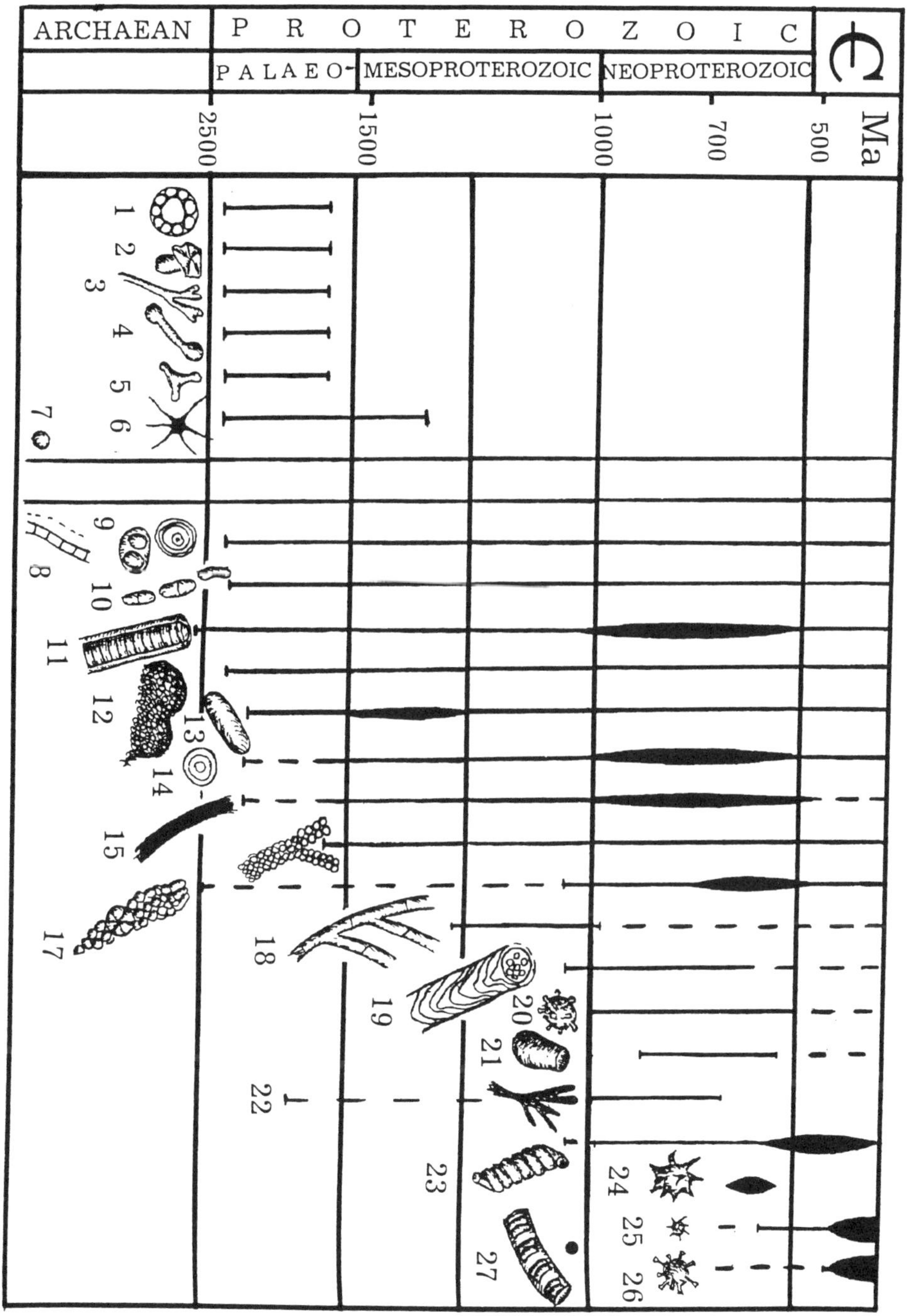
ARCHAEAN
P R O T E R O Z O I C
PALAEO-
MESOPROTEROZOIC
NEOPROTEROZOIC
Ma
2500
1500
1000
700
500

Figure 2. The distribution of the main morphological groups of microfossils in Precambrian. 1-6. The morphologically complex prokaryote microorganisms from the iron-banded formations of Gunflint type. 7. Small (less than 10 μm in diameter) single spherical microfossils. 8. Thin (less than 10 μm in diameter) filamentous microorganisms. 9. Unicellular chroococcalean *Gloeocapsa*-like cyanobacteria (Genus *Gloeodiniopsis*). 10. Unicellular chroococcalean *Synechococcus*-like cyanobacteria (Genus *Eosynechococcus*). 11. The remnants of oscillatoriacean cyanobacteria (genera *Siphonophycus, Oscillatoriopsis* and *Palaeolyngbya*). 12. Enthophysalidacean cyanobacteria (Genus *Eoentophysalis*). 13. Akinetes of nostocalean or stigonematalean cyanobacteria (the ellipsoidal microfossils of Genus *Archaeoellipsoides*). 14. Large spherical *Chuaria*-like microfossils. 15. Remnants of large filamentous microorganisms of possibly eukaryote origin. 16. Remnants of *Stigonema*-like cyanobacteria.. 17. Remnants of *Pleurocapsa*-like cyanobacteria. 18. Branching filaments of Genus *Palàeosiphonella*. 19. Stalked cyanobacterium *Polybessurus*. 20. Acanthomorphic acritarchs of Genus *Trachyhystrichosphaera*. 21. Vase-shaped microfossils of Genus *Melanocyrillium*. 22. Branching filaments of green algae (genera *Proterocladus* and *Ulophyton*). 23. Spiral cylindrical microfossils of Genus *Obruchevella*. 24. Large acanthomorphic acritarchs from assemblages of *Pertatataka* type. 25. Small acanthomorphic acritarchs of Genus *Micrhystridium*. 26. Acanthomorphic acritarchs of genera *Skiagia* and *Baltisphaeridium*. 27. Remnants of red bangiophycean algae.

THE MESO-NEOPROTEROZOIC TRANSITION

The principal biological event at Meso-Neoproterozoic (Middle-Late Riphean) transition was the explosive radiation of eukaryote microorganisms and consequent the appearance of morphologically more complex microfossils in Neoproterozoic deposits, including acanthomorphic acritarchs, branched filaments of green algae, vase-shaped protists, thalli of bangiophyte red algae, *etc.* (Fig. 2). This radiation about 1000 Ma ago is now named the "Neoproterozoic revolution". The transition was however not instantaneous. Some latest Mesoproterozoic microbiotas contain microfossil remnants typical of Neoproterozoic, such as red algae in the Hunting Formation of Canada, and even spiny acritarchs in the Sukhaya Tunguska Formation, Siberia [16] and the Ruyang Group of China[29]. In this article, I am not going to analyze this "eukaryote revolution" (for discussion of this problem see [9, 11, 19]). I would rather try to analyze the changes in the cyanobacterial communities. In spite of the evolutionary conservation of blue-green algae, the composition of Neoproterozoic cyanobacterial communities was different from earlier assemblages. There were at least two new morphological types of cyanobacteria now known from Neoproterozoic or latest Mesoproterozoic: the cylindrical spiral form of Genus *Obruchevella* and the stalked cyanobacterium *Polybessurus*. At least the appearance of the spiral cyanobacteria in the Neoproterozoic can be explained from the data of molecular biology, because the modern counterpart of this form, Genus *Spirulina*, is one of the two morphologically complex cyanobacterial taxa, whose systematic position based on morphology does not coincide with sequences of the 16 ribosomal RNA [26].

The general composition of the conservative microcommunities from peritidal Neoproterozoic carbonates has changed too. Most silicified microbiotas of this type are dominated by empty sheaths of the hormogonian cyanobacteria Genus *Siphonophycus* and different kinds of coccoidal microfossils. Because of morphological similarity in the primitive eukaryote and prokaryotes, it is difficult to decipher the nature of these morphologically simple coccoidal microorganisms. At least part of them are probably remnants of green algae [4]. So, the "hidden" expansion of eukaryote microorganisms is probably at least partially responsible for the change in composition of the conservative communities of microorganisms from restricted environments. However, neither the idea of competing eukaryotes nor that of taphonomic bias can completely explain the observed decline of *Eoentophysalis* at the Mesoproterozoic-Neoproterozoic transition. *Eoentophysalis*-dominated assemblages are widely distributed in Palaeo- and Mesoproterozoic strata [6, 15, 20, 23, 30], but they have not been reported from Neoproterozoic marine rocks, where *Eoentophysalis* occurs as scattered colonies in assemblages dominated by other organisms [13, 19]. This seems anomalous, as the entophysalid cyanobacteria are widespread mat builders in recent intertidal environments [2]. On the Kotuikan and Yusmastakh tidal flats, *Eoentophysalis* colonies preferentially colonized hard substrates. The distinctive nature of Palaeo- and Mesoproterozoic peritidal environments, with their widespread sea-floor cements, may have facilitated *Eoentophysalis* for getting to dominance [11].

CONCLUSION

The Mesoproterozoic microbiotas from peritidal carbonates were dominated by akinetes of Genus *Archaeoellipsoides*, associated with the short trichomes and enthophysalidacean cyanobacteria of Genus *Eoenthophysalis*. However, there are other types of Mesoproterozoic microbiotas which were dominated by coccoidal and filamentous cyanobacteria. This taxonomic difference in composition of the microbiotas is related to specific environments. The eukaryotes probably were widespread in Mesoproterozoic time, but morphologically complex forms are missing in most representative microbiotas of this age. The main event at the Meso-Neoproterozoic transition was the radiation of eukaryote microorganisms and appearance of morphologically complex forms, especially the acanthomorphic acritarchs. But this transition between Meso- and Neoproterozoic is gradational, with many microfossils typical of Neoproterozoic age first appearing in late Mesoproterozoic. However, the two eras were also different in composition of preserved cyanobacteria, perhaps reflecting evolving substrate conditions and the ecological effects of newly evolved eukaryote benthos.

Acknowledgements

I thank the Geohost Organization Committee of the 30th International Geological Congress for the financial support of my trip to the Congress. I thank Professor A.H. Knoll of Harvard University, USA, and Professor M.A. Semikhatov, Academician of the Russian Academy of Sciences for critical reading of the manuscript. Preparation of this article was supported by Grant of Russian Fund for Basic Research (RFBR) and by Grant of PalSIRP of the Paleontological Society of America.

REFERENCES

[1] V. K. Golovenok and M. Yu. Belova. Riphean microbiotas in cherts of the Billyakh Group on the Anabar Uplift, *Palaeontologicheskyi Zhurnal*, **4**, 20-30 (English version) (1984).

[2] S. Golubic and H. J. Hofmann. Comparison of Holocene and mid-Precambrian Entophysalidaceae (Cyanophyta) in stromatolitic algal mats: cell division and degradation, *Journal of Palaeontology* **50**, 1074-1082 (1976).
[3] I. M. Gorokhov, M. A. Semikhatov, E. P. Drubetskoi *et al.* Rb-Sr and K-Ar vozrast osadochnyh geochronometrov nizhnego rifeya Anabarskogo massiva (Rb-Sr and K-Ar ages of sedimentary geochronometers from the Lower Riphean deposits of the Anabar Massiv), *Izvestia AN SSSR, seriya geologicheskaya* 7, 17-32 (1991).
[4] J. W. Green, A. H. Knoll and K. Swett. Microfossils from silicified stromatolithic carbonates of the Upper Proterozoic Limestones-Dolomite 'Series', Central East Greenland, *Geological Magazine* **119**, 527-551 (1989).
[5] J. P. Grotzinger. New vies of old carbonate sediments, *Geotimes* **38:9**, 12-15 (1993).
[6] H. J. Hofmann. Precambrian microflora, Belcher Island, Canada: significance and systematics, *Journal of Palaeontology* **50**, 1040-1073 (1976).
[7] H. J. Hofmann and C. D. Jackson. Shelf-facies microfossils from the Uluksan Group (Proterozoic Bylot Supergroup), Baffin Island, Canada, *Journal of Palaeontology* **65**, 361-382 (1991).
[8] R. J. Horodyski and J. A. Donaldson. Microfossils from the Middle Proterozoic Dismal Lakes Group, Arctic Canada, *Precambrian Research* **11**, 125-159 (1980).
[9] A. H. Knoll. The Early Evolution of Eukaryotes: A Geological Perspective, *Science* **256**, 622-627 (1992).
[10] A. H. Knoll, A. J. Kaufman, and M. A. Semikhatov. The carbon isotopic composition of Proterozoic carbonates: Riphean successions from north-eastern Siberia (Anabar Massif, Turukhansk Uplift*), American Journal of Science* **295**, 823-850 (1995).
[11] A. H. Knoll and V. N. Sergeev. Taphonomic and Evolutionary changes across the Mesoproterozoic-Neoproterozoic Transition, *Neues Jahrbuch für Geologie und Paläontologie, Abhandlungen* **195**, 289-302 (1995).
[12] A. H. Knoll and K. Swett. Carbonate deposition during the Late Proterozoic Era: an example from Spitsbergen, *American Journal of Sciences* **290-A**, 104-132 (1990).
[13] A. H. Knoll, K. Swett and J. Mark. Paleobiology of a Neoproterozoic tidal flat/lagoonal complex: the Draken Conglomerate Formation, Spitsbergen, *Journal of Palaeontology* **65**, 531-570 (1991).
[14] I. N. Krylov and V. N. Sergeev. Rifeiskie mikrofossilii Yuzhnogo Urala v raione goroda Kusa (Riphean microfossils from the vicinity of town Kusa, southern Ural Mountains). In: *Stratigraphia, litologia i geochimia verchnego dokembria Uzhnogo Urala i Priuralia* (Stratigraphy, lithology and geochemistry of the Upper Riphean of the southern Urals and Near Ural). pp. 95-109. Ufa. Izdatelstvo Bashkirskogo filiala AN SSSR (1986).
[15] D. Z. Oehler. Microflora of the middle Proterozoic Balbirini Dolomite (McArthur Group) of Australia, *Alcheringa* **2**, 269-310 (1978).
[16] P. Yu. Petrov, M. A. Semikhatov and V. N. Sergeev. Development of the Riphean carbonate platform and distribution of silicified microfossils: the Sukhaya Tunguska Formation, Turukhansk Uplift, Siberia, *Stratigraphy and Geological Correlation* **3**, 79-99 (1995).
[17] J. W. Schopf and C. Klein (Eds.). *The Proterozoic Biosphere*. Cambridge University Press, Cambridge (1992).
[18] V. N. Sergeev. Microfossils from transitional Precambrian-Phanerozoic strata of Central Asia, *Himalayan Geology* **13**, 269-278 (1989).
[19] V. N. Sergeev. *Okremnennye mikrofossilii dokembrya i kembrya Urala i Sredney Azii* (Silicified microfossils from the Precambrian and Cambrian deposited of the southern Ural Mountains and Middle Asia). Nayka, Moscow (1992).
[20] V. N. Sergeev. Silicified Riphean microfossils of the Anabar Uplift, *Stratigraphy and Geological Correlation* **1**, 264-278 (1993).
[21] V. N. Sergeev. Microfossils in cherts from the Middle Riphean (Mesoproterozoic) Avzyan Formation, southern Ural Mountains, Russian Federation, *Precambrian Research* **1**, 231-254 (1994).
[22] V. N. Sergeev, A. H Knoll, S. P. Kolosova and P. N. Kolosov. Microfossils in cherts from the Mesoproterozoic Debengda Formation, Olenek Uplift, Northeastern Siberia, *Stratigraphy and Geological Correlation* **2**, 23-38 (1994).
[23] V. N. Sergeev, A. H. Knoll and J. P. Grotzinger. Paleobiology of the Mesoproterozoic Billyakh Group, Anabar Uplift, Northeastern Siberia, *Palaeontological Society Memoir* **39** (1995).
[24] M. L. Sogin, G. H. Gunderson, H. J. Elwood, R. A. Alonso and D. A. Peattie. Phylogenetic meaning of the kingdom concept: An unusual ribosomal RNA from Giardia lamlia, *Science* **243**, 75-77 (1989).
[25] T. V. Yankauskas *et al. Mikrofossilii dokembrya SSSR* (Precambrian microfossils of the USSR). Nauka, Leningrad (1989).

[26] A. Wilmotte. and S. Golubic. Morphological and genetic criteria in the taxonomy of Cyanophyta/ Cyanobacteria, *Algological Studies* **64**, 1-24 (1991).
[27] M. S. Yakschin. *Vodoroslevaya mikrobiota nizhnego rifeya Anabarskogo podnyatia* (Algal microbiota from the Lower Riphean deposits of Anabar uplift). Novosibirsk, Nayka, Sibirskoe Otdelenie. **29** (1991).
[28] P. Zhang. and S. Gu. Microfossils from the Wumishan Formation of the Jixian System in the Ming Tombs, Beijing, China, *Acta Geologica Sinica* **60**:13-22 (1986).
[29] Y. Yan and G. Liu G. Discovery of acanthomorphic acritarchs from the Baicaoping Formation in Yongli, Shanxi, and its geological significance, *Acta Micropalaeontologica Sinica* **9**, 278-281 (1992).
[30] Y. Zhang. Proterozoic stromatolite microfloras of the Gaoyuzhuang Formation (Early Sinian: Riphean), Hebei, China, *Journal of* Palaeo*ntolology* **55**, 485-506 (1981).
[31] Y. Zhang. Stromatolitic microbiota from the Middle Proterozoic Wimishan Formation (Jixian Group) of the Ming Tombs, Beijing, China, *Precambrian Research* **30**, 277-302 (1985).

Proc. 30th Intern. Geol. Congr., Vol. 1, pp. 187-199
Wang *et al.* (Eds)

Histological Study on the Neoproterozoic Organism's Fossil Remains: Implications for Origin of Multicellularity and Sexuality

ZHANG YUN
College of Life Science , Peking University, Beijing, 100871, P. R. China

Abstract

The histological study on the three dimensionally well-preserved multicellular fossils embedded in the phosphorites of the Neoproterozoic Doushantuo Formation in central Guizhou Province, south China, provides some details about the major evolutionary innovations that occurred prior to Ediacaran radiation and Cambrian "explosion". A comparative histological study on the fossil forms in different grades of multicellular organization have offered the information that may be conductive to understanding the evolutionary transition and development of multicellularity. The study have also provided the evidence for sexual reproduction modes of the early multicellular organisms. It was possible that, the sessile thallophytes (*e.g. Thallophyca*), with polarized and directional growing thallus, possibly evolved from a prostrate colonial ancestral form, by changes of habit and growth pattern. The benthic photoautotrophic organisms adapted to a turbulent habitat by increasing stability, this promoted the evolutionary change of habits from prostrate to sessile, and spurred on tissue differentiation and structural complication. The evolutionary transition from prostrate colonial forms to sessile thallophytes had been possibly also promoted by competition for light. The prostrate colonies might expand their surface for light acceptance by upward growth and branching, thus resulted in polarization and differentiation. The study demonstrates that, some of the thallophytes possess the structures that are comparable to female and male reproductive organs. These thallophytes showed both sexuality and multicellularity, they have a certain level of tissue-differentiation and possibly a diploid-dominant sexual life cycle. Development of sexual reproduction in multicellular organisms led to an increase of genotype and phenotype variability, this preludes the remarkable diversification of the metaphyte and metazoans at the beginning of Phanerozoic.

Keywords: *multicellularity, sexuality, histological study,* Neo*proterozoic, macroscopic fossils.*

INTRODUCTION

Attainment of multicellular organization and sexuality, as well as complex poly-phased sexual life cycle, are among the most important evolutionary events in the early life history, and are the essential prerequisites for the early Phanerozoic "explosion" of metazoans and metaphytes. Although Proterozoic macroscopic fossils have been discovered and described recently in China and other areas of the world, however, so far we still know little about the evolutionary transition from the microscopic, unicellular to macroscopic, multicellular organizations, because most of the discovered Precambrian macrofossils are two-dimensionally preserved compressions or impressions [1, 2, 4-6], lacking microsctructural information, or occasionally having degraded microstructures [14, 19]. The study on the three dimensionally well-preserved fossils embedded in the phosphorites of the Neoproterozoic Doushantuo Formation in central Guizhou Province, south China [16-18] have offered the possibility to understand some details about the major evolutionary innovations that occurred prior to Ediacaran radiation and Cambrian "explosion".

The Doushantuo fossil assemblage embraces diverse organism's remains with well-preserved micro- structures, the fossils harbor morphological information from cellular to tissue-organ organizations. The fossil assemblage consists of varied forms of plankton and benthose, which show different levels of morphological complexity. The comparative study on these fossil forms that shown different grades of multicellular organization, may help to understanding evolutionary transition and development of multicellularity. Some of the fossils possess distinctive structures that are recognized as possible sexual and asexual reproductive organs, comparable to that of the modern counterparts. The study provided evidence for the sexual reproduction modes of the early thallophytes, and the information about relationship between multicellularity and sexuality in the early evolution. The Doushantuo fossils were preserved in the huge Neo-proterozoic phosphate deposits, which formed just after the late Neoproterozoic glacial epoch (called as Nantuo glacial epoch in China) and before the Ediacaran radiation of metazoans. The fossil-bearing phosphorites of the Doushantuo Formation might harbor profound mystery and meaningful information about the global environmental changes and the related biological events during the terminal Proterozoic, that await for bringing to light. In addition, the Doushantuo fossils are also valuable for palaeoecologial and taphonomical studies, because of their diverse morphology and unique preservation.

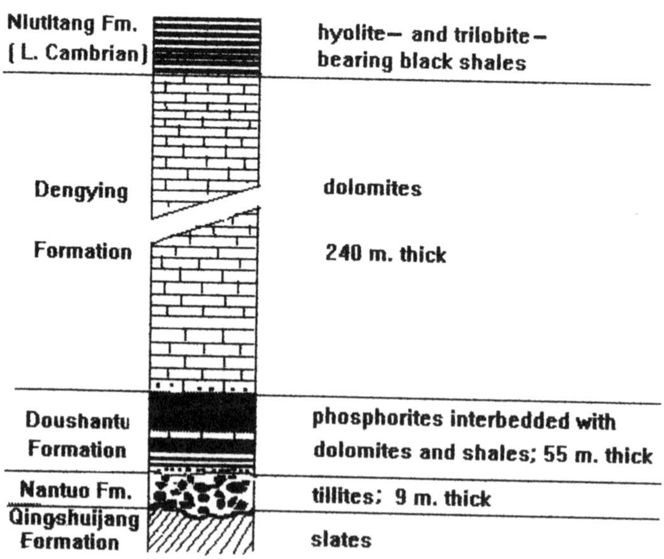

Figure 1. Stratigraphic column of the Neoproterozoic sequence in Wengan Phosphate Mining area, Guizhou Province, south China. The studied fossils came from the phosphorite beds of the Doushantuo Formation, which underlie the Dengying dolomites and overlie the Nantuo tillites. The former has been regarded as Ediacaran in age, as some Ediacaran fossils were found from the equivalent horizon in the area of Yangzi Gorges.

THE DOUSHANTUO PHOSPHORITES AND THE FOSSIL PRESERVATION

The samples for this study were collected from the phosphorite beds of the Doushantuo Formation in the northeast part of the Wengan Phosphate Mining Area (latitude: 27° 05' N; longitude: 107° 25' E) in central Guizhou Province, south China. The stratigraphy of the Neo- Proterozoic rocks of this area have been described previously [16, 17]. The fossil-bearing phosphorites of the Doushantuo Formation underlies the Dengying Dolomites and overlays the Nantuo tillites (Fig. 1). The lower part of the Dengying Formation has been regarded as deposits of Ediacaran age [3, 11]. The underlying Nantuo tillites has been considered as the deposits of late Neoproterozoic glacial epoch with an estimated age of 700 Ma [12]. The Doushantuo phosphorites have been measured as 620 to 690 Ma [15], and the fossils from the phosphorites have an estimated age older than 570 Ma (dating for Ediacaran fauna) and younger than 700 Ma.

GRADES OF MULTICELLULARITY

The macroscopic fossils from the Doushantuo phosphorites showed different levels of morphological complexity and different grades of multicellularity: from the simple colonial forms to the thallophytes with differentiated tissues and organs.

Colonial Forms

The colonial forms are the most simple multicellular organisms, composed of aggregated non-differentiated cells. Two types of colonial organisms have been observed. The first type represented by the colony consisting of irregularly aggregated monomorphic cells, with indefinite shape and size (Fig. 2A), The colonies possibly propagated by asexual way: a new colony produced from a "mother" colony by segregation, as seen in Fig. 2B. The second type of colonial forms is represented by *Wengania* [16]. The colony of *Wengania* consists of regularly arranged monomorphic (rectangular) cells, with a definite shape (rounded shape) and certain range of size (Fig. 3). The colonies possibly conducted asexual reproduction: a small doughter colony might be produced and separated directly from a large "mother" colony, as seen in Fig. 4.

Thalloid Forms with Differentiated Tissues

The thalloid forms of the Doushantuo phosphorites show relatively higher level of organization, they have differentiated vegetative tissues and reproductive organs.

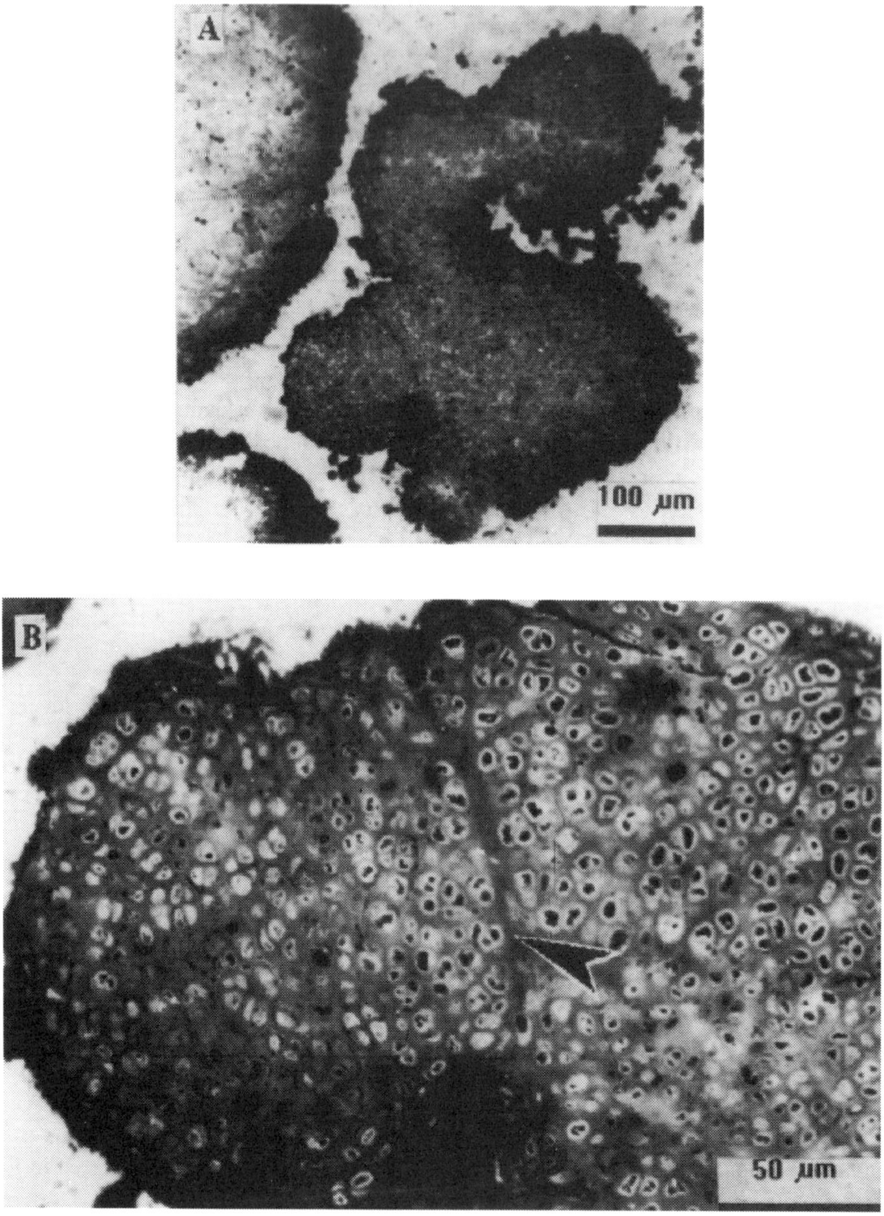

Figure 2. The colonial multicellular organisms from the Doushantuo phosphorites. The colony with indefinite shape and size, consisting of aggregated cells (A and B); it was possibly propagated by segregation: a new colony might be produced by constriction and segregation of the "mother " colony along the furrow (arrow in B).

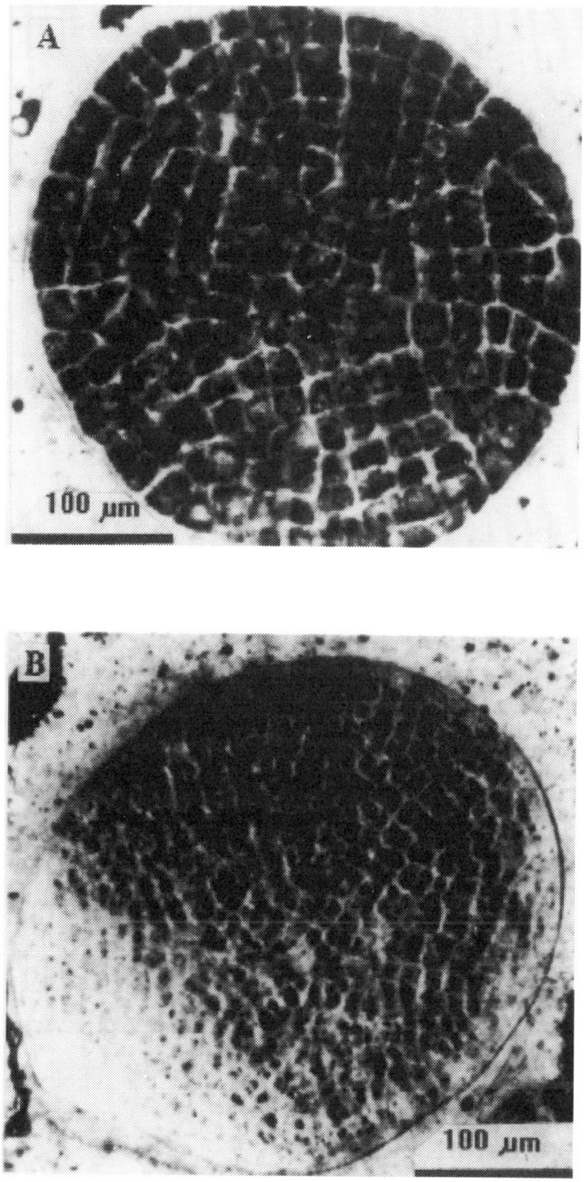

Figure 3. *Wengania globosa,* the colonial form of Neoproterozoic multicellular organisms, the colony consists of regularly arranged cells, with a definite shape and a certain range of size. The rounded shape and smooth surface of the colonies are possibly adaptive to floating and moving in sea water.

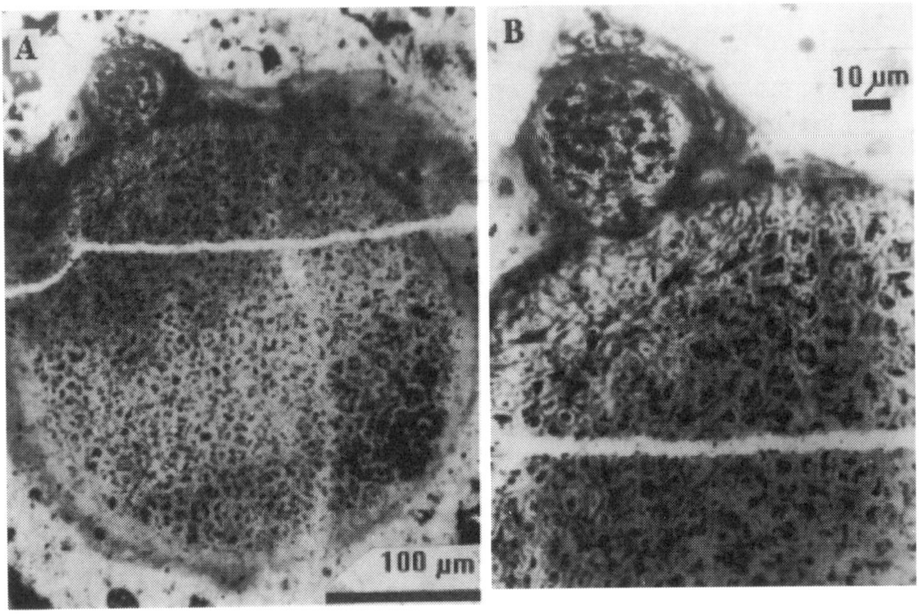

Figure 4. Asexual reproduction of *Wengania globosa*: a small "daughter colony" produced from the large "mother colony" (B is the enlarged view of A). The fossils were observed in thin section of the Neoproterozoic Doushantuo phosphorites.

Differentiation of cortical tissue *vs.* medullary tissue have been observed in the thallus of *Thallophyca* (Fig. 5A); The cortical tissue is composed of spherical, elongated or fibrous cells, which arranged in palisade or parallel layers. The medullary tissue consists of spherical, spheroid, elongated, rectangular or polyhedral cells, which regularly arranged in following patterns:

(a) Pseudoparenchyma: cells vertically arranged in cell-rows, which are in turn arranged in parallel bunches (Fig. 5B).

(b) Parenchyma: cells regularly or irregularly arranged in a compact tissue (Fig. 5C).

REPRODUCTIVE STRUCTURES

Distinctive structures comparable to reproductive organs exist within the vegetative tissue of the thallophytes. Carposporangium- and conceptacle-like as well as spermatangia-like reproductive structures have been observed in *Thallophyca* and other thallophytes from the Doushantuo phosphorites.

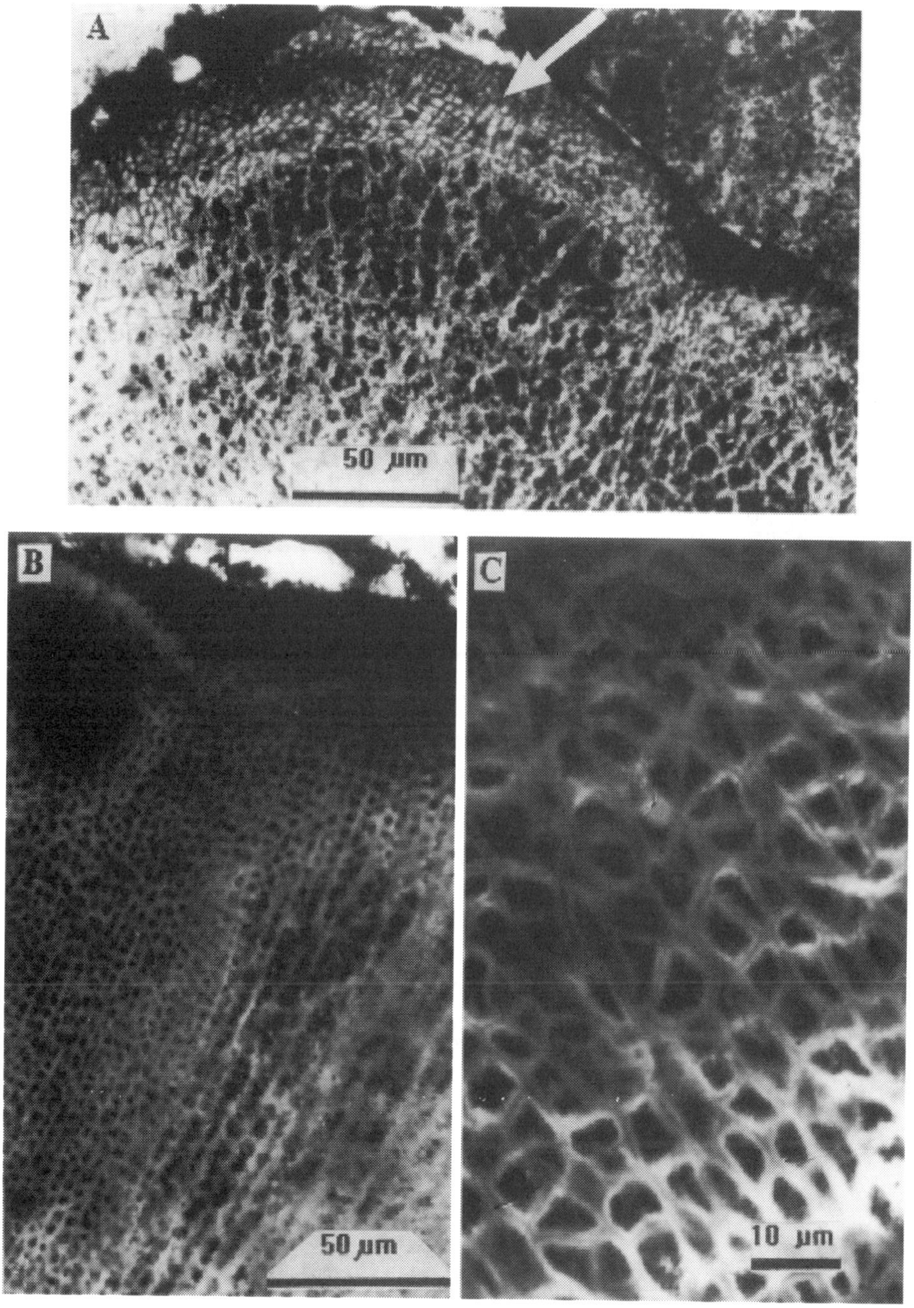

Figure 5. Tissues of the thalloid fossil form *Thallophyca* (in thin sections of the Doushantuo phosphorites). A. The tissues of *Thallophyca corrugata*. Note the differentiated cortical tissue (arrow) surrounding the medullary tissue. B. The pseudoparenchyma tissue of *Thallophyca corrugata*. C. The parenchyma tissue of *Thallophyca phylloformis*.

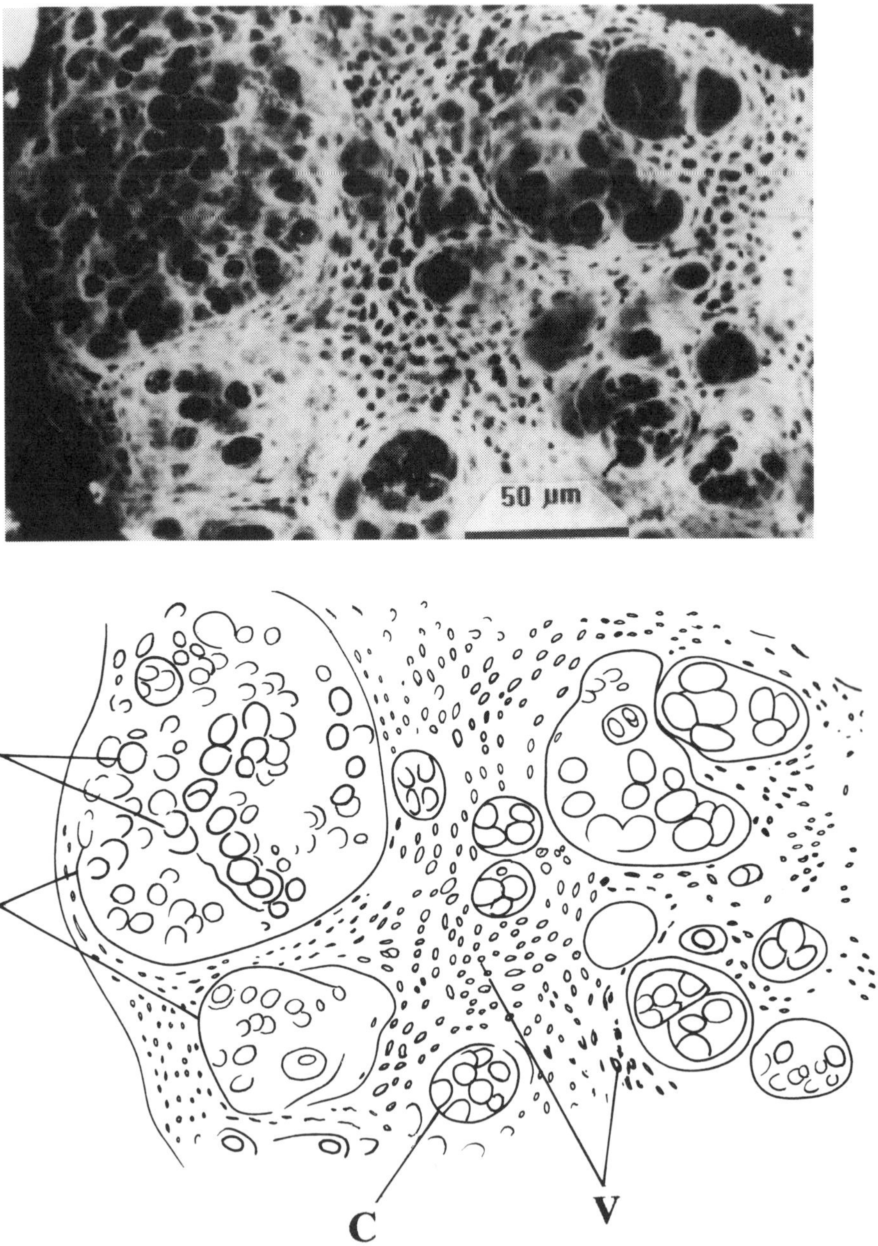

Figure 6. Microphotograph and its sketch of carposporangium-like reproductive structures of *Thallophyca*: the structures consist of a few to numerous ovoid cells surrounded by vegetative tissue. The large structure near the left edge seem to be a matured carposporangial conceptacle containing numerous carpospores.

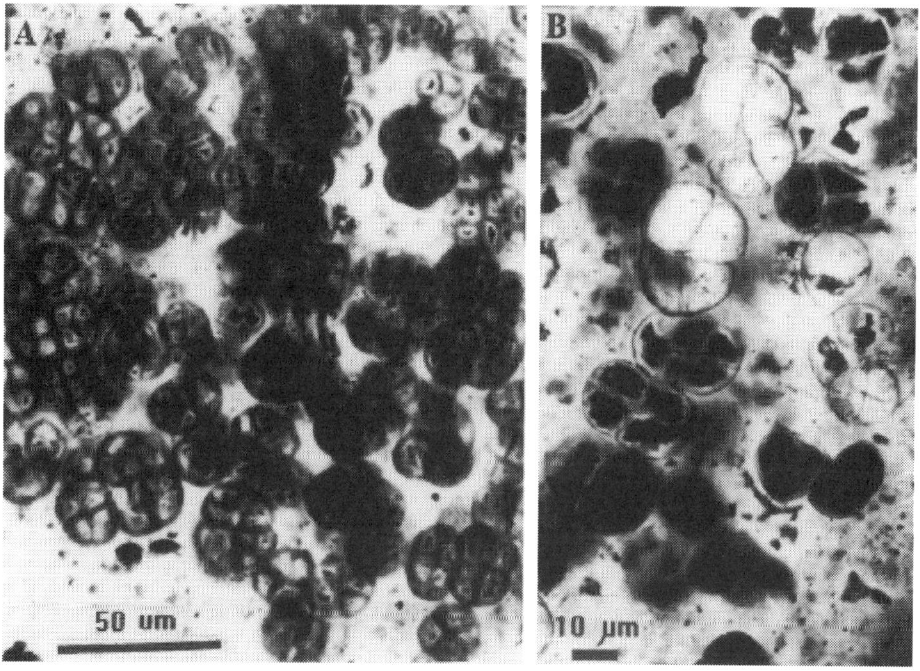

Figure 7. Microphotographs of the crucial tetrads in thin section of the Doushantuo phosphorites, each consists of 4 or 8 compactly arranged cells, they are similar to carposporangia that are often observed in the reproductive portion of thallus in *Porphyra* (a modern red algae).

The thalli of *Thallophyca* have distinctive structures within the vegetative tissue, each structure consists of a few to numerous dark ovoid cells surrounded by a few layers of curved vegetative cells (Fig. 6). The structures commonly occurred in the marginal portion of the thalli, and often protruded from the surface. The dark ovoid units within the structures are supposed to be carpospore-like reproductive cells, which distinctly differ from the surrounding vegetative cells in shape and size. Numerous crucial tetrads, each consisting of 4 - 8 compactly packed cells, have been observed in the thin sections of the phosphorites. The cells of the tetrads are about 8-12 μ m in diameter, with distinct mono-layered wall, with or without dark inclusions (Fig. 7).

The tetrads are distributed in patches, and are morphologically similar to spermatangia or carposporangia which are often seen in the thallus of modern bangiophycidae red algae, *e.g. Pophyra*. The population of the tetrads possibly represents a reproductive portion of the thallus.

DISCUSSION AND CONCLUSION

The comparative histological study on the macroscopic organism's remains that was three-dimensionally preserved in the phosphorites of the Neo-proterozoic Doushantuo

Formation in South China has demonstrated that, some of the organisms achieved multicellular organization with differentiated tissues and organs, which possibly conducted sexual reproduction. The study has also offered information about the development of multicellularity among the colonial and thalloid plants.

Development of Multicellularity among the Early Organisms

The Doushantuo fossil assemblage contains both typical planktonic forms represented by acanthomorphic acritarchs, and typical sessile benthos, *e.g. Thallophyca,* which has polarized thallus. The colonial forms include two modes of adaptation. The colony consisting of a lump of aggregated cells (Fig. 2) was possibly a prostrate benthic form, colonized on the substrate of sediments (Fig. 8A). The colonies of *Wengania globosa* was possibly adapted to floating. Buoyancy of the aquatic organisms depends mainly on the ratio of the body surface area to body volume.

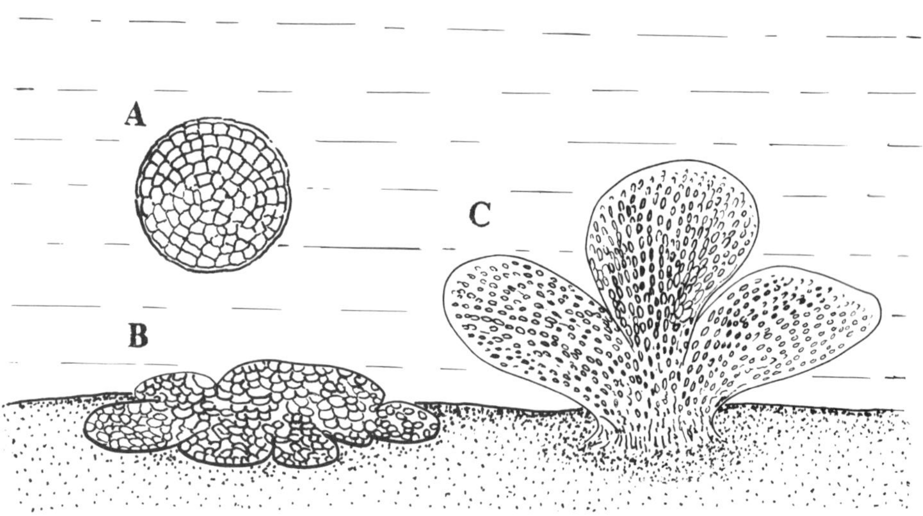

Figure 8. Illustration of the life habits of the Doushantuoan colonial and thalloid organisms.
A. The prostrate colonial form . B. The floating colonial form (*Wengania*). C. The benthic thalloid form (*Thallophyca*).

The Doushantuo acritarchs, with numerous processes resulted in a relatively large surface area and higher buoyancy, were adapted to floating in sea water. However, the colonies of *Wengania* were solid cell-aggregation with rounded shape and smooth surface, thus having relatively large volume and smaller surface area that resulted in a buoyancy relatively lower than that of acanthomorphic acritarchs. *Wengania* may be neither a benthos nor a typical plankton. It probably lived in the lower horizon of the sea water, where the water might have a relatively larger gravity and viscosity due to phosphate concentration. The colonies of *Wengania* could sink or float near the bottom, and moved along the bottom driving by current or turbulence (Fig. 8 B).

The sessile thallophytes, *e.g. Thallophyca,* with polarized thallus differentiated into an upper and a basal part resulted from directional growth pattern, possibly evolved from a prostrate colonial ancestor. The benthic photoautotrophic organisms adapted to the turbulent environment by changes of habit and growth pattern, from prostrate to sessile and from irregular growth to directional upward growth, thus leading increase of stability, which in turn promoting tissue differentiation and morphological complication. In addition, the evolutionary transition from prostrate colonial form to sessile thallophyte might have been promoted by competition for light. The prostrate colonies might have expanded their surface for light-acceptation through upward growth and branching (like some algal mats), thus resulting in morphological polarization of the colonies, *i.e.* the differentiation of the upper and basal portions. Evolutionary changes of habits (from prostrate to sessile) and growth patterns (from irregular growth to directional upward growth) were possibly promoted by natural selection.

Size of the Doushantuo Thallophytes

Although the Doushantuo thallophytes attained macroscopic size of individual body, but comparing with their modern counterparts, *e.g.* some modern seaweeds (rhodophytes and phaeophytes), they were relatively small. For example, *Thallophyca ramosa,* the largest thallophyte in the Doushantuo fossil assemblage, has a thallus size of 1 - 20 mm wide and 0.5 - 4 mm high, but the thalli of modern *Corallina* (rhodophyte) are about 30 - 50 mm high, and the thalli of *Pophyra* are more than 100 mm tall. Two possible reasons for the limited size of the Doushantuo thallophytes: (1) they were in low level of multicellularity; (2) they sufferred from environmental stress of the phosphorus-rich reduced conditions.

Sexuality of the pre-Ediacaran Organisms

As to the origin of sexuality, two questions have been presented and discussed in recent years. First, in what geological time and in what kind of organisms did sex first appear ? Second, what biological and physical factors drove the evolution of sexual reproduction, and what adaptive advantage of sexual reproduction had made it dominant in the multicellular eukaryotes. As sexual reproduction exists in some protists, it is possible that sexuality evolved at the time early before emergence of metaphytes and metazoans. Until recent years, there are not any definite palaeontological evidence for the earliest sexual reproduction in the Precambrian unicellular eukaryotes. Thus we have no definite answer on the question of " in what geological time did sexuality arise". However, among modern organisms, sexual reproduction associated with diploid-dominant multi-phased life cycle exists mostly in multicelluar eukaryotes, and it possibly evolved relatively later. The Doushantuoan thallophytes showed both sexuality and multicellularity, and the plants had a certain level of tissue-differentiation and possibly had a diploid-dominant sexual life cycle. This indicates that sexual reproduction possibly had become common among the multicellular thallohytes in the terminal Proterozoic.

Among modern higher organisms, only a small number of species conduct asexual reproduction, the majority being sexual. This leads to the deduction that sexual reproduction must have large adaptive advantage over asexual. The conventional argument is that sexuality has a long-term advantage to the species, because it greatly increases genetic variation and accelerates rate of evolution by genetic recombination that favor new adaptive combination of genes [13]. This argument has been challenged by the deduction that the diploid sexual genotype must pay 50% cost for meiosis [9, 13]. How can sexual reproduction possess an adaptive advantage large enough to overcome the 50% lost of fitness per generation? One of the possible solution is from Williams' statement, as he pointed out that if organisms live in and adapt to a constant, homogeneous environment, natural selection will favor the asexual reproduction, because it reproduces offsprings that are genetically most like their parent. But the situation will be the reverse if the organisms live in a changing and heterogeneous environment. Natural selection will then lead toward to sexual reproduction, as it will reproduce genetically varied offsprings that may adapt to the changing and heterogeneous habitats. The recent world-wide geological and geochemical investigation has demonstrated the changeable environments of Neoproterozoic time , and the most dramatic changes of global environment occurred during the time between 800 Ma to 550 Ma [7, 8]. The tillites and large-scale phosphate-manganese deposits of the Nantuo and Doushantuo formations in south China implicate the unstable and distinctive environments during the Terminal Proterozoic. In this changeable and distinctive environment, sexual reproduction possibly evolved and became dominant among the multicellular thallopytes, because it could expressed its large advantage that may overcome the cost of meiosis.

In addition, as soon as sexual reproduction become dominant, reproductive isolation comes to be the key mechanism of speciation; population and species, instead of individual or clone, come to be the evolutionary units. Natural selection will favor the sexual population with higher genetic diversity, which could occupy changeable and heterogeneous habitats, and become wide-spread. This will in turn accelerate the rate of speciation.

It is reasonable to infer that sexual reproduction associated with diploid-dominant sexual cycle could evolve and possibly become common among the early multicellular organisms in the late Neoproterozoic changeable heterogeneous shallow marine environments. Sexual reproduction and development of diploid-phase in sexual life cycle, associated with the diverse changeable environments in the terminal Neoproterozoic time, might greatly accelerate the rate of speciation, and consequently brought about a dramatic increase of diversity in the beginning of Phanerozoic. Development of multicellularity of the Neoproterozoic organisms might have been controlled by the global environmental events indicated by geochemical records. Development of sexuality and sexual reproduction in multicellular organisms had possibly led to an increase of speciation rate, which would in turn promote morphological diversification and complication. These may partly interpret the remarkable bio-diversity increase of metaphytes and metazoans at the beginning of Phanerozoic.

remarkable bio-diversity increase of metaphytes and metazoans at the beginning of Phanerozoic.

Acknowledgements

I thank Dr. A. H. Knoll, Dr. Yin Leiming and Xiao Shuhai for helpful discussions. Thanks are due to Mr. M. K. Pang for his financial support to my research work in recent years.

REFERENCES

[1] Chen Menge, Xiao Zhongzheng and Yuan Xunlai. A great diversification of macroscopic algae in Neoproterozoic, *Scientia Geologica Sinica* **4**:**3**, 295-308 (1995).

[2] Ding Lianfang, Li Rong, Hu Xiaosong, Xiao Yaping, Su Chunqian and Huang Jiangcheng. *Sinian Miaohe Biota Of China*. Geological Publishing House, Beijing (1996).

[3] Ding Qixiu and Chen Yiyuan. Discovery of soft metazoan from the Sinian System along eastern Yangtze Gorges Hubei, *Journal of Wuhan Geological Institute* **2** , 53- 57 (1981).

[4] Du Ruling, Tian Lifu and Li Hanbang, Discovery of megafossils from the Gaoyuzhuang Formation, Changchengian System, Jixian, *Acta Geologica Sinica* **60:2**, 115 - 120 (1986).

[5] M. F. Glaessner. *The dawn of animal life*. Cambridge University Press (1984).

[6] H. J. Hofmann. Precambrian carbonaceous megafossils. In: *Palaeoalgology: contemporary research and application*. D. F. Toomey and M. H. Nitecki (eds.). pp. 20 - 33. Springer Verlag, Berlin (1985).

[7] A. J. Kaufman , S. B. Jacobsen and A. H. Knoll. The Vendian record of Sr and C isotopic variation in sea water: implications for tectonics and *palaeo*climate, *Earth and Planetary Science Letters* **120**, 409-430 (1993).

[8] A. H. Knoll. Neo-proterozoic evolution and environmental change. In: *Early Life On Earth*. S. Bengston (ed.). pp. 439-449. Nobel Symposium No. 84, Columbia U. P. New York (1994).

[9] M. Ridely. *Evolution*. Blackwell Science, Cambridge, MA (1993).

[10] W. J. Schopf, R. E. Haugh, D. F. Molnar and Satterthwait. On the development of metaphyta and metazoans, *Journal Of Palaeontology* **47**, 1-9 (1973).

[11] Sun Weiguo. Late Precambrian Pennatulids (sea pens) from the eastern Yangtzi Gorges, China: *Paracharnia* gen. nov., *Precambrian Reassert* **31**, 361-375 (1986). .

[12] Wang Yuelun, Lu Zongbin, Xing Yusheng, Gao Zhenjia, Lin Weixing, Ma Guogan, Zhang Luyi and Lu Songnian. Subdivision and correlation of the Upper Precambrian in China. In: *Sinian Suberathem in China*. Wang Yue Lun et al. (eds.), pp. 1-30. Tienjin Sci. & Techn Press, Tienjin (1980).

[13] G. C. Williams. *Sex and Evolution*. Princeton U. P., Princeton (1975).

[14] Yan Yuzhong. Discovery and preliminary study of megascopic algae (1700 Ma) from the Tuanshanzhi Formation in Jixian, Hebei, *Acta Micropalaeontologica Sinica* **12:2**, 107- 126 (1995).

[15] Ye Lianjun et al. *The Phosphorites of China*. Science Press, Beijing (1989).

[16] Zhang Yun. Multicellular thallophytes with differentiated tissues from Late Proterozoic phosphate rocks of South China, *Lethaia* **22**, 113-132 (1989).

[17] Zhang Yun and Yuan Xunlai. New data on multicellular thallophytes and fragments of cellular tissues from Late Proterozoic phosphate rocks, South China, *Lethaia* **25**, 1-18 (1992).

[18] Zhang Yun and Yuan Xunlai. Discovery of sexual reproductive structures of the Terminal Proterozoic multicellular rhodophytes, *Science In China* (B) **25:7**, 749-754 (1995).

[19] Zhu Shixing and Chen Huineng. Megascopic multicellular organisms from the 1700 million-year-old Tuanshanzhi Formation in Jixian area, north China, *Science* **270**, 620-622 (1995).

Proc. 30th Intern. Geol. Congr., Vol. 1, pp. 201-213
Wang et al. (Eds)

Climatic and Geodynamic Significance of Cenozoic Land Surfaces and Duricrusts of Inland Australia

HELMUT WOPFNER
Geological Institute, University of Cologne, D- 50674 Koln, Germany

Abstract

Cenozoic duricrusts of interior Australia comprising, in younging order, silcretes, ferricretes, calcretes and gypcretes are products of surficial processes within pedogenic and shallow phreatic systems. They are related to climate and to specific land surfaces and demonstrate a systematic reduction of precipitation from Eocene times to Recent. Early Tertiary columnar silcretes of the Cordillo Surface of interior Australia reflect warm, pluvial conditions at a time, when Australia was still proximal to Antarctica. The inundation of a vast peneplain caused deep leaching of underlying rocks by acid groundwaters in a reducing environment, resulting in kaolinization of feldspars and other silicate minerals. Reduced precipitation in the Oligocene combined with structural differentiation caused reworking of columnar silcrete and the formation of silcrete breccias cemented by ferruginous silica in an oxidizing environment. Seasonally arid conditions leading to fluctuating groundwater levels in Miocene times are indicated by the formation of extensive ferricretes succeeded by cherty dolomites. A further increase of aridity in the Late Tertiary caused widespread formation of gypcretes and chert breccias. The final stage of this development is characterized by precipitation of halite and locally by subhalite formation of sulphides. The development of Australian duricrusts and intervening sediments thus reflects the separation of Australia from Antarctica and its northward drift over some 27 degrees of latitude during the last 55 million years, and also a dramatic change of global climatic belts during the course of the Cenozoic.

Keywords: Cenozoic, regolith, silcrete, ferricrete, gypcrete, climate

INTRODUCTION

Continental Australia was never affected directly by the Alpine Orogeny. This is especially evident in the interior of the continent where quasi-stability reigned after withdrawal of the Cretaceous sea in Late Cenomanian-Turonian times. In regions of Alpine tectonism thousands of meters of sediments were accumulated and subsequently piled up to form mountain chains in the course of the Cenozoic. During the same time span interior Australia received less than 500m of sediments and except for regional epeirogenic uplift, tectonism was restricted to some normal faulting and the formation of broad domal upwarps and intervening synclines [15].

For most of the past 60 million years interior Australia assumed a flat and stable surface with intermittent events of terrestrial deposition, but for most of the time was dominated by chemical weathering processes. Episodes of sedimentation generally followed periods of mild epeirogenic movements which created sufficient morphological relief for effective erosional gradients and depositional basins. For these reasons a great variety of deep weathering profiles together with their products of neomineralisation and associated landforms have developed, some of them covering thousands of square kilometres. Their excellent preservation, especially within the arid part of Australia, provides us with a unique opportunity to reconstruct the events leading to the formation of individual weathering profiles and duricrusts, so as to decipher the large scale climatic framework in the Cenozoic.

The present paper is concerned essentially with the northern parts of South Australia (Lake Eyre Basin; eastern Officer Basin) and adjoining regions in western Queensland and the Northern Territory (Fig. 1).

REGIONAL SETTING

At the time of the Cretaceous-Tertiary boundary, about 68 Ma ago, Australia was still connected with Antarctica, although a major rift system more or less paralleling the southern coast of Australia had already developed between the two Gondwanan portions. Continental tilt, probably in response to the formation of this rift system, had expelled the Late Cretaceous marine and lacustrine regimes from the area under consideration [19] and broad meandering streams had gouged their beds into the late Cretaceous sediments of the Winton Formation. The Large scale chaccel and pointbar deposits of the Turonian-Coniacian Mt. Howie Sandstone [16, 6, 9] bear witness to this final depositional episode of the Mesozoic megasequence.

The Cenozoic sequence of events was initiated in the Palaeocene by the deposition of fluviatile siliciclastics of the Eyre Formation [23]. Braided to highly sinuous stream systems led to the sedimentation of mature to clayey sandstones, but there were also frequent interludes of overbank and paludal sedimentation, producing carbonaceous siltstones and lignites. These stream systems extended from the northwestern Barrier Ranges in New South Wales and the Channel Country in western Queensland right across to the western margin of the Simpson Desert and to the SW of Lake Eyre (*e.g.* Willalinchina Sandstone, [9]). Time equivalents even abut Ayers Rock in the southern Northern Territory [12].

The gradients required for the formation of the stream systems were produced by epeirogenic uplifts in northwestern New South Wales, the Gawler Block in South Australia and in the southern part of the Northern Territory. These movements are her referred to as Tibooburra Movements. The depositional phase of the Eyre Formation lasted well into the Eocene. Alley *et al.* [1] reported well preserved micro- and macrofloras from a locality south of Lake Eyre, indicating a Middle Eocene age. This depositional event which corresponds to "sedimentation phase 1" of Krieg *et al.* [8] was terminated by epeirogenic uplift and surface modifications. These movements commenced in Late Eocene and extended well into Oligocene [19, 20]. They are here referred to as Nappamilkie Movements.

The Eyre Formation provides an important time control as all Cenozoic regoliths postdate the onset of it. The oldest Cenozoic duricrusts developed during the later half of this depositional cycle. These are the silcretes of the Coudillo Surface.

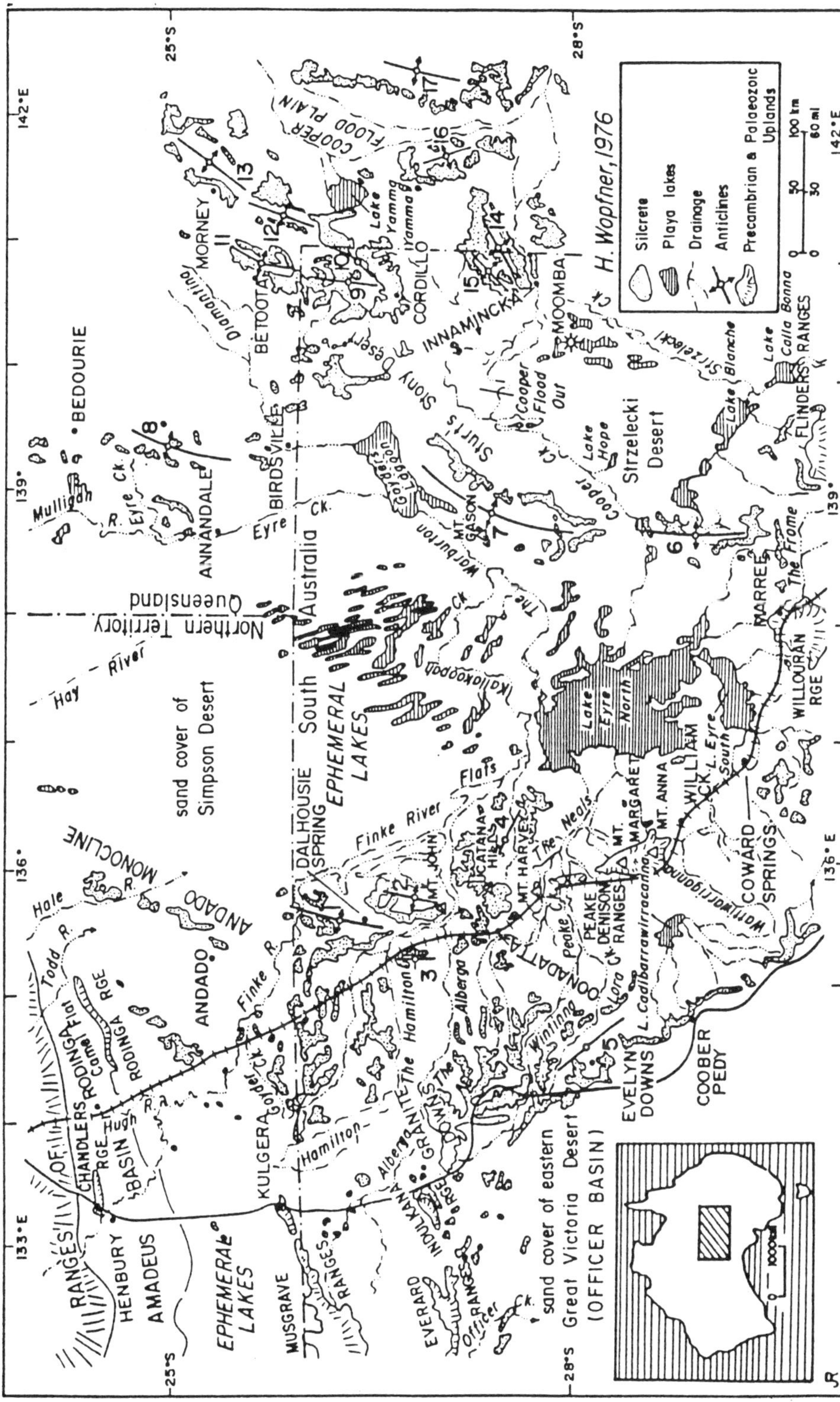
Silcrete
Playa lakes
Drainage
Anticlines
Precambrian & Palaeozoic Uplands
H. Wopfner, 1976
Northern Territory
Queensland
South Australia
sand cover of Simpson Desert
sand cover of eastern Great Victoria Desert
(OFFICER BASIN)
Strzelecki Desert
Sturt's Stony Desert
EPHEMERAL LAKES
Lake Eyre North
Lake Eyre South
BEDOURIE
BIRDSVILLE
BETOOTA
INNAMINCKA
MOOMBA
DALHOUSIE SPRING
COOBER PEDY
COWARD SPRINGS
MARREE
ANDADO
KULGERA
FLINDERS RANGES
WILLOURAN RGE
AMADEUS BASIN
MUSGRAVE RANGES
EVERARD RANGES
ANDADO MONOCLINE
COOPER FLOOD PLAIN
Finke River Flats
Diamantina
Warburton
Hay River
142°E
139°
136°E
133°E
25°S
28°S

Figure 1. Map of study area showing schematic distribution of early Tertiary silcretes of the Cordillo Surface. Silcretes are most prominently developed on anticlinal structures and uplifted table lands. Major structures of folded silcrete: 1) Dalhousie, 2) Mt. John, 3) Mt. Sarah, 4)Ucatanna, 5) Evelyn Downs, 6) Dulkaninna, 7) Mt. Gason, 8) Birdsville, 9) Nappamilkie, 10) Mt. Howie, 11) Betoota, 12) Curallee, 13) Morney,14) Innamincka, 15) Packsaddle, 16) Durham Downs; 17) Grey Range.

Barnes *et al.* [2], and to a certain extent also Krieg *et al.* [9] argued that sandstones containing reworked silcrete pebbles should be separated from the Eyre Formation. This phenomenon had already been recognized by Freytag *et al.* [5] while mapping the Oodnadatta 1:250 000 sheet area, where they identified and mapped the Macuvba Sandstone as an Early Cenozoic unit. Wopfner *et al.* [22] and Wopfner [21] discussed the relation of multiple silcretes, whereby tectonically tilted and subsequently eroded silcretes became "unconformably" overlain by a newly formed silcrete. This indicates the intimate interactions between depositional events and chemical weathering processes and their relation to tectonics. These interactions are vital for the understanding of landform development and duricrust formation. For this reason the author regards the formation of these Early Tertiary sandstones as a dynamic process and considers all these sandstones as part of "sedimentation Phase 1".

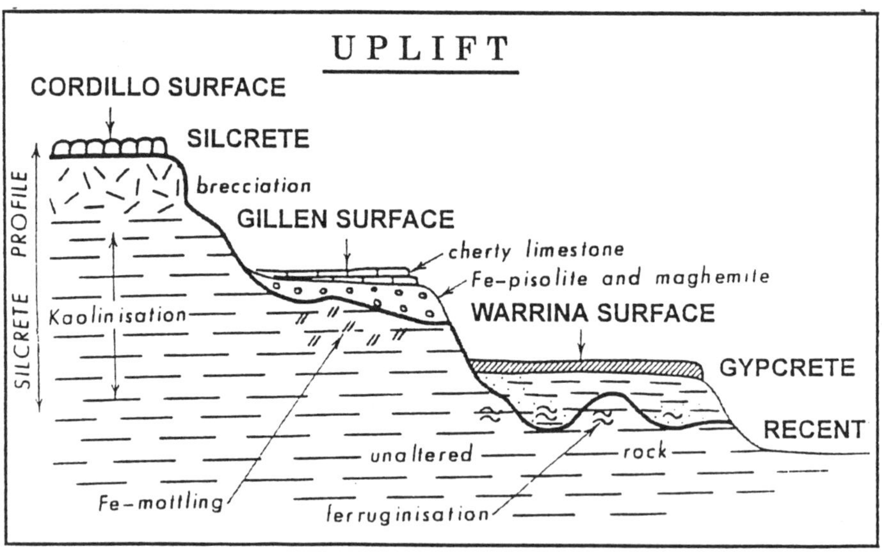

Figure 2. Schematic relationship of Cenozoic land surfaces and their duricrusts in the interior of Australia.

LAND SURFACES AND DURICRUSTS

Tectonic disturbances, though very modest on a global scale, controlled the patterns of erosion, denudation and deposition (*i.e.* surface dynamics), whereas the chemical processes for a given state of surface dynamics were governed by the prevailing climate, primarily by the amount of precipitation. Thus successive prominent land surfaces were developed, each can be identified by a specific weathering profile, specific duricrust and in certain cases associated deposits.

The identification of the fossil land surfaces is best achieved in areas of episodic regional uplift where the oldest surface stands highest and the younger ones occupy successively downstepped positions (Fig. 2). This is well demonstrated along the western and northwestern watershed of the Lake Eyre drainage and in the uplifted regions of northeastern South Australia and adjoining regions in Queensland. On the other hand, in areas dominated by regional subsidence, duricrust formation was preceded by or was concomitant with sedimentation events. In this case a duricrust may have been buried or even substituted by its erosional debris derived from more elevated areas.

In the interior of Australia, especially within the area under consideration, three Cenozoic land surfaces can be recognized. In order of decreasing age they comprise the Cordillo surface, the Gillen Surface and the Warrina Surface. The relative position and the nature of their respective duricrusts are shown in figure 2. A major depositional event in the Miocene intervened between the Gillen and the Warrina Surfaces.

Cordillo Surface

The oldest Cenozoic land surface is the Cordillo Surface. It is characterized by a thick silcrete which generally displays a pronounced columnar texture, the sides of the columns exhibiting a wax drip pattern. Quartzitic silcretes consisting of homoaxially intergrown quartz and pseudo-stromatolithic textures are also associated with the silcretes developed on that surface. The thickness of the actual silcrete crust averages about 1.6m to 2.0m, but much thinner and considerably thicker silcretes (up to 5m) have been observed [21].

The silcretes developed on all types of rock exposed on the Cordillo surface at that time. The host material ranges from Mesoproterozoic granites to the Eocene sandstones of the Eyre Formation. The latter was the most frequent host and provided the environment for the silcrete formation in its terminal stage.

The silcretes of the Cordillo Surface are invariably associated with a deep kaolinization profile sometimes exceeding 25m thick. Much confusion has arisen from the failure of some researchers to recognize polygenetic overprinting of weathering profiles in the discussion on silcrete formation. Monogenetic profiles developed on a reasonably homogeneous host rock are exposed near Tieyon Station on the northwestern margin of the Great Artesian Basin, where silcretes are developed on Mesoproterozoic granites and tonalites of the Kulgara Suite. One such profile has already been described [21], and a nearby profile was sampled and reevaluated by Wopfner and Walther in 1995 for further checking. The detailed analyses which confirmed the results published by Wopfner *et al.* [21] will be published elsewhere.

The profiles near Tieyon exhibit fresh granite at the base, a thick kaolinized zone with everything changed to kaolinite except quartz, a pedogenic transition zone and a columnar silcrete at the top. The kaolinized zone is pseudomorphic after granite indicating that the original granite texture has been perfectly preserved. despite the complete transformation of all silicates to kaolinite. Of special interest is the concentration of phosphate minerals crandallite and gorceixite together with rare earth elements at the base of the pedogenic zone, which is indicative for an acropetal element transport. Mass balance calculations show that only about 25 to 30% of the total released silica is required to produce the silcrete at the top of the profile. The rest of the silica apparently went out of the system (Dr. H. Walther, pers. com., 1995). Therefore there is probably a causal connection between the release of large amounts of continental silica and the event of the "silica window" in the Late Eocene deposits of the South Pacific.

Silcretes assigned to the Cordillo Surface are widely distributed (shown in Fig. 1). They are recognized over the whole Lake Eyre Basin, from southwestern Queensland to the Everard Ranges and northward into eastern Amadeus Basin. They are also present in the Officer Basin, where they cap amongst others the Observatory Hill. They are also known in typical association on the northern shore of the Wyola Lake, 170km N of Cook on the Trans-Australian Railway Line.

Towards the end of the Eocene and probably in the Oligocene, the cordillo Surface was warped and differentiated into large dome shaped anticlines and broad synclines. This phase of deformation, probably equivalent to the Pyrenean Phase, is here called Nappamilkie Movement. During this phase multiple silcretes were formed, uplifted silcretes were eroded, and the material was emplaced in younger silcretes down slope [21]. After deformation and further erosion of the Cordillo Surface, the second Cenozoic land surface, the Gillen Surface began to develop.

Gillen surface

This surface is a multiple surface and probably diachronous from W to E. The regolith of this surface was first described from exposures around the Observatory Hill in the Officer Basin [17]. The profile of the Gillen regolith comprises a ferricrete with an high content of maghemite (F_2O_3) at the top and an underlying zone of kaolinized rocks with profuse brick-red to purplish-red mottling. The mottles consist of fine grained aggregates of hematite and maghemite displacing kaolinite. In some instances, as in the Arckaringa Hills, some 90km WSW of Oodnadatta, the aggregates are formed by minute, concentric spherules, measuring 0.10mm to 0.15mm in diameter. They consist of a central grain of hematite and an outer rim of radially textured hematite and maghemite needles. This may indicate that the precursor mineral was lepidocrocite. The displacement of kaolinite by hematite and observations on the iron distribution along fissures and microfractures show that the iron mottles were emplaced by downward percolating fluids postdating kaolinization.

The most prominent expression of the Gillen surface exists along the western margin of the Lake Eyre drainage system and extending W thereof into the Officer Basin. Large parts of the escarpment which delimits the Lake Eyre drainage as well as isolated table hills in front of the erosional breakaway are capped by the ferricrete and the indurated mottled crust of the Gillen Surface. Mt. Gillen, the type locality of this land surface, is situated just east of this escarpment, about 95km NNW of the opal town Cobber Pedy. The mottled kaolinitic claystones are a prominent feature around Cobber Pedy and an excellent example of the regolith of the Gillen Surface. Here, the mottled claystones were formed by alteration of the originally montmorillonitic mudstones of the Aptian Bulldog shale. Further westward, in the Officer Basin, the alteration has affected Palaeozoic rocks indicating the Cambrian Observatory Hill Beds [17]. A pronounced bevel observed on some slopes of the Everard and Mann Ranges covered by maghemite pebbles probably also represents the Gillen Surface, but no systematic observations have been made in that region.

To the east of the type locality, the Gillen Surface covers Arckaringa Hills, a large tract of tableland between Arckaringa Creek and The Neales. The easternmost point where the surface is still clearly recognizable is at Ucatanna Hill, some 40km ENE of Oodnadatta. East of this longitude the older Cenozoic land surfaces (both Cordillo and Gillen Surfaces) dip rather rapidly towards the depression of Lake Eyre and the Simpson Desert. In the Lake Eyre bore hole 20, drilled in the central Madigan Gulf, about 3m of a ferrugineous granule conglomerate with abundant reworked grains of goethite and hematite underlies unconformably the Miocene Etadunna Formation [7]. The ferruginous components of the conglomerate are considered to be reworked material from the Gillen Surface, thus linking the elevated areas on either side of the Lade Eyre depression [22, 13].

Apparently maghemite is more resistant to weathering than the other components of the ferruginous crust. After denudation of the profile well rounded fragments of maghimite were incorporated as water-laid pebbles to granular layers in younger calcretes or as deflation residues on recent surfaces. The amount of stripping of the Gillen Surface is thus also reflected by the grain size distribution of reworked maghemite on the present land surface. It is ubiquitous west of the escarpment bounding the Lake Eyre drainage system. To the E of the escarpment the occurrence of maghimite diminish both in volume and in grain size. There, fine grained maghemite can be observed frequently as thin, aeolian deposited layers and stringers within the red sand dunes between the escarpment and the old Central Australian Railway line. N and E of Lake Eyre maghimite is a common accessory mineral of the aeolian sands, but coherent layers are rare.

East of Lake Eyre and the Simpson Desert, the morphological equivalent of the Gillen Surface forms broad and shallow erosional channels within the dome structures of the deformed Cordillo Surface. The land surface is evidenced by the Doonbara Formation, a sequence of brick-red ferruginous and aluminous pisolitic ferricretes, interspersed with current bedded, ferrugineous, coarse grained sandstones. The pisolitic portions show loaf-shaped structures 50 to 80cm in diameter, frequently with concentrically bleached rims. The rocks underlying the formation are invariably pedogenically brecciated with interstices filled with ferruginous matter [20].

There is no direct correlation between the Gillin Surface and the surface occupied by the Doonbara Fornmation. But both postdate the Nappamilkie diastrophism and precede

Miocene carbonate deposition and development of the Warrina Surface [20, 8].

Miocene carbonate deposition

This event which has been termed Cenozoec Phase 2 [8] coincides with the mid-Miocene transgression which inundated wide reaches along the southern and southeastern coast of Australia. Within the subsiding regions of Lake Eyre, Strzelecki Desert and Simpson Desert, large shallow and slightly alkaline water bodies were formed in which the dolomites and dolomitic marls of the Etadunna Formation and its equivalents were deposited. In the marginal and possibly isolated lakes, cherts were formed apparently later than magadiite (*e.g.* Cadelga Limestone, [20]). It was described [8] as the transgressive phase which overlapped the Eyre formation and extended into the Northern Territory and SW Queensland in a "giant lakes" environment. However, it is not clear whether these lakes were entirely land-locked or was somewhat connected with the transgressive seas. Occasionally foraminifera were described [7, 8] from the Etadunna formation, also observed by the author in a dolomite NW of Oodnadatta, which are not confirmative for a marine connection. The dolomite formations reach a thickness exceeding 120m in the depocentres but are much thinner in the marginal regions where channel erosion alternated with deposition and chemical weathering. In notheastern South Australia and neighbouring Queensland the cherty carbonates (Cadelga Limestone) rest directly on ferricretes of the Doonbara Formation [20]. Identical profiles also exist at Wyola Lake in the southernmost Officer Basin.

Warrina Surface

The Warrina Surface was originally called "Gypsite Surface" [17, 18, 22] and mapped as such on the geological sheet Oodnadatta 1:250 000 [5]. As the original name indicates, the Warrina Surface is characterized above all else by the presence of one or several gypsum crusts. Due to its widespread occurrence around the Peake and Denison Ranges [18] and especially within the area of the Warrina 1:250 000 sheet map area, the morpho-stratigraphic term Warrina Surface was introduced [21].

The constitution of the gypsum crust varies over a wide range from finely granular and sometimes crudely bedded gypsum layers to large intrasediment crystal aggregates of butterfly twins with 10 to 15cm diameters. The most common occurrences are gypcertes with sand, and silty siliciclastic particles cemented by a matrix of fine to medium-grained butterfly twinned gypsum, mixed with various amounts of powdery gypsum. Generally, the gypcrete shows signs of early gypsum emplacement by evaporative pumping. In the proximity of active fault scarps the siliciclastic fraction may increase to pebble, or even cobble-size. The thickness of the gypsum crusts varies between 30 and 200cm. Multiple gypsum crusts are not uncommon, usually separated by 3 to 5 m beds of green clay to silty mudstones. Occasionally these contain pauperate foraminifera of Plio-Pleistocene age (Ludbrook, quoted by [22]).

The bedrock below the gypsiferous succession was affected by intense ferruginisation and formation of "iron stones". A very special modification is present where the base of the Warrina Regolith affected the limestone concretions of the Early Cretaceous Bulldog Shale. These concretions in fresh conditions are invariably pyritic and changed completely to iron oxide, whereas the calcium apparently combined with sulphate to form thick sheets of selenite, occupying joints and bedding planes. This process has even produced gypsum pseudomorphs after the belemnite *Dimytobelus*. This indicates that SO_4 was released from the weathering of pyrite and subsequently combined with Ca to form gypsum and released CO_2.

Frequently the underlying, kaolinized and partially iron stained Cretaceous mudstones have been changed to porcellanite. Where the Warrina Surface abutted onto older hill slopes (Fig. 2) lateral infiltration of opaline silica took place. Invasion of pedogenetically fractured rocks by silica, genetically changing the host rock to opal-CT but preserving the finest details of the original texture, is also common and leads to the formation of red and white silcrete breccias [3, 21].

The Warrina surface has a very wide distribution within the Lake Eyre drainage basin. It is also widespread in the eastern Officer Basin, where gypcretes up to 3m thick form prominent escarpments along desiccated drainage channels around Observatory Hill [17]. thick gypcretes and gypsum beds interbedded with gypsiferous sands underlying vast tracts of land extend from Emu Field for at least 160km to the north. Significantly, the shallow ground waters of that region are highly saline. In many cases gypcretes have been stripped, leaving only thc porcellanitic and ferruginous bedrock as witness of its existence. As the development of the surface was polyphased, its formation near the margins probably occurred earlier than near the depressions of Lake Eyre and the Simpson Desert, where the Warrina Surface shows a complicated interplay with diminishing depositional area of the desiccating precursor of Lake Eyre (Lake Dieri).

CLIMATIC DEVELOPMENT

The mineral distribution within the silcrete profiles of the Cordello Surface indicates an environment, where the solubility of Al_2O_3 exceeded that of SiO2 near the top of the profile. This required a permanently high water table with Ph-values around 3.5, suggesting a very humid, subtropical to tropical climate. Such an environment is supported by the presence of a prolific macro-flora associated with the silcretes [1]. The floras are of mid-late Eocene age which is also the time of major silcrete development. Multiple silcretes were formed in the Oligocene.

The ferricretes of the Gillen Surface and especially the invasion and displacement of kaolinite by hematite indicate a fluctuating water table and thus a somewhat less humid condition than in the Early Tertiary. Maghemite is most commonly formed by dehydration of lepidocrocite. This latter is a typical mineral of bog iron accumulations, and suggests a seasonal humid and warm climate with fluctuating water table. The latter would account for the downward percolation of iron oxides to produce intense mottling. The formation of the Gillen ferricretes and the Doonbara Formation reflect thus a gradual change from semi-humid to a semi-arid climate in Oligocene to Early Miocene times.

The dolomites of the Miocene Etadunna Formation and its equivalents were deposited in shallow, slightly alkaline water bodies. Plant and animal fossils together with sedimentary features suggest permanent water cover with open woodlands surrounding the lakes [8], and indicate a seasonal semi-arid climate. However, local evaporating conditions are indicated by the magadiite-type cherts.

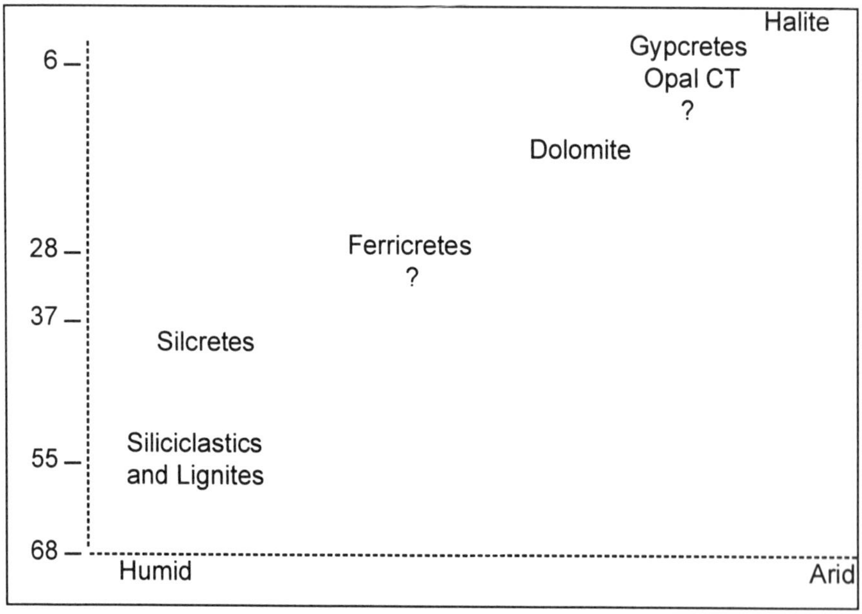

Figure 3. Duricrusts and sediments of interior Australia plotted against the Cenozoic time scale show a constant change from humid to arid conditions. Scale on left hand column in Ma.

The Warrina Surface, like many other land surfaces, is a hybrid resulting from interaction of clastic deposition, and chemical precipitation from both free surface water and evaporative pumping, and from chemical interaction with the underlying bedrock. Undoubtedly, some of the gypsum was produced by chemical interaction and new mineralization processes within the weathered profile. But the origin of enormous mass of gypsum found on the surface is difficult to interpret from purely intracratonic sources. Under Lake Eyre and adjoining depocentres the Etadunna Formation is overlain by sediments with increasing gypsum content. The widespread gypcretes and gypsum deposits on the Warrina Surface may have been formed in the terminal stage of the "giant lakes" phase, during which large playas and sabkhas developed owing to the increasing aridity. Reliable dating is scarce, but the increasing aridity probably occurred in Plio-Pleistocene.

Renewed tectonism in the Plio-Pleistocene considerably modified the landscape and created the present Peake and Denison Ranges [18]. This rejuvenated erosion and caused dissection and stripping of the Warrina surface. The change to an arid climate as manifested today was initiated at the end of this phase.

The present climate of the studied area is arid and evaporative, most of the region having an annual precipitation of less than 125mm. This is expressed by precipitation of halite and subhaline formation of sulphides, mainly melnikowite, within the subhaline ground waters (*e.g.* Kallakoopah).

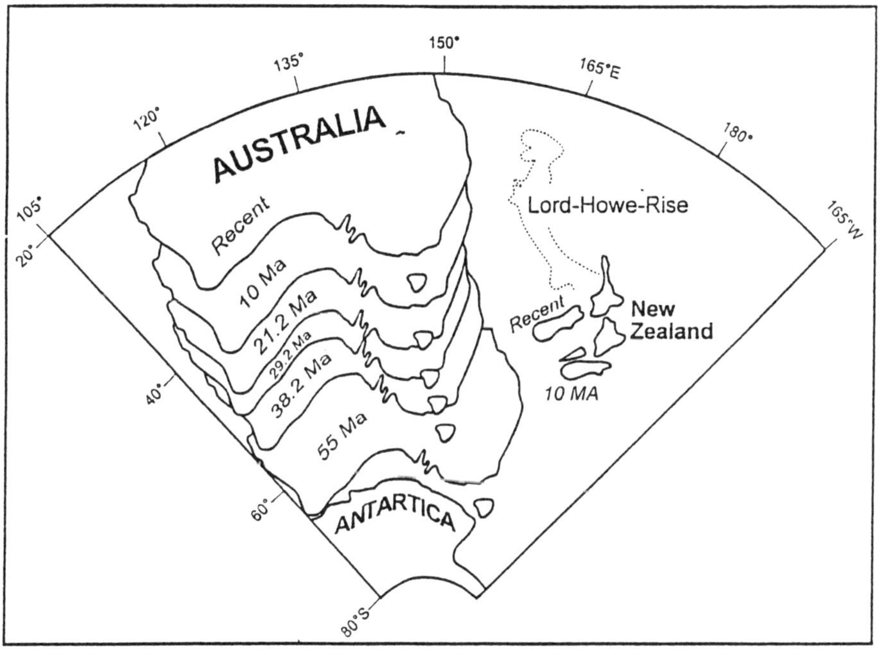

Figure 4. Migration path of Australia from its location adjacent to Antarctica at the beginning of the Tertiary to its present low latitude position. Modified after [5].

CONCLUSIONS

Figure 3 shows the relation of the Cenozoic duricrusts and the sedimentary deposits of the study area against time. It should be noted that the specific age of some of these units is approximate. Dating of the land surfaces and regoliths is also problematic, as shown by the papers of the Australian Regolith Confercnce "94" [10]. The main purpose of figure 3 is to show the temporal relationship of duricrust development and depositional events in the Cenozoic, which demonstrates a gradual change from a humid climate at the beginning of the Tertiary to the arid climate prevailing today, while the temperature experienced little change and remained fairly warm throughout.

This gradual climatic change in the interior of Australia is explained by the interaction of plate tectonics and the cooling of the south polar region since Oligocene. At the beginning of the Tertiary Australia was still attached to Antarctica, although substantial rifting existed between these two Gondwanan fragments since Late Jurassic (Fig. 4).The evidently warm, pluvial climate in the Palaeocene/Eocene of interior Australia must have prevailed also in the adjoining parts of Antarctica. Ocean currents flowed around

Australia in a clockwise direction [4]. It may be assumed that northwesterly air currents transported moist air across the continent, causing exceptional high rainfalls. Morphologic effects of Antarctic mountain ranges may have played an additional role.

Australia was separated from Antarctica in Oligocene [4, 14], thus permitting the establishment of the circum-Antarctic current system. The climate over interior Australia changed from pluvial to semi-humid as indicated by the maghemitic ferricretes of the Gillen Surface. At about the same time the lowering of average temperatures in the South Polar region commenced [11], and caused a rearrangement of climatic belts. As Australia moved into lower latitudes, the intensity of insolation increased, thus compensating the continued lowering of mean annual temperatures at the pole.

When Australia reached about 40 degrees latitude (Fig. 4), it came under the influence of the westerly anticyclonic system, which was established in response to the new climatic belts. The unusually pronounced continental climate caused by the great longitudinal extension of the continent controlled the climatic development thereafter. As Australia continued to move northward, the intensity of insulation increased. The rising temperature over the interior was responsible for the less frequent saturation of moisture, and led ultimately to the arid situation today. Climatic fluctuations during the Pleistocene were, like other episodic changes, only second order cycles superimposed on the long term Cenozoic trend.

REFERENCES

[1] N. F. Alley, G. W. Krieg and R. A. Callen. Early Tertiary Eyre Formation, lower Nelly Creek, southern Lake Eyre Basin, Australia: palynological dating of macrofloras and silcrete, and palacoclimatic interpretations, *Austral. Journal Earth Sciences* **43**, 71-84 (1996).

[2] L. C. Barnes and G. M. Pitt. The Mount Sarah Sandstone, *South Aust. Geol. Survey, Quart. Geol. Notes* **62**, 2-8 (1977).

[3] R. A. Callen. Late Tertiary "grey billy" and the age and origin of surficial silicifications (silcrete)in South Australia, *Journal Geol. Soc. Australia* **30**, 393-410 (1981).

[4] R. M. Carter, L. Carter and I. N. McCave. Current controlled sediment deposition from the shelf to deep ocean: The Cenozoic evolution of circulation through the SW Pacific gateway, *Geol . Rundschau* **85**, 438-451 (1996).

[5] I. B. Freytag, G. R. Heath and H. Wopfner. Oodnadatta 1:250 000 Geological Map, Geol. Atlas of South Australia, Sheet SG 53-15, zone 5. *South Aust. Geol. Survey* (1967).

[6] B. G. Forbes. Possible post-Winton Mesozoic rocks northeast of Maree, South Australia, *South Aust. Geol. Survey, Quart. Geol. Notes* **41**, 1-3 (1972).

[7] R. K. Johns and N. H. Ludbrood. Investigation of Lake Eyre, *South Aust. Geol. Survey, Rept. Invest.* **24**, 104 pp (1963).

[8] G. W. Krieg, R. A. Callen, D. I. Gravestock and C. G. Gatehouse. Geology. In: *Natural history of the north east deserts.* M. J. Tyler, C. R. Twidale, M. Davies and C. B. Wells (Eds.). pp. 1-26. Royal Soc. South Aust., Occas. Publ. **6** (1990).

[9] G. W. Krieg, P. A. Rogers, R. A. Callen, P. J. Freeman, N. F. Alley, and B. G. Forbes. Curdimurka, South Australia 1:250 000 Geological Map, Explanatory Notes. *South Aust. Geol. Survey*, 60 pp (1991).

[10] C. F. Pain and C. D. Ollier. Regolith stratigraphy: principles and problems, *AGSO Journal of Austral. Geology & Geophysics* **16**, 197-202 (1996).

[11] N. J. Shackleton. Palaeogene stable isotope events, *Palaeogeography, Palaeoclimatology, Palaeoecology* **57**, 91-102 (1986).

[12] C. R. Twidale and W. K. Harris. The age of Ayers Rock and the Olagas, central Australia, *Transact. Royal Soc. South Australia* **101** (1977).
[13] C. R. Twidale and H. Wopfner. Dune fields. In: *Natural history of the north east deserts*. M. J. Tyler, C. R. Twidale, M. Davies & C. B. Wells (Eds.). pp. 45-60. Royal Soc. South Australia, Occas. Publ. **6** (1990).
[14] J. K. Weissel and D. E. Hayes. Magnetic anomalies in the south-east Indian Ocean, *Antarctic Research Series* **19**, 99-165 (1972).
[15] H. Wopfner. On some structural developments in the central part of the Great Artesian Basin, *Transact. Royal Soc. South Australia* **83**, 179 - 193 (1960).
[16] H. Wopfner. post-Winton Sediments of probable Upper Cretaceous age in the Central Great Artesian Basin, *Transact. Royal Soc. South Australia* **86**, 247-253 (1963).
[17] H. Wopfner. Some observations on Cenozoic land surfaces in the Officer Basin, *South Aust. Geol. Survey Quart. Gelo. Notes* **23**, 3-8 (1967).
[18] H. Wopfner. Cretaceous sediments on the Mt. Margaret plateau and evidence for neo-tectonism, *South Aust. Geol. Survey Quart. Geol. Notes* **28**, 7-11 (1968).
[19] H. Wopfner. Depositional history and tectonics of South Australian sedimentary basins. *Proceedings 4th Symposium on the development of petroleum resources of Asia and the Far East- Mineral Resources Development Series* **41:1**, 251-267, United Nations, New York (1972).
[20] H. Wopfner. Post-Eocene history and stratigraphy of north-eastern South Australia *Transact. Royal Soc. South Australia* **98**, 1-12 (1974).
[21] H. Wopfner. Silcretes of northern South Australia and adjacent regions. In: *Silcretes in Australia*. T. Langford-Smith (Ed.). pp. 93 - 142. New England University Press, Armidale (1978).
[22] H. Wopfner and C. R. Twidale. Geomoprphological history of the Lake Eyre Basin. In: *Landform studies from Australia and New Guinea*, J. N. Jennings and J. A. Mabbut (eds.). pp. 118-143. Aust. National University Press, Canberra (1967).
[23] H. Wopfner, R. A. Callen and W. K. Harris. The Lower Tertiary Eyre Formation of the south-western Great Artesian Basin, *Journal Geol. Soc. Australia* **21**, 17-51 (1974).